Praise for *Lonely Hearts of the Cosmos*:

"Overbye's personal narrative . . . conveys the tremendous excitement at the front lines of modern cosmology."
— *Publishers Weekly*

"A compelling history of the science of cosmology . . . witty, intelligent, and highly readable."
— *Kirkus Reviews*

"Essential reading for anyone wishing to understand modern cosmology."
— *Booklist*

"Overbye has discovered a fiendishly clever way of tricking readers into understanding cosmology . . . [and] reveals [it] to be a human and passionate enterprise."
— *Time*

"Written with such wit and verve that it is hard not to zip through it in one sitting."
— Martin Gardner, *Washington Post*

"Overbye spins this complicated yarn with eloquence and wit. . . . Suspenseful, often poignant human dramas. . . . Compelling."
— John Carey, *Business Week*

"Clear and direct. . . . Overbye describes one of the greatest quests of all times."
— David Graham, *San Diego Union*

"Overbye seems determined to break the formulaic pattern of science writing. . . . The characters who walk the immense stage of *Lonely Hearts* are wonderful."
— Richard M. Preston, *Los Angeles Times*

"First-rate science journalism. . . . Written with intellectual sophistication and dazzle."
— Jeanne K. Hanson, *Minneapolis Star Tribune*

LONELY HEARTS OF THE COSMOS

The Scientific Quest for the Secret of the Universe

Dennis Overbye

HarperPerennial
A Division of HarperCollins*Publishers*

Photographs in the photo insert are credited as follows. Page 1—top: Caltech; bottom: J. R. Eyerman/*Life*. Page 2—top left: Frank Bello/*Fortune*; top right, bottom right and left: Caltech. Page 3—top right: Julian Calder; top left: Joe McNally; bottom right and left: Dennis Overbye. Page 4—top: P.J.E. Peebles; center: Princeton University; bottom: Yale University. Page 5—top and bottom: David Schramm. Page 6—top and bottom: Fermilab; center: Ohio State University. Page 7—top: Alan Lightman; bottom: University of Pennsylvania. Pages 8 and 9—Allan Sandage/Carnegie Observatories. Page 10—top: Marianne Kun; center: James Cornell/Center for Astrophysics; bottom: Caltech. Page 11—top: Carnegie Institution; center: Steven Seron; bottom: Harvard-Smithsonian Center for Astrophysics. Page 12—top: Dennis Overbye; bottom: Samuel Thaler. Page 13—top and bottom: Fermilab. Page 14—top: Caltech; bottom: Fermilab. Page 15—top: Doug Cunningham; bottom: Allan Sandage/Carnegie Observatories. Page 16—top: University of Hawaii; bottom: P.J.E. Peebles.

Page 425: Song lyrics © Nancy Abrams. Reprinted by permission.

Pages 183, 185: Excerpts from interview of Allan Sandage by Spencer Weart, 1977, American Institute of Physics, New York City.

Pages 185, 189: Excerpts from *My Daughter Beatrice: A Personal Memoir of Dr. Beatrice Tinsley, Astronomer*, Edward Hill. The American Physical Society, 1986.

Page 199: "Cosmic Gall" copyright © 1960 by John Updike. Reprinted from *Telephone Poles and Other Poems* by John Updike, by permission of Alfred A. Knopf, Inc. Originally appeared in *The New Yorker*.

A hardcover edition of this book was published in 1991 by HarperCollins Publishers.

First HarperPerennial edition published 1992.

Designed by Cassandra J. Pappas

The Library of Congress has catalogued the hardcover edition as follows:

Overbye, Dennis, 1944–

    Lonely hearts of the cosmos/Dennis Overbye.—1st ed.
      p.   cm.
      Includes index.
      ISBN 0-06-015964-2
      1. Cosmology.   2. Astronomers—Biography.   I. Title.
QB981.096   1991
523.1—dc20   89-45700

75176

ISBN 0-06-092271-0 (pbk.)
92   93   94   95   96   NK/RRD   10   9   8   7   6   5   4   3   2   1

For my mother and father

CONTENTS

II   FERMILAND

III   THE SHADOW UNIVERSE

IV   THE LAST GENTLEMAN

Prologue

DELEGATES TO ETERNITY

In 1954 *Fortune* magazine ran an article on the state of American science. Accompanying it was a photo essay featuring ten promising young scientists. One of them was an astronomer, who had been photographed leaning against the base of a famous 200-inch telescope on Palomar Mountain. He looked lean and Jimmy Stewartish, wearing a bomber jacket and grinning with dimpled cheeks, a spit of curl hanging over his high forehead. His eyes sparkled. He seemed both cocky and serious, like an ace bound for a battle with the Red Baron. All that was missing was the cigarette dangling from the lower lip. His name was Allan Sandage.

Sandage had a right to look cocky, and eager. At the time of the article, he was only twenty-eight, just a year past his Ph.D., and one of a handful of humans who had access to the 200-inch telescope, the most famous instrument of its time. Among that privileged few, he owned the darkest nights, the purest skies, the heaviest burden. As the magazine delicately put it, "He is helping to define the age and structure of the universe."

It is a poignant image, this jaunty young man with dimples and his new telescope, and a poignant moment, one that crystallized the hope that science, with these new tools and bright eager men, could unravel the mystery of the cosmos. Allan Sandage had become the first person in history whose job description was to determine the fate of the universe.

Five years later and half a world away in India, a wandering religious scholar named Huston Smith had a strange encounter with the Dalai Lama, the spiritual and governmental head of Tibet who had fled into India after the Chinese invasion of his country. According to the traditions of Tibetan Buddhism, the Dalai, whose real name was Tenzin Gyatso, was the fourteenth reincarnation of the Buddhist god of compassion; in practice he was a curious fellow.

As Smith recounted it years later, an audience with God was usually a brief, formally scripted affair: few words were spoken; you said your lines, paid your respects, bowed, and left. "Where are you from?" asked the Dalai as part of the game. Smith, a slender, prematurely bald man, explained that he taught philosophy and religion at the Massachusetts Institute of Technology, in Cambridge, Massachusetts. The Dalai Lama, an educated man, promptly abandoned ceremony, somewhat to Smith's surprise. "Stick around for a few minutes," he said. "I want to talk to you."

The Dalai went on to ask if Smith could tell him what was going on in the raging debate about the origin of the universe. Recently astronomers had coalesced into camps espousing two opposing versions of cosmic history. The first theory—called the big bang—held that the universe had begun in a fiery cataclysm 10 billion years ago or so, and might end in an equally spectacular crash billions of years in the future—a notion splendidly redolent of the cycles of destruction and re-creation of the world that the Tibetan Buddhists called *kalpas*. The other theory—known as the steady state—maintained that the universe was infinite and was forever the same. The Dalai listened attentively as Smith told him that, according to the latest words from Palomar, the balance was perhaps shifting toward the big bang. The Dalai Lama nodded and smiled ironically. "Of course," he said, "we have a position on that."

It's traditional for a book about cosmology to start by recounting the colorful creation myth of some ancient or primitive society, perhaps partly to show how far we've allegedly come. More important, it reminds us that the questions of who we are, where we come from, why we die, and why there is something rather than nothing at all are in our bones. I grew up in the sputnik generation of the fifties, maybe the first generation raised with a creation myth that was supposedly scientifically certifiable. The Palomar astronomers were then just beginning what would *surely* be the definitive exploration of the cosmos. My friends and I grew up reading—like the Dalai Lama, but perhaps with less personally at stake—about the debate between the big bang and the steady state, about curved space-time, the mystery of the expanding universe, and energy and matter transforming themselves into each other.

We were raised on science fiction movies and the presumption that mankind would conquer space and discover amazing mystical things about the origin of the universe *in our lifetimes*. My childhood was the age of the bomb, transistors, the arrival of television, frozen food, and tail fins. Even LSD was first presented as a triumph of technology. Scientists were heroes, and science, still giddy from its Faustian triumph with the bomb and the discovery of DNA, was ready to claim jurisdiction over the ultimate questions. Science was heady stuff by itself. It was jazzy, and our elders—brought up to believe that vacuum-tube radios were miraculous—weren't hip to it, which reinforced the generational prejudice that they and their dusty ways were irrelevant. Why *shouldn't* the Dalai Lama defer to astronomers?

The surprising launch of sputnik in 1957 was a defining moment for my generation. Afterward, science and technology became national obsessions, matters of national security. The best and the brightest students of the late fifties and early sixties were drawn into science, or at least thought about it—just as ten years later they would go to law school and twenty years later, alas, to business school. Because the road to the frontier was long and arduous, most of them wound up doing something else, whether selling computers or dipping candles. A handful made it all the way to becoming cosmologists. This is a book about them.

This is a book about what it's like to be on the cosmological quest in the second half of the twentieth century, about how men and women armed with computer chips, underground particle accelerators, 10-ton hunks of aluminized glass, radio telescopes, humor, and pride are still grappling with the issues that tantalized me in my boyhood. They are the priests and the mythmakers of our technological age.

What could be closer to the flavor of myth than the notion that the universe did in fact appear, perhaps out of nothing; that the atoms in our bones and blood were forged in stars light-years away and billions of years ago; or that the even more ancient particles of which those atoms are composed are fossils of barely comprehensible energies and forces that existed during the first microsecond of creation? We are all artifacts of the universe, walking reminders of the ultimate mystery. We are walking dust, waking stardust.

It's a great story, the modern version of the history of the universe, and maybe it's even true. It is probably part of the human condition that cosmologists (or the shamans of any age) always think they are knocking on eternity's door, that the final secret of the universe is in reach. It may also be part of the human condition that they are always wrong. Science, inching along by trial-and-error and by doubt, is a graveyard of final answers. But at least cosmologists are always wrong in different ways, like Woody Allen, who described the effects of increased fame by saying that he now fails with a better class of woman.

So, in a way, this is a book about failure. But also guts, hope, stubbornness, pride, genius, and luck. I've tried to tell this story through the eyes of the participants. There are always distortions in a narrative history; nobody can be everywhere or do everything. Every discovery or theoretical breakthrough stands on a pyramid of anonymous contributions. The closer you look at some events in science, the harder it can be to figure out with whom or where an idea really originated; authorship often seems to vanish like some quantum uncertainty. In general, when there was a choice, I have chosen to write about people who stayed with a particular subject and committed themselves to an idea, rather than those who made a suggestion and moved on. I apologize to those who may feel their contributions were left out and hope they will at least understand why. Any misattributions are, of course, my own error.

The stories told here are meant to be representative, not all-inclusive. Some cosmologists, not all cosmologists, are in the book. I've tried to follow mainstream science as I see it (not always an easy call), on the grounds that orthodox cosmology is strange and miraculous enough without invoking new mysteries. I apologize to the reader in advance for the physics. I've tried to include enough science so that the reader can catch the flavor of how cosmologists actually work and talk, but not so much that the prose reads like that of a textbook. Undoubtedly I have erred in both directions more than once. Take heart, for the physicists themselves, it seems, often don't understand what their colleagues are saying.

It should probably come as no surprise that strife and controversy, as well as love and loyalty, characterize this story. Cosmologists rarely speak with one voice, no matter how seamless whatever version of cosmic history temporarily in vogue appears to be. Scientists ask questions of nature in accordance with a set of formal rules. Nature in some sense is "out there," but the questions come from "in here," and the answers are received back in that same heart of confusion, mystery, and hope.

There are many voices, even in this abbreviated version of cosmological history, but one voice speaks louder, one heart, it seemed to me, harbored more of that cosmological confusion and hope, had harbored it longer than a normal mortal could bear.

I first met Allan Sandage at an astronomical conference in Tucson in January 1985, but I had heard of his legend long before that. For thirty years Sandage had operated the 200-inch telescope on Palomar Mountain, the most famous scientific instrument of the century, as if it were his backyard spyglass, measuring and remeasuring the universe, scraping from the shadows of photographic plates and enigmatic spectra and mathematical drudge-work clues to the size and fate of the universe. Measuring the universe was a man's work, and took a corresponding toll on Sandage's psyche. It is the kind of work that attracts critics. You weren't anybody in astronomy if Sandage hadn't stopped speaking to you at one time or another.

When I first met him he was engaged in a feud with several other groups of astronomers about the size and age of the universe. They disagreed by a seemingly irresolvable factor of two. For several years it seemed that every few months *The New York Times* was announcing that one group had corrected the other, and that the universe was now 20 billion years old instead of 10, or vice versa. I had heard that he had ripped his telephone out, and didn't talk to his colleagues, let alone the press.

One night during the conference Sandage and I drove out into the desert looking for a dinner party that was supposed to be held on an old movie set. I told him that I thought there were two histories to the universe. One was the sequence of physical events that went from the putative big bang that started it all, to the galaxies, to the sun lighting up, to the building of personal computers. The other was what I call the secret history, the story of the pain and imaginations and disagreements among the cosmologists. I wanted to know the secret history of the universe. I wanted to explain to people why their universe as reflected in the pages of the *New York Times* was such a yo-yo: first big, then small, infinite one day, doomed to collapse the next.

"Astronomy is an impossible science," Sandage said, laughing, as we drove over sand dunes, lost in the desert somewhere near Tucson. "It's a wonder we know anything at all."

Ten months later, after many postponements, we met in San Diego, where Sandage had rented a beach house for a sabbatical from the Mount Wilson and Las Campanas observatories, where he had worked his whole adult life. For two weeks we met twice a day, usually in a restaurant bar on the beach in La Jolla and talked, laughed, fumed, complained, teased, joked, and drank coffee. He has kept no journal, no diary, no record of his life beyond the logbooks, plates, notebooks, technical papers, and reports—no record, that is, except the science itself. He began every conversation by trying to convince me that he didn't remember anything about his life, and, moreover, that he knew nothing about galaxies and the universe. But his nothing could be another man's encyclopedia.

In the months that followed, my travels and his sabbatical wanderings found us crossing paths often. It became a ritual for me to walk into his office—whether in Honolulu, Baltimore, or Pasadena—unannounced. He would look up and cover his face in mock horror and theatrically groan "Oh shit!"

Sandage, the young man in the bomber jacket, haunts this narrative simply because he has been doing cosmology, trying to solve the universe, so much longer and more intensely than anyone else. Scientifically, he was the father of a generation of cosmologists. The universe of theorists who dream of quantum bubbles in their fertile imaginations is mostly the universe that Sandage has compiled from dry numbers in yellowed logbooks and squinty blurs through the great Palomar telescope.

Almost everything that can happen to someone on the road to knowledge has happened to Sandage. The longer I worked on this project, the more I felt that the story of Allan Sandage was a paradigm for the cosmological quest, indeed for science itself.

For the last five years I have been a cosmological camp follower. Besides haunting Sandage, I have haunted cosmology conferences in all the glamorous parts of the world (no fools, the cosmologists) in which they are held. I sat through weeks of workshops at the Aspen Center for Physics, where the theorists congregate every summer to argue without telephones or students. I climbed mountains and ate too many French dinners with the astrophysicists. I spent a week crawling with a strep throat around CERN, the European Centre for Nuclear Research, and sat in the control room at Fermilab while protons were slammed together at the highest energies yet attained on earth. I walked behind Stephen Hawking's whining wheelchair and helped lift him onto podia. And I rode in the prime focus cage of the 200-inch Palomar telescope, the vantage point from which quasars were discovered.

Many physicists and astronomers, many of whom subsequently did not even appear in the final version, gave me time and hospitality beyond all reasonable bounds during the preparation of this book. I cannot begin to mention them all. Special thanks must go to Allan Sandage, Stephen Hawking, John Wheeler, David Schramm, Gustav Tammann, Jim Peebles, Brent Tully, Alex Szalay, Vera Rubin, Joel Primack, Alan Guth, Jim Gunn, Gary Steigman, John Huchra, Maarten Schmidt, John Schwarz, Marc Davis, Kip Thorne, and the late Marc Aaronson. Invaluable logistical aid was supplied by Sally Mencimer of the Aspen Center for Physics, Dennis Meredith then at the California Institute of Technology, the late Margaret Pearson of Fermilab, John Gustafson of Lick Observatory, and Jean Hrichus and Spencer Weart at the American Institute of Physics. Michael Turner patiently read the entire manuscript and suggested many helpful corrections. Some of the quotations by Allan Sandage and James Peebles were taken from interviews conducted by the American Institute of Physics for its History of Physics project, whose director, Spencer Weart, kindly allowed them to be reproduced here.

Many people—far more than can be thanked in one meager paragraph—provided emotional support and encouragement during the long years this book required. Among them were my brother Gordon Overbye,

Tom Franzel, Kalia Doner and the Amazons of Phoenicia, Tom Dworetzky, James Polk, Joan Munkacsi, Conrad Fenwick, Gary Greene, Pat Sims, Bil Jaeger, Kathy Cahill, Ron Kriss, Suzanne Richie, and the gang at Misty's. Dan and Alex Hafner made a strategically important loan of a computer at a crucial moment. The crucial efforts of my plumber, William Heckeroth, and my accountant, Bernie Becker, must also be acknowledged. I would like to thank my editors at Sky and Telescope and Discover—the late Joseph Ashbrook, Lief Robinson, Leon Jaroff, and Gil Rogin—for encouragement and stimulation.

I owe a special thank-you to Gary Taubes, who introduced me to Kris Dahl, who became my agent. She weathered every crisis with patience, faith, and a crack of the whip. Equally invaluable has been Richard Kot, my editor, about whom I'm afraid to say too much for fear that some other publisher will scoop him up; I would prefer to keep him for myself.

Natalie Angier spent selfless days and months helping me to rescue my manuscript from chaos. She was always available, dispensing advice, criticism, and encouragement with her characteristic cheerful crispness; without her this book would not exist.

Lula and Rebecca Blackwell-Hafner shared their lives and their home with me, fetching aspirin and flowers through difficult years. To all those who never gave up on me, my debt is beyond repayment.

I

THE MAN IN THE CAGE

*To undertake executions for the master executioner
[Heaven] is like hewing wood for the master carpenter.
Whoever undertakes to hew wood for the master
carpenter rarely escapes injuring his own hands.*
— *Lao-tzu*

1

HIS MASTER'S VOICE

Few men are handed the keys to heaven, but Allan Sandage was one.

As long as he could remember, he said, the "Hounds of Heaven" had been pursuing him. He was born in Iowa City in 1926, an only child. Two tendencies, the worldly and the otherworldly, clashed in his background and in his nature. One of his grandfathers, Moses Sandage, was a Missouri dirt farmer. The other was a college president and an elder in the Reorganized Church of Latter Day Saints—Mormons who didn't move west. His father was a professor of advertising at the University of Miami, in southern Ohio. His mother was born in the Philippines, where her father had been sent by President Taft to reform that country's educational system.

Sandage described himself as a religious child. As a youngster he was either blessed or cursed with a spiritual itch he just couldn't figure out how to scratch. His earliest impressions of nature were charged with awe. Existence, he later remembered bitterly when the feeling seemed forever lost, was full of wonder; it was a miracle. The woods were full of magic. "Daddy, Daddy, look look—a flower! Isn't that something?" His voice took on a screeching girlish glee as he re-created his boyish self. The youngster felt alone in his appreciation. He recalled going to church by himself while his parents slept late on Sunday.

Meanwhile, the other half of him grew up toughened by the

Depression, intense and driven. He was raised in an intellectual household in which ambition and achievement were the norm. College and graduate school were always assumed. Early on he somehow absorbed the lesson that whatever he did would never be good enough, never complete enough, never right enough. There would never be any margin for error, never enough time to do everything. It didn't help that he came of age in the Depression years. There was a lesson in the men who came to the back door looking for any sort of work. The world of men seemed like a grim place, a morass of competition and doubt. Failure was always waiting like a vulture. The only remedy was to keep running, to work harder and harder. It was his duty, he felt, to learn everything.

"From the beginning," he recalled, "back into my childhood, the setting of goals and the completion of those goals seemed to me to be where all the action of the process of living was. I thought everyone was made that way."

When Sandage was nine, his father took a temporary assignment with the IRS, and the family moved to Philadelphia for a couple of years. On a trip to Washington, D.C., young Allan looked through a friend's backyard telescope one night and was "hooked." A whole new arena of wonder, beyond the sky, beckoned to him.

Up until then he had been interested in numbers but knew little about science. It was the perfect outlet, the coalescence of wonder and drive, a process that never ended because there was always something new to learn, a task of endless worship. He embraced astronomy with more than the usual fervor of the backyard stargazer. On returning home to Ohio, Sandage started to build himself a telescope. He got as far as grinding the mirror for a 6-inch-diameter model—a respectable backyard size—but he was no mechanic. His father bought him a commercial instrument. At night young Allan watched the Milky Way wheel across the midwestern sky, distracting his neighbors who were trying to sleep. By day he tracked the sun. During one four-year period he kept a chart of the number of sunspots he counted every day.

Who can explain why the stars are beautiful to us? They look like jewel chips, faithful in their seasonal shiftings and wanderings, but the cold terrible distances between them should make us want to scream with insignificance. To young Sandage that was exactly the grandeur of it all. The stars didn't care. The universe was immense, it was out there, independent of what he calls "the human morass." There was something greater and more permanent beyond the whims of man, time, and circumstance. Nature had its own rules, remote, impersonal, and terrible, but with science you could find them. After one look through the telescope it was clear what he had to do with his life. The stars were magic. He couldn't wait for them to come out each night.

He began to devour science and math books, especially astronomy, teaching himself the stars, constellations, and elementary celestial mechanics. As a teenager he read *The Realm of the Nebulae,* published in 1936, by Edwin Hubble, the great American astronomer. Sandage also absorbed the popular writings of Arthur Stanley Eddington, an English astronomer and friend of Einstein's. He learned that he was living in a period of a revolution in cosmological consciousness unparalleled in history.

For centuries most astronomers and philosophers had assumed that the flattened cloud of stars we know as the Milky Way constituted the entire universe, but they argued about the nature of the fuzzy little clouds of light, called *nebulae* after the Latin for clouds, mixed in with the stars. They had first been pointed out by a French comet hunter, Charles Messier, and the brightest ones were known by the numbers he had given them. On closer inspection some turned out to be interstellar gas clouds lit up like lampshades by stars inside them, and while others were tight clumps of stars, the "whirlpool" nebulae remained enigmas until the first decades of this century when large reflecting telescopes began to be constructed on dark, clear mountaintops of the American Southwest.

With the largest of these, on Mount Wilson overlooking Pasadena, California, Hubble had managed to gauge the rough distances to a handful of nebulae as hundreds of thousands of light-years. The puny-looking nebulae suddenly stood revealed as vast assemblages, tens of thousands of light-years across, of billions of stars—island universes comparable to the Milky Way. Hubble discovered that galaxies were the citizens of the cosmos. The farther he looked the more he found, strewn like dust across space.

Hubble had followed up with an even more remarkable discovery: These nebulae seemed to be all rushing away from our own as if they had been blown outward, like the shards of a grenade. It turned out that such a strange circumstance could be predicted—or explained—by Einstein's then-infant theory of general relativity, which ascribed gravity to warped space. Space and time were expanding like a balloon; the galaxies, Einstein's surfers, were just along for the ride. The universe was in flight.

From what and toward what?

If Hubble's observations, which amounted to barely more than a preliminary reconnaissance, and Einstein's theory were right, the universe had a beginning—back when the galaxies were all together. It was written in the stars for everybody to see. That realization jolted the impressionable young Sandage. "It was a shock," he recalled, "I mean, how does one feel in the presence of reality? Well, the whole thing is a miracle, in a sense."

If Hubble was rather dry and empirical about it—in his book he devoted four out of two hundred pages to relativity theory—Eddington in his own work was semimystical, as in this passage from *Space, Time, and Gravitation.*

We have found that where science has progressed the far-thest, the mind has but regained from nature that which the mind put into nature.

We have found a strange footprint on the shores of the unknown. We have devised profound theories, one after another, to account for its origin. At last we have succeeded in reconstructing the creature that made the footprint. And lo! it is our own.

Sandage didn't understand half of what he read, but he heard the Hounds baying. He recognized another seeker across the generations. His blood was stirred. He longed to be able to write like Eddington, to penetrate the mystery of mind with the precision with which Einstein and Newton had penetrated the mystery of matter.

He enrolled at Miami University in Ohio, where his father taught, and for two years majored in physics and minored in philosophy, reading Spinoza and Nietzsche. Then he was drafted into the navy where he spent eighteen months doing electrical maintenance in Gulfport, Louisiana, and Treasure Island in San Francisco. The experience left him with a taste for Polynesian restaurants and tropical rum drinks and a distaste for elec-tronics.

In the meantime Sandage's father moved to the University of Illinois and convinced Sandage to resume his education there, at Urbana, and live at home. Illinois was a bigger place than Miami; it had both physics and astronomy. Sandage made a typical Sandage choice: He chose physics because astronomy was easy and physics was hard, and physics was what he needed.

To keep up astronomically, he volunteered to work at the uni-versity observatory. The Harvard astronomer Bart Bok had organized a nationwide network of observers to photograph the sky and count stars of different brightnesses. Sandage was assigned the constellation Perseus. Years later he was asked what his first taste of professional observing was like. "Well, it was cold," he answered typically. "Those Midwest winters are incredibly cold, and Perseus is in the winter sky. I had to learn how to develop plates and measure magnitudes by myself. I knew the future would take care of itself if I trained myself sufficiently."

In those days a young aspiring astronomer naturally hoped that his future was in Southern California, home of Hubble, land of the big telescopes. Once, when his father was on a sabbatical at Berkeley, Sandage had visited Mount Wilson, which overlooks Pasadena from the front of the San Gabriel mountains, and seen the telescopes Hubble had used to discover the expansion of the universe. It had fired his imagination. Mount Wilson was a name that, like Einstein's face, was recognized in every dusty lane of the world. It was a legend; it crackled with the mystery and authority of science. Sandage couldn't think of anything more wonderful than being part of Mount Wilson.

For the last two decades—most of Sandage's life—the California Institute of Technology and the Carnegie Institution, which owned Mount Wilson Observatory, had been building an even bigger instrument, a 200-inch telescope, on Palomar Mountain, 50 miles to the south.[1] The press had followed the planning and construction of the device, the drama and heartbreak of the casting of the giant glass blank from which the mirror would be ground at the Corning Glass Works in New York, the slow odyssey of the blank across the country while millions lined the railroad tracks, the finished mirror's triumphant journey up the mountain in a rainstorm. Books had already been written about it: the instrument by which the Mount Wilson astronomers would investigate the deepest questions science knew to ask: Was the universe really expanding? Did the universe really begin? Where was it going? Would it ever end?

Hoping to get as close to the action as possible, Sandage applied to graduate school at Caltech. He applied in physics, but was accepted into the first class of Caltech's brand new astronomy program. He arrived in Pasadena in 1948, virtually on the eve of the inaugural of the 200-inch telescope, one of the very select, child and heir to the new era.

Pasadena was a far cry from Urbana. For one thing the winters weren't cold, and you could see Perseus or Orion without killing yourself. The California Institute of Technology, a collection of hacienda-style buildings with thick walls, arches, and tiled courtyards set in a garden, rested easily in a nexus of rose blossoms, Hollywood society, wealth, sunshine, smog, and the emerging high technology that was about to transform California into a world power. The undergraduate and graduate student bodies combined were smaller than that of an average high school. Caltech was known as a sociable place to be a scientist—a land of tolerance, shirt sleeves, and barbecues—and a hellish place to be a student. Behind the veil of sunshine, of course, was the genius and drive that produced Nobel Prize winners. In stepping into Pasadena astronomy, Sandage was entering into an arena of personalities and institutions whose tangled re-

---

1. The light-gathering power of a telescope is proportional to the square of the diameter of its primary mirror. The 200-inch would be able to see galaxies and stars four times fainter—1.5 astronomical magnitudes—or twice as far away as the 100-inch model.

lations would cast shadows the length of his career. Feuds begun in Pasadena before 1920 were still being played out in the 1980s by participants who had forgotten the original issues or protagonists.

The partnership between Carnegie and Caltech to build Palomar had been a shotgun marriage. Mount Wilson Observatory, Carnegie's ward, had been founded by George Ellery Hale, a brilliant and indefatigable fund-raiser and promoter as well as solar astronomer, but given to periodic nervous breakdowns. Hale had dreamed of a giant telescope to dwarf the 100-inch-diameter Hooker telescope on Mount Wilson, and had sold his dream to the Rockefeller Foundation way back in 1928. The Rockefellers weren't prepared to hand over money to the Carnegies, however, and gave $6 million to Caltech, instead. The deal was that Caltech would own Palomar, and Carnegie, Mount Wilson, but they would run them jointly as the Mount Wilson and Palomar Observatories. (In the 1960s the consortium became known as Hale Observatories.)

The Mount Wilson astronomers were appointed Caltech professors, but they got neither offices nor salaries. No money was supposed to change hands between the institutions. They each kept up their own staffs and telescopes. The principle was observed even in small transactions when Caltech bought a van to transport astronomers to the mountain, Carnegie agreed to hire the driver.

To begin with, then, Caltech had the prospective telescope while Mount Wilson had the astronomers. The observatory offices were housed in a two-story white limestone neoclassic building at 813 Santa Barbara Street, two miles from campus. It was known throughout the astronomical world simply as Santa Barbara Street. There were about a dozen gentlemen and no women inmates of Mount Wilson, and they had the cosmos pretty well divided up among them. Part of the gentleman's code of Mount Wilson was that you didn't intrude into somebody else's field unless you thought he was an idiot. And Mount Wilson gentlemen didn't call each other idiots.

The biggest slice of the sky, of course, had belonged to Hubble, the polestar of Pasadena astronomy. He referred to astronomy as a "calling." Born in Missouri, he had been a boxer, Army officer, Rhodes scholar, and high school teacher and basketball coach in Indiana before succumbing to the lure of the nebulae. He now lived in the exclusive old-money enclave of San Marino and entertained the Aldous Huxleys and the Igor Stravinskys. He had been on the cover of *Time* magazine and nominated for the Nobel Prize. Einstein made a pilgrimage to see him.

Hubble's personality shone through in his papers—grave, aloof, impersonal, and grand. Just as test pilots and eventually all pilots learned to speak in the hillbilly aw-shucks understatement of Chuck Yeager, cosmologists, particularly in Pasadena, wound up sounding like Hubble. His voice became the voice of cosmology. He reported the news about the universe as if it had been handed to him on stone tablets. A typical sentence was, "Nebulae are found both singly, and in groups of various sizes up to the occasional, great compact clusters of several hundred members each." No word about *who* found the nebulae. Nobody had done the work, the work had just been done. The galaxies *were*. You could hear the intergalactic winds creaking through his prose.

Hubble was hurt when, after the Second World War, he was passed over for the directorship of the joint observatories. The post went instead to a spectroscopist named Ira Bowen. That decision had been made in part because Hubble was perceived as not being interested in administration and in part because the same aloofness—some called it arrogance—that characterized his papers characterized his relations with certain other astronomers. Particularly Harlow Shapley, who had been the star of Mount Wilson before Hubble arrived and who had invented most of the methods Hubble subsequently used. Shapley felt uncredited. Once Hubble had scrawled "of no consequence" across a draft of one of Shapley's papers and the comment had been set in print in a journal. Shapley made the untimely mistake of leaving Mount Wilson to become the director at Harvard just before Hubble made his great discoveries, and he spent the rest of his life sniping at the West Coast astronomers.

Hubble had also slipped slightly from the forefront of the profession. During the war he had been recruited to run a ballistics school in Aberdeen, Maryland, and when he returned to Pasadena he had lost a step. He and his assistant, Milton Humason, had already pushed as far outward into the realm of the galaxies as they could go with the 100-inch on Mount Wilson, and were reduced to waiting for the 200-inch to become operational.

He had spent the last twenty years devising a grand program of observations that would begin to answer the questions posed by the expansion of the universe, the flight of the nebulae. At an informal early meeting of future users of the Palomar telescope, Hubble asked for half the available observing time on the upcoming telescope for this work. He was gently turned down. Hubble took it like the English gentleman he pretended to be. One of the participants later said there was a "sense of personal tragedy" in the air.

When Sandage got to Pasadena, the cutting-edge work at Mount Wilson was being done by Walter Baade, a little German who had joined the observatory in the thirties. In the photographs of the cosmological founders that line the halls at Santa Barbara Street he looks like a banker, with a hawk nose and carefully parted hair. As a German, he had been restricted from any role in the war effort and, in fact, from traveling more than five miles from his house. The observatory managed to get his range extended far enough to reach the domes of Mount Wilson, however, and during the long blacked-out California war nights he mastered the 100-inch Hooker telescope as no astronomer ever had. With it he had made fundamental discoveries about the basic kinds of stars in the universe and their relationship to the structures and histories of galaxies. It was the beginning of a revolution that within a decade would culminate in an understanding of the nature of stars, their life cycles, and the evolution of the Milky Way galaxy. Sandage was destined to play a pivotal role in that revolution.

After the war, realizing that the graduate students were about to start beating down its doors to get at the big telescopes, Caltech had hired Jesse Greenstein, a promising young astronomer, away from Yerkes Observatory in Chicago, to set up an astronomy program. Greenstein, a gravelly voiced ex-New Yorker with slicked-back hair and a black moustache, was the son of a wealthy furniture manufacturer; he had interrupted his education during the Depression to save the family fortune. His specialty was stars, especially old ones whose fires have failed or are about to. He liked to call himself a stellar mortician. He was also an amateur psychologist. He got to Pasadena, looked around, and sighed.

His only charge, the sole Caltecher doing astronomy, was a tall, irascible naturalized Swiss named Fritz Zwicky, who had been trained as a solid state physicist. Zwicky was brilliant, but had so many ideas it was almost impossible for other astronomers to sort the good from the off-the-wall. One of his major discoveries—that 90 percent of the matter in the universe seems to be invisible—was not to be taken seriously for forty years. He invented an intuitive system that he called morphological astronomy, with which he tried to guess all the possible types of galaxies and stars in the universe. He also proposed shooting artillery bursts over Palomar to make the air more transparent.

Zwicky and Baade had collaborated in the thirties, but by the time of the war, Zwicky hated Baade because of his German ancestry and called him a Nazi. Once Zwicky threatened to kill him if he caught him on campus.

In 1948 when Sandage and the other first students arrived, Greenstein was in a quandary. The Mount Wilson staff had no teaching experience, they were rather stuffy, and they were two miles away, although they did come down to campus for lunch at the Atheneum, the faculty club, once a week. And because Zwicky couldn't be trusted not to turn students into slaves on his obscure projects, Greenstein taught the astrophysics courses himself.

His students stayed up all night rewriting Greenstein's notoriously disorganized astrophysics lectures, and in the process, absorbed the material more thoroughly than they would have otherwise. They also had to flail through Caltech's tough physics courses. The road to Palomar was intentionally a hard one.

Sandage later claimed that the wonder and enchantment with which he had always viewed the world disappeared after two weeks at Caltech, replaced by a near mystical reverence for the mathematical equations that nature mysteriously seemed to know how to obey. He had a tough time that first year. His appointment came with a fifteen-hour-per-week laboratory job that paid the princely sum of $900 a year. There was little time for sleep. He often referred to himself disingenuously—as if he were not descended from a line of college educators—as a hick who had just fallen off the turnip truck. Outwardly, he struck Greenstein and the others as shy and withdrawn, but they recognized tremendous determination beneath. Greenstein knew from the start that he was special.

"From the time he was a student, Sandage wanted to solve the mysteries of the scale and fate of the universe; he kept to it," Greenstein, semiretired, recalled one summer day in his office. His hair was gray, and he was wearing a Hawaiian shirt with its tails hanging out. "Sandage had a hard time because he set high goals for himself. He wanted to do difficult problems and do them best. He was born that way—or his parents made him. He had to work hard at physics. He was a very hard self-driver. I don't think I've ever seen someone so intense who succeeded."

Greenstein himself felt like a papa to his students. "I was subfather for a generation," he said with a paternal gleam in his eye. He felt that Sandage was his greatest accomplishment. He had worked the hardest and come the furthest.

Sandage, according to Greenstein, had the cosmological fever. All the cosmologists had a special intensity, and it was something that had always intrigued the psychologist in him. He wasn't particularly interested in doing cosmology himself, although he acknowledged that it was important, and he supported Hubble and the rest in their voracious requests for telescope time. He thought cosmology was a slightly childish preoccupation, a quirk more psychological than scientific. What was this drive to get back to the first moment of time? Greenstein said he thought it had something to do with their fathers.

The 200-inch telescope was formally dedicated in the summer of 1948, but, to the dismay of the astronomers, Bowen kept tinkering and testing the mirror. (In the East, Shapley was leaking stories to the press that the 200-inch was a failure.) Hubble fretted. Finally, in January 1949 Hubble went to the telescope. The outer 18 inches of the mirror were still unfinished, but he could tell the telescope was everything he had hoped it would be. The Great Campaign swung into action at last. Hubble observed steadily until late April, clicking off photographic plates of the galaxies brought to new full-blooded starry brilliance by the giant mirror. In May the mirror was removed from the telescope for its final finishing. Hubble took stock, and feeling the pressure of the oncoming observational campaign, he decided he needed an assistant; he asked Greenstein to recommend someone. Knowing that Sandage had practical observing experience from his undergraduate days, Greenstein sent him to Hubble.

Up to then Sandage had had little contact, other than weekly colloquia, with the Mount Wilson astronomers. To the graduate students, Mount Wilson sat across town like a sort of unapproachable colossus. The observatory offices had the atmosphere of a Victorian gentlemen's club. The astronomers wore suits. On the mountain they dressed for dinner and ate on real linen. The halls were carpeted and paneled with oak. The sign on the men's room door read Gentlemen. Sandage went to see Hubble feeling tremulous. It was like an audience with God.

Hubble, of course, did not disappoint: He looked exactly like somebody who should be sending teams of astronomers out to measure the universe. He was a tall, handsome man with a noble boxer's jaw in which a pipe was usually clamped. He dressed like an English lord, and an Oxbridge accent had somehow grown over his native Missouri tongue. He had a lawyer's knack for making a visitor feel like a jury that had the full focus of his attention. One of his tricks was to sweep all the papers off his desk into the wastebasket with an imperious wave of the arm when a visitor walked in. After the visitor had left Hubble would be observed picking through his trash looking for his papers again.

Hubble explained to Sandage one of the programs he had devised, once the telescope was operational: to determine the nature and degree of any Einsteinian warp in the universe. With the new equipment at Palomar he would be able to make precise counts of galaxies at different levels of faintness. Fainter galaxies, Hubble presumed, would be on average farther away from earth. If galaxies were distributed uniformly in the universe, counting them in this fashion was a crude way of measuring the volume of space, the way that the number of trees in a forest is a rough gauge of its acreage. You would expect a thousand galaxies to take up about ten times as much space as a hundred galaxies. But what if they somehow only encompassed twice as much space? You would conclude that space itself has a warp.

If, as Einstein and the relativists contended—and Hubble was by no means convinced they were correct—the outrush of the nebulae was a cosmological fact, due to the expanding of warped space, the direction and degree of warp determined the future of the universe. If it were warped one way, the universe would eventually collapse back on itself like an accordion; the nebulae would all rush in. If it were warped the other way, it would just go on expanding forever. The sensitivity of the huge 200-inch mirror would finally allow him to see far into the depths of space and time. The way the galaxy counts increased with faintness (or distance) could reveal which way space was actually bent and what destiny had already been inscribed thereon. In short, Hubble was trying to find the fate of the universe.

Sandage was already familiar with all this, although he was still years from a true mastery of the mathematics involved. It was a small, but crucial part he was now being offered.

Most of the galaxies that Hubble planned to count were so far away that they appeared on the plates as no more than soft dim dots barely larger than stars. To make his plan work, Hubble needed a quantitative measure of the relative faintnesses of all those galaxy dots; his idea was to compare them to stars that appeared in the foreground of the observations. Sandage's job was to find suitable stars for the comparison. His duties didn't require observing, but catalog work and perusing plates until his eyes wouldn't focus. It was technical and arduous and really the backbone of Hubble's whole scheme.

During the summer of 1949 Sandage began plugging away on Hubble's magnitude sequences in an office in the basement of Santa Barbara Street while Hubble went fishing at his ranch in Colorado. One day in July Sandage ran into one of the older astronomers, Rudolph Minkowski, and asked when Hubble was coming back. Just then the phone rang. Minkowski picked it up, listened awhile, and hung up. "Hubble won't be back for a long time," he told Sandage. "He just had a heart attack."

Hubble spent a month in the hospital out in Colorado and then returned to Pasadena and spent another month in bed. His doctor forbade him from observing. It would be more than a year before he returned to Palomar. The new era was getting off to a gloomy start.

Sandage didn't see much of Hubble during this period. Without guidance, he quickly got bogged down in technical problems with his star sequences and set the project aside. He looked around for something else to do. The contact with Hubble had whetted his appetite for research.

Sandage and a fellow student, Halton "Chip" Arp, who was a year behind him at Caltech, decided they wanted to do some serious observing together. They were an unlikely pair. If Sandage was a "hick," Arp was an antihick. The son of an artist, he had grown up in Greenwich Village and Woodstock and attended Harvard, where he had become one of the best fencers in the United States, and had also been involved in Bok's star-counting network; he and Sandage had met one summer at the Harvard Observatory. At Caltech they became best friends.

Greenstein sent them to Baade, the finest observer in Pasadena.

Baade agreed to give them a job measuring stars in globular clusters. Globular clusters, as the name implies, are spherically shaped clouds of starlight, reddish in color, almost minigalaxies of up to a million stars, that form a halo around the Milky Way like bees around a saucer of honey. Very little was known about the stars that inhabited globular clusters, partly because they were so far away, mostly concentrated toward the center of the galaxy tens of thousands of light-years distant, and partly because the stars in them seemed inherently dim and squashed together and, as a result, hard to see. It was likely that globular clusters had something important to say about the formation and history of the galaxy and the mystery of the evolution of stars; they were at the core of Baade's research.

Sandage and Arp were thrilled to be handed work on the cutting edge. It was also daunting work that would test the resolve and skill of the most experienced astronomer: taking high-resolution photographs of globular clusters and then measuring the intensity of thousands of tiny star images with a microscope.

They did, in fact, need Baade's skill, so Baade himself took them up Mount Wilson. As their learning instrument he chose the oldest telescope on the mountain, a 60-inch-diameter reflector that had been built in 1905, back when a mule train was the only way up the mountain and the astronomers used to hike up for their observing stints.

Veering steeply upward from the back of Pasadena, the San Gabriels form an abrupt wild wall to the cultivated sprawl of Los Angeles. Coyote, sagebrush, scrub oak, sequoia, and black-cone Douglas fir predominate on the knobby peak of Mount Wilson, which at 5700 feet breaks through the Los Angeles smog layer into clean steady air. On a good night, Pacific fog fills the basin and valleys below, completely blanking city lights. The domes, towers, and other buildings of the observatory are strung along a narrow ridge. The machine shop and library—wood-frame sheds with sloping tin roofs to keep off the snow—run along one edge of the mountainside joined by a narrow balcony and walkway overlooking a thousand-foot drop. Not a place for a misstep on an ink-black night. The dome of the hallowed 100-inch Hooker telescope, the one that had discovered galaxies and the purported expansion of the universe, sits slightly apart, at the end of a long wooden footbridge that leads over a shallow gorge. At the far end of the footbridge from the telescope is a small shed in which the cook would leave a midnight lunch for the 100-inch observer.

Life on the mountaintop was a constant contrast between such civilities and the rugged vicissitudes of observing. The astronomers dressed each night for dinner at the Monastery (so-called because no women were allowed), the dorm at one end of the ridge. Hubble even wore a tie to the telescope. At the dinner table they were seated in a precise and unchanging hierarchy. Whoever was scheduled to observe that night on the 100-inch telescope got the place of honor. Next to him sat the 60-inch observer, and so forth, down to the students and assistants. For each Mount Wilson staff member the dining room kept a wooden napkin ring with his name on it; students and visitors were assigned clothespins. One of the great rites of passage in a young astronomer's life was when he went from a clothespin to a wooden napkin ring with his name on it.

These rituals were carried out against a background of wilderness, cold, and industrial waste. The observatory domes were unheated, their concrete floors littered with steel and grease; they smelled like machine shops. The great Hooker telescope was driven by a pendulum clock drive through a nearly incomprehensible set of gears one of which was wider across than a man could reach. Every dome was a living museum of turn-of-the-century technology, including huge copper double knife-blade electrical switches like you see in movies such as *Frankenstein*.

The 60-inch was Sandage's first big telescope, and he would remember it as one of the most complicated to operate. For one thing, it was of a Newtonian design: At the bottom a big concave mirror gathered starlight and sent it to a tiny mirror suspended overhead, which then deflected the converging beam of light sideways out of the telescope to an eyepiece, camera, or spectrograph. The focus, where the observer had to station himself, was on the side of the device, way up in the air.

Paradoxically, Sandage, like others before him, found that he was more comfortable up there during actual observing situations in the dark, when he couldn't see the precariousness of his perch, than he was in the light climbing up to tinker with instruments. To make sharp photographs the telescope had to track the stars as they moved across the sky. Unfortunately, you couldn't trust even the finest telescope to track the stars faithfully, let alone an antique like this. As the telescope heeled over, it would flex, minute imperfections in its gears either speeding it ahead or dropping it behind the stars it was following; also differential refraction would shift the apparent location of the stars in the sky. So you had to stand on the moving platform in the sky with your eye to the separately moving eyepiece, with a control paddle in your hands hitting buttons to goose the telescope or slow it down.

Guiding could be literally painful. On cold nights tears would freeze one to the eyepiece. When the telescope moved, the dome had to move, to keep its slit opening in front of the telescope, and the observer's platform had to move, as well. Over the course of a night these movements, all independent of each other, could gradually scrunch an astronomer's neck and back into celestial pain.

There were other complications. The guiding eyepiece also sampled the light far off the optical axis of the telescope, where aberration turned the stars into a teardrop shape called a coma. To Baade, that little blur spoke volumes, and he taught Sandage and Arp how to read it too. They learned how to diagnose the shape of the mirror from it, how to recognize the trembling striations in the faint part of the fan that presaged a change in the seeing conditions, how to adjust the focus when that happened, *during an exposure,* how to keep the starlight on a photographic plate concentrated into crisp specks.

When Baade in his gentle German had said everything there was to say and shown everything there was to show, he ordered Sandage to the precarious observer's platform. Then Sandage discovered there was one more trick that Baade had withheld. Because of an error in one of its gears, every eighty seconds the telescope leapt ahead of the stars it was supposedly tracking, and then fell behind. "Baade wanted to see whether we would discover that," said Sandage, "whether we'd push the East button, then the West button, then the East button, then the West button. He could tell by listening to the relays click and the time period whether we were guiding well or not."

Night after night the trio took their posts over the skies of Pasadena. When it was cloudy they sat inside and talked about stars. Sandage and Arp came down after seven days thoroughly indoctrinated, both in the intricacies of big-telescope observing and in Baade's love-detective affair with the stars. And with their names on clothespins.

Baade had directed his students to two particular clusters, named M92 and M3 (*M* is for Messier). Sandage and Arp managed eventually to get decent photographs of M92. In the negative image on a photographic plate M92 looked like a pile of pepper. They spent the next year taking turns measuring the intensities of the individual grains with a special microphotometer in the basement of Santa Barbara Street. Sandage eventually took on the second cluster, M3, by himself, and the work developed into his Ph.D. thesis.

It had become clear to Hubble that while the Great Campaign had to go on, he could no longer do the observing himself. In photographs from the period he looks cadaverous and hollow eyed. Humason had been his faithful lieutenant on Mount Wilson, but Humason too was getting old. He sent for Sandage again in 1950, a year after their first go round. This time more was at stake. Hubble gave him what Sandage later called "a series of strange tests," hunting on photographic plates in the M31 and M33 spiral nebulae for variable stars—stars that undergo cyclic changes in brightness. He must have passed; shortly thereafter Humason took him to Palomar, to the 200-inch.

Palomar Mountain, 5600 feet high, is about two hours south and east of Pasadena, in ranching and orange grove country. The white dome of the 200-inch rises from a meadow tufted with pine, its fourteen-story height disguised by its classic proportions.

Sandage and Humason walked through a columned portico to the ground floor of the dome, a vast room that looked like an industrial staging area, littered with stacks of steel, barrels of oil, and forklifts. One floor up they came on the telescope itself, painted battleship gray, sprawling in the perpetual twilight of the cavernous dome. Its most noticeable feature was a thick steel horseshoe, as wide as a tennis court, inside which the 200-inch-diameter concave mirror, 15 tons of glass, swiveled demurely like a child in a swing. The skeleton of a tube pointed skyward toward the dome. At its upper end, suspended roughly in the center of the vaulted space, halfway between heaven and the hard concrete, hung a small mesh cage resembling the gondola of a hot-air balloon: the prime focus cage. That was their destination.

They ascended a series of stairs, and elevators, past the library and darkrooms, the kitchen, the control room, and offices that ringed the great inner space of the dome to a catwalk, up a ramp, and climbed into a sort of cherry picker basket. With a whine that echoed through the great chamber Humason drove them out across empty space, with the big reflector looking up at them like a shaving mirror, to the prime focus cage. They scrambled over a railing and squeezed into the cage together between red-lighted instruments and the hollow pier that contained the beam of light coming up to a focus from the big mirror.

Sandage later recalled that the seeing—the astronomer's technical term for the quality of star images, which is affected by atmospheric turbulence—was very, very good that night. He was already a professional and his excitement at being at Palomar was outweighed by his relief that it was a much easier telescope to use than the Mount Wilson telescopes. Sandage spent most of the first night learning to focus the telescope. Since everyone's eyes are a little different, each observer at a big telescope has to find for him- or herself the relationship between what he sees as the focus and the "true" focus. Sandage took a series of test exposures, which showed that he was dead on. Later when he told Hubble about it, Hubble was angry. "If the seeing was so good, why didn't you observe," he asked, "instead of frittering away your time on something you knew would come out right?"

And so Sandage began to commute between Palomar and Pasadena three times a month as Hubble's surrogate observer. Sandage's cup ran over. "Oh, it was fabulous, so much was happening," he said. "It was an opportunity that was beyond imagination—first of all, observing with the 200-inch, and, second, working on the long-range program of cosmology with Hubble. And at the same time being a graduate student, trying to pass the courses in physics and astronomy. So it was a very high pressure atmosphere. The work on the mountain was an escape, but you knew that your sins would catch up with you, because you were four days away from campus and courses, and you were pretty well swamped."

In the fall of 1950 Hubble started returning occasionally to Palomar, where he slept in a room rigged with a bell so that he could call for help if needed. More and more he sent surrogates to observe and ran the show from his huge office on Santa Barbara Street. Every time Sandage went to the mountain he went with three sets of instructions, depending on the observing conditions. Sandage brought the exposed plates back to the master like a schoolboy handing in homework. Hubble would examine them, comment on them, and then take them to analyze.

The most important and exhausting project was to photograph galaxies and search for variable stars within them so that the distances to these galaxies could be measured. The whole world knew that this was how Hubble had found the range to the famous Andromeda Nebula M31 and a handful of others in the twenties. Those had all been part of the "local group," a small clustering to which the Milky Way belonged. Now he was trying to reach out of the local group, to the next nearby clusters. It could take a year, Hubble figured, just to get the distance to the next group of galaxies; it could take decades or a lifetime to follow the entire observational path that he had laid out, to gather the evidence that would give definitive answers to the questions of the size, age, and fate of the universe. Hubble became Sandage's personal tutor in cosmology.

Hubble took Sandage under his wing in other ways, too. He invited him home to dinner with the likes of the Huxleys and the Stravinskys. "I was like a son to them for a year or two," Sandage said. It left an indelible mark on his personality. Hubble, Sandage thought, interacted with people as a god might, a noble man. He spoke to Sandage about philosophy, art, the universe, religion. Sandage, who considered himself a functional illiterate, worked hard to keep up, to grow out of his hickness. He tried to be like Hubble, a man seemingly dedicated to truth. Hubble gave Sandage the impression that he was the great man he seemed to be. He was a dedicated empiricist. He would not twist the data or go beyond them. He would not tolerate dishonesty.

At the same time Hubble was very formal; he shared very little personally with Sandage. He read widely and hungrily in religion and philosophy—whatever conclusions or connections he drew from those realms to the known universe he kept to himself. Even with his chosen successor Hubble was aloof, always correct, polite—but cold.

Sandage felt the gulf, and in some respects it tempered his worship. On the one hand, he told me, "Hubble was the greatest scientist of the last four hundred years." On the other, Sandage made a point of digging out a historical article showing that Hubble had been teaching school in Kentucky when his official résumé said he had been practicing law.

During the remainder of his years at Caltech, Sandage went back and forth between Hubble and Baade, the two poles of Pasadena cosmology. Next to Hubble, Baade looked like Walter Mitty but, Sandage realized, he was a warmer person. He was gregarious and told jokes on himself. And he was the better observer. "Baade and Hubble were kind of at odds with each other," Sandage explained. "Hubble wanted the great vast things that you've talked about, the geometry of space-time, the distance scale, cosmology. Baade said that you cannot really understand all of that unless you, okay, approach this problem microscopically instead of macroscopically by looking in detail at the structure of a given galaxy. Only after both Hubble and Baade died did the confluence of those two trains of thought come together."

The third great influence on Sandage over at Santa Barbara Street and in the red-lighted domes through the long observing nights on Palomar and Mount Wilson was Humason, a dreamy-looking man with wire spectacles and fat, boyish lips. A wisecracker and eighth-grade dropout who had worked his way up from mule skinner to janitor to night assistant to astronomer by asking questions and being clever, Humason was Hubble's leveler as well as his chief observer. "Humason was of a different kind, very meticulous, simple in some ways, but streetwise," said Sandage fondly. "He was very streetwise. He knew how to get along with Hubble, and he knew how to get along with the rest of the staff. He was the intermediary between Hubble and the rest of the staff. Humason taught me to fish and chew tobacco and how to do all sorts of other things, and how to get on with people. Hubble taught me how to classify galaxies."

Sandage spent more and more time at Santa Barbara Street. Every time he went in he felt like he was entering Valhalla. It was a privilege to walk the halls and handle the plates, plates from the 100- and 200-inch telescopes; he likened them to "the plates of Moses." The formality of the prewar years was beginning to dissolve in the presence of graduate students, but it was still a place where, as he put it, "You were never more than ten feet from a gentleman." Giants, it seemed to him, strode the oak-paneled halls in those days. He fell in love with the library, a sanctuary with a sunken carpeted floor surrounded by floor-to-ceiling dark, polished bookshelves and light filtering through the curtains of one tall corner window, where an astronomer could cloister himself and drift through centuries of starry parchments undisturbed.

In 1952, when Sandage was almost finished with his degree, Bowen offered him a post at Mount Wilson, a seat in Valhalla with the gods. It was all he had ever wanted in life, but the stewardship of those great telescopes was also an awesome responsibility. He thought of the immense programs that had been carried out by Hubble and Humason and Baade and Shapley. Nowhere else would they have been possible. He thought of the immense programs still to be carried out, the extension of Hubble and Baade's work. "The whole world knew that the 200-inch had been built to do those tasks," Sandage said in a dry voice. "By accepting the Mount Wilson post you already knew what was expected of you."

Sandage accepted, but delayed his appointment a year to go to Princeton and study the theory of the evolution of stars, an outgrowth of his Ph.D. thesis.

In September 1953, not long after Sandage had returned, Hubble had another heart attack and died. His wife, Grace, called the observatory with the news. "It was such an incredible shock," Sandage recalled. "I walked out the door of Santa Barbara Street and walked around Pasadena by myself for two or three hours."

There was no funeral that Sandage or any other astronomers knew about, and no one knows what happened to his body. Rumors circulated that Grace had cremated it. Sandage would continue to go to the house on Woodstock Road in San Marino to see Grace Hubble for the next seventeen years. For some strange reason, he confesses, puzzled now, he never asked her what had happened to her husband.

The full impact of Hubble's death did not hit Sandage immediately; he was immersed in work on stellar evolution, but he knew what was coming. Someone had to carry on the tradition of Mount Wilson and complete the work Hubble had planned to plumb the size and destiny of the cosmos.

It was both scary and the chance of a lifetime, both an opportunity and a sentence, and he still recalls it both ways, depending on his mood. "My responsibility is that Hubble died too young," Sandage said in a low voice one rainy afternoon. "Taken that way it's an awful statement. Hubble died too young and left me with a burden, an incredible burden, to carry out his program. And that burden, I still feel it, but my love and interest is the stars.

"It was all laid out by him," Sandage recalled with an almost desperate tone in his voice. "It would be as if you were appointed to be copy editor to Dante. If you were assistant to Dante and then Dante died, and then you had in your possession the whole of the *Divine Comedy*, what would you do? What actually would you do?

"It was all laid out, and I was the only one left after Hubble died. It had kind of been in the back of my mind all along, to do his program. So not only did I think I had to because that was my job," he explained. "It was laid on me, but it was also what I kind of wanted to do." He paused.

"If I didn't grab that opportunity I would be nuts."

2

THE CANDY MACHINE

The next few years—the early fifties—were among the happiest of Sandage's scientific life. He became the bomber-jacketed ace flying the 200-inch telescope deeper and farther than even its builders had dreamed. The thirties and forties had seen revolutions not just in cosmology but in every aspect of astronomy and astrophysics, including a whole new understanding of the nature of stars, how they are born, burn, and die. With the war over and with new telescopes, it was possible for the first time to test these new ideas.

The Hale telescope was affectionately known as the "Big Eye" to the Pasadena astronomical crowd, but Sandage liked to think of it as a giant candy machine. "I was a kid in a candy store that was so magnificent, full of everything that you wanted, that it was life's greatest carnival. I was the only one with a key to the store, and somehow the candy kept miraculously appearing."

Sandage inherited Hubble's generous allotment of observing time—thirty-five nights a year on the 200-inch telescope alone—as well as the plates and data Hubble had already collected in his campaign to measure the universe. He spent most of half of every month, the half centered on the new moon, when the sky is darkest and the blue-water extragalactic sailors of astronomy come out, yo-yoing between Pasadena and Palomar.

Some astronomers preferred to stay in the control room, if they were observing as part of a team or using the telescope in the Cassegrain mode, in which the light was reflected back down from the prime focus cage through a hole in the center of the 200-inch mirror. (Particularly heavy or bulky instruments could be hung behind the hole, on the back of the mirror, to dissect starlight, but lengthening the focal path that way made the telescope optically less efficient.) It was warmer observing that way, and the astronomers could joke with the night assistant—the man who actually operated the telescope—eat cookies, drink coffee, go to the bathroom, catch up on paperwork, or even sleep during lulls or long exposures.

Sandage liked being in the prime focus cage.

Night after night he rode the lift through darkness, the dome rumbling and whirring like the secret machinery of the cosmos itself as the night assistant positioned the dome slit for the first of the observations that Sandage planned with military precision. He scrambled eagerly the few inches across the void and settled onto a tractor bench, lugging his sky charts, notebooks, and photographic plates whose emulsions had been soaked in hydrogen or baked in nitrogen until the already sensitive grains were hysterical for the light that had left some star or galaxy before the human race was born. The telescope pier, a hollow barrel, made a chest-high table in the center of the cage. At that point, the prime focus, a portion of the universe about one and a half times the apparent size of the full moon was splayed out over the area of a postcard for inspection and investigation. Sandage was alone, at the center of the universe.

Einstein's relativity theory taught that the center of the universe was everywhere and nowhere. It was the present, wherever, surrounded by concentric shells of the past—history racing at him in the form of light rays at 186,282 miles per second, the speed of light, the speed of all information. The prime focus cage of the 200-inch was the cockpit of a time machine, the finest and largest ever constructed, its yawning mirror pearled with starlight pointing in the only direction any of us could ever face: backward. The moon he saw was an image composed of light that left its surface a second and a half ago, Mars glowing martial red was half an hour away, the center of the galaxy, hiding behind the thick star clouds of Sagittarius, 30,000 years. The cold eyepiece by which Sandage, sighting a star on cross hairs, could correct the fine movements of the telescope, steer it to his precise target, and keep it locked on the same celestial object, all night if need be—which was often the case— was an eyelash, a tenth of a nanosecond ago.

Sandage bragged that he had iron kidneys; he could go up to the prime focus cage and not have to come back down for fourteen hours, an entire long winter's observing night. That, he joked, was the main reason for whatever success he had as an observer, whatever candy he brought down—that and the wonderful instruments that the Mount Wilson staff had built. Discomfort was part of the romance. "You sit there," Sandage chuckled dryly, "straddling the pier with your privates nestled up against the cold of the universe." In the winter he donned an electrically heated jumpsuit of the sort that bomber pilots used to wear.

Music of the observer's choice—Sandage was partial to opera—was piped up through the intercom. Sometimes Sandage would pop the neoprene cover on top of the cage and look out with his own eyes on the unamplified universe. By peering over the side he could look down on the mirror. By daylight its face, the size of a Manhattan studio apartment, was pitted and chipped from accidents over the years. When I saw it, there was one crater the size of a dinner plate where a workman had dropped a hammer once. By nighttime, when the dome slit was open to the heavens, the mirror was a bowl of stars floating in front of his face like fireflies.

The view was great, but it was the solitude more than anything else that Sandage prized. Up there he could think about the data coming in, meditate on them. What if this plate worked? What if the seeing went bad? What was the next step? The ideas seemed to come more easily up in the quiet Zen-like intensity of the cage when it was just him and his plates, the universe breathing light in and out on them.

Theory, Einstein had once remarked sarcastically with reference to epistemology and quantum mechanics, determines what observations can be made. Sandage agreed: The mind had to be prepared.

His own, he thought, was more receptive up there to the whisperings of the subconscious. He liked to imagine its deep recesses as a cauldron where ideas, sensations, emotions, and dreams were slowly, slowly stewing in an endless ferment. Once in a while there was a volcanic outburst into the consciousness, when some new idea, the solution to a problem, burbled up flecked with the detritus of its birth—pride, rage, ambition, prejudice, all the freight of human creativity.

You couldn't control the volcano, he knew, but you could be ready for the explosion. You had to prepare the mind if you wanted to solve problems. Truth was all around you, there was only one universe, but all theory and interpretation were a jungle of subjectivity. You saw only what you wanted to see, what you were prepared to see.

In the first years that Sandage returned to Santa Barbara Street, he dutifully continued to collect the plates for Hubble's campaign, but didn't really do anything with them. There simply weren't enough data; neither he nor they were ready, and they went on accumulating in his office.

He continued the work he had started with Baade, using the great telescopes as if they were geologists' hammers to break apart the globular clusters and get at the nature of the stars inside. Stars had been his first love. They were also the primal astronomical mystery, but now that the war was over that mystery was yielding to a great wave of theory and data. The stars were being unraveled, and thirty years later Sandage would still say that his contribution to that process was his proudest moment.

The bonfires of the universe came in a staggering diversity: blue stars, red stars, yellow, mud brown, white, green, orange. Bright stars; dim stars; double, triple, and quadruple stars; stars in ragged gangs sparkling through dust and gas; and great clouds of stars packed so tightly into galaxies and globular clusters that their glow combined into one smooth smoosh of light. Stars barely larger than the earth. Stars into which a million suns would fit. But there was a secret order in the chaos of the sky. The stars were not random.

The first clues to this secret order had come at the end of the last century, when astronomers realized that the color, or spectral type of a star, was an indication of the temperature at its surface, the outermost layer of its boiling gases.

In 1910 the astronomers Ejnar Hertzsprung and Henry Norris Russell independently got the idea to graph the magnitudes of stars against their colors, or temperatures. When they corrected the apparent magnitudes[1] of stars to account for the variations caused by distance (or looked at a whole bunch of stars at the same distance, in a cluster), a simple relation emerged: The hotter a star was, the more luminous it was. So plotted, the vast majority of stars formed up in a swaybacked line called the main sequence. The sun, a so-called yellow dwarf with a surface temperature of 5500°K, was right in the middle of the main sequence. Down at one end were dim stars, dull red in color. At the other extreme were giant blue stars, ten times as hot, and thousands of times brighter than the sun.

What was the main sequence? That was the big question in astronomy.

The prevailing idea early in the century was that the main se-

---

1. The astronomical system for ranking star brightness derived from the ancient Greeks, who divided the visible stars into five classes or "magnitudes." First-magnitude stars were the brightest in the sky, like Sirius or Vega, the real sparklers. The dimmest stars that the eye could discern were fifth magnitude. After the telescope brought even dimmer stars into view in the seventeenth century, the system was standardized. Each magnitude step represents a ratio of about 2.5 in brightness. First-magnitude stars are thus 2.5 times as bright as second, which in turn are 2.5 times brighter than third-magnitude stars. Five magnitudes is a factor of 100 in brightness. A further step in standardization is to correct the magnitudes of the stars as seen or measured on the sky—their *apparent* magnitudes—for the effects of the different distances of the stars. Astronomers define the *absolute* magnitude of a star as the magnitude it would have as seen from a distance of 10 parsecs (32.6 light-years). Unfortunately, the distances of relatively few stars are known accurately.

quence charted the life path of a star. Like sparks sputtering down a fuse, stars would burn their way along the main sequence, starting out blue and luminous at the top, ending up cool and dim at the other end. Like athletes they had brilliant youths and then faded away. This theory remained vague, because astronomers had no idea how stars generated their prodigious energies. Moreover, other mysterious features decorated the H-R diagram, as it was called: anomalous clumpings of stars with unusual qualities, for example, red stars that were also very bright—so-called supergiants—and blue stars that were dim. What were they? How were they related to main sequence stars? What *was* the main sequence? Astronomers used the diagram as a diagnostic tool. They could plot on it each new clump of stars they found and notice that this or that part of the usual pattern was missing—perhaps there were no bright blue stars in some cluster—but astronomers didn't know what it meant. It was as if doctors studying the people of some town found that the population was normal in all respects—except that there were no fat, blue-eyed people. Clearly understanding the reason for this anomaly would tell them something new about medicine or people. Similarly, the H-R diagram was some sort of hieroglyph awaiting translation.

The second clue to cosmic order came in the next two decades with the growing suspicion that thermonuclear fusion was the energy source of the stars. In 1938 Hans Bethe of Cornell worked out the details for the sun (and later got the Nobel Prize). The sun, he demonstrated, is just a hydrogen bomb held together and fed by gravity. At its center, where the temperature is estimated to be 15 million degrees K, 600 million tons of hydrogen are fused into about 596 million tons of helium every second. The other 4 million tons, about 0.7 percent, are transformed into energy; they become sunshine and constitute the mortgage payment for life in the solar system and the violent winds of Jupiter.

The first crude nuclear models of stars gave impetus to a new interpretation of the main sequence on the H-R diagram. Because gravity was responsible for stoking the furnace, the theory predicted that the luminosity and temperature of a star should depend on its mass. The most massive stars should burn brighter and hotter and exhaust themselves more quickly. The main sequence was, therefore, really a mass sequence, with 100-solar-mass stars on top and 0.1-solar-mass stars on the bottom. In this scenario, as stars aged they didn't move along the main sequence— they mostly stayed put and then, at the end of their lives, moved *off* it, puffing into red giants or cooling to white dwarf cinders. As a group of stars aged, its main sequence would slowly disappear. The stars in the top of the main sequence would disappear first. Then the middle and finally the bottom of the main sequence would vanish. The older a group of stars was, the shorter would be its main sequence and the lower and redder would be the top of the chain.

Enter Baade with his virtuosity and his wartime monopoly of Mount Wilson. Baade discovered that there are two different "populations" of stars in the galaxy and in the universe: regular stars, like the sun, which contain hydrogen, helium, and heavier elements and another, older set of stars, which are basically all hydrogen and helium. The latter were redder and dimmer than the former and were found almost exclusively in the round central bulges of galaxies (as opposed to their bluish, twisting spiral arms) and in the spherical clouds of stars called globular clusters.

In the East, and particularly at Princeton where Martin Schwarzschild, the guru of the H-R diagram, lived, theorists realized immediately that Baade's discovery was confirmation of the idea that the main sequence was a mass sequence. Galactic bulges and globular clusters—what Baade termed "Population II" stars—lacked the massive blue stars because they were older than other parts of the galaxy. The blue luminous end of their main sequence had already disappeared. (In fact, there had already been reason to believe that Population II stars and globular clusters were older than the rest of the galaxy: Both were distributed in the shape of a halo, as if they had formed when the galaxy was still a round cloud of gas just beginning the collapse that would result in the modern flattened disk known as the Milky Way.)

Globular clusters, then, were a test tube in which to study the evolution of stars, how they changed when they ran out of hydrogen. It dawned on Schwarzschild and the others that they could calculate the age of a globular cluster (or any other group of stars) and, by extension, the age of the galaxy and perhaps the universe, if they knew exactly how far down from the top this process of disappearance from the main sequence had proceeded, where it was cut off. Massive stars evolved faster and thus had shorter lifetimes, so the main sequence was also a sequence of life expectancies. As time went on and the main sequence of a cluster got shorter, the cutoff point represented stars whose projected lifetimes were the same as the age of the cluster. The more massive stars that had been above the cutoff had shorter lifetimes and had already gone away; stars below it still had billions of years to burn. Just find the mass of the stars at the cutoff and, using the new theories of stellar structure and evolution, calculate their life spans—that was all.

Of course, those were almost impossible observational and theoretical tasks. The samples of Population II stars—globular clusters and galactic nuclei—were all so far away, and the stars themselves were so dim, that the properties of the individual stars in them—their magnitudes and colors—had never been assayed and plotted on an H-R diagram. That was where Sandage came in.

In 1950, as Sandage soldiered through the observations and analysis of M92 with Arp and then M3, a bigger cluster in Canes Venatici, alone for his thesis, he was largely unaware of these new ideas. He thought stars still evolved *along* the main sequence. Baade had initiated Sandage and Arp into the mystery of the H-R diagram when he took them to Mount Wilson the first time, but Baade was just an observer. As an observer himself, Sandage felt that whatever was going on *inside* the stars was beyond his purview or his competence. His job was to measure colors and magnitudes. He wanted to connect the dots on the H-R diagram, and his intuition told him that his globular clusters were important. He didn't know how important.

Even in the early fifties, the big discoveries in cosmology happened not at the telescope but on a piece of graph paper when a table of data from nights or from years on the mountain was put into a form the eye could read. Calvinist work, worthless if not performed with picky precision, and work that was, therefore, perfect for young Sandage. For all its glory, it was still tedious work. In practice the process began with at least two plates being exposed for each cluster—one with a blue-sensitive emulsion, another with red-sensitive. Two piles of pepper, and the same stars would have to be located and measured on each plate.

Sandage measured about a thousand stars and plotted the main sequence of the globular cluster M3, but it was a funny sort of diagram he wound up with. The bottom half looked like the H-R diagram of any star cluster in the galaxy, but the top half of the main sequence was missing. All the yellow to blue luminous stars were simply gone. In their place, a funnel of dots ran off the top of the abbreviated main sequence off to the right of the diagram, where the stars known as red giants lived.

One day in 1952 Sandage walked upstairs from the basement at Santa Barbara Street with the H-R diagram of M3 in his hands and ran into Baade and Schwarzschild, who visited Pasadena regularly, probably hoping for just this kind of moment. "When Schwarzschild saw the main sequence cutoff he got very excited," Sandage recalled.

Sandage had found the missing link. In his hands were the data that Schwarzschild needed to solve the problem of stellar evolution not just in principle but in numerical detail. A few days later he came around to Sandage and asked him to come to Princeton for a year.

Sandage didn't get it at first, but finally it sank in: the sky was alive with connections. The stars fell into his hands like candy from a busted piñata. The thrill has never left him.

"You only had to make one simple statement, that stars as they changed their magnitudes moved *off* the main sequence instead of *up* the main sequence," he said. His eyes danced and his fingers played an imaginary piano along the edge of the tablecloth as he recalled the glory days of stellar evolution. "All this stuff was just lying out there in the fifties, ready to be taken. I can remember the tremendous excitement when we first found the main sequence cutoff of the globular clusters and the turn-off point. Suddenly everything fell into place."

Schwarzschild's plan was to turn Sandage into a theoretical physicist for a year. Together they would follow the trail of those dots on the H-R diagram with mathematical computations of nuclear interiors of stars, trying to match their results to the colors and magnitudes that Sandage had so studiously measured. At Princeton, Sandage traded his telescope for a mechanical calculating machine. He hated it.

To "solve" the interior of a star meant that you had to solve four complicated differential equations of physics at every point outward from its core. The calculator they were using was just a glorified adding machine. All the elaborate theory and calculus had to be reduced to an endless series of multiplications. *Punch punch punch,* turn a crank to perform the mechanical computation. It was not only time-consuming—it was noisy. *Punch punch punch. Ka-chunk. Punch punch punch punch ka-chunk.* The theory was terribly difficult, which was perfect for Sandage's Calvinist psyche, of course, because it had to be done.

It took hundreds of hours for each calculation, and it took dozens of calculations to map a significant part of the life of a star, even the seemingly boring eons when it was burning hydrogen into helium in its core and barely changing its outer appearance. Through it all Sandage thought of himself as Schwarzschild's clerk, just turning the crank.

After a year Sandage had finally reached the point in his work at which the core of the star begins to run out of hydrogen, its fires bank, and the core collapses with a jolt that puffs out the rest of the star. The star cools and reddens, and leaves the main sequence. It becomes a red giant, with a radius maybe the size of the orbit of Venus and a core of helium. For a star as massive and as hot as the stars at the top of what was left of the main sequence in M3, Sandage calculated that this transition would occur at an age of 3.2 billion years. That meant that the cluster itself had to be at least 3.2 billion years old.

Sandage and Schwarzschild had originally hoped to get much further. All the wild and interesting times in the life of the star were still ahead of it, all the loops and countertracks on the H-R diagram, but it would be twenty years before Schwarzschild and his theoretical models made it to the end of the line. A journey begun on mechanical adding machines would end on air-conditioned mainframes.

Nevertheless, the research was a triumph. Sandage could calculate the ages of stars. It worked, the H-R diagram worked, stellar evolution was explicable. Physics ruled. The stars were old but not impossibly old, the galaxy was old but not infinitely old. The universe could be plumbed.

"The world had always been rational," he explained, "in the sense that the equations of physics were always real, in the sense of Plato's archetypes; the world is not irrational, and I think the brain can understand the logic of the universe. But I felt an elation, that something so understandable as the H-R diagram was really understandable by the laws of physics, by chemical compositions of the stars and just giving the stars enough time to age."

The importance of Sandage's work went beyond, far beyond, stellar evolution and the niceties of the H-R diagram. It added a new dimension to the infant field of cosmology. The universe couldn't be younger than its oldest stars, which meant that the universe itself was at least 3.2 billion years old. That was an amazing thing to be able to say from looking at a relative handful of stars in a corner of the galaxy. Allan Sandage had measured the age of the universe. In 1954 it got him his first *New York Times* headline.

There was another way to estimate the age of the universe. Although the data for the expanding universe were still sketchy, it was possible to calculate backward from the rate at which the galaxies were speeding away from each other to find out how long ago they had all been together—the hypothetical moment when the expansion started. The answer was 4 billion years.

Here were two independent clocks. One, based on the embryonic science of nuclear reactions, said the universe was at least 3.2 billion years old. The other, based on the equally embryonic science of distancing galaxies, said the universe was no more than 4 billion years old. Neither number was accurate enough to take seriously as a final answer, but to Sandage the fact that they were so close bordered on revelation. It helped forge an almost unshakable faith in his soul that the expanding universe must be right. Within the generous limits of astrophysical error the two answers were the same. One could have turned out to be thousands of years, the other quadrillions. It was a miracle, or it was science's witness to a miracle.

Or it was an embarrassment. Few astronomers liked to talk about the idea that the universe had to have a beginning. Hubble and Einstein both had thought it was nutty. That was the weakest, most incomprehensible aspect of the expanding universe theory. It wasn't science; it was theology.

Not Sandage. He called it the Creation Event. The Hounds of Heaven had never stopped chasing him. In the fifties he read a popular book about existentialism called *The Outsiders,* by Colin Wilson. The outsider, said Wilson, drawing from the works of Sartre and Dostoyevsky, is someone who lives in a hell of too much consciousness, bounding from gloom to exaltation, insect lust to angelic sacrifice, while all round him others seem to sleepwalk. His only hope, Wilson suggested, is religious acceptance. Sandage recognized himself—he was the outsider—and he pressed the book on his friends.

Few scientists up till then had been willing to confront the beginning seriously enough to try to tackle scientifically the issue of what it had been like. One notable exception was a Russian-born astronomer and nuclear physicist named George Gamow. Gamow suggested that the early universe, in addition to being squeezed together and dense, was also very, very hot—millions or billions of degrees. He conceived of the beginning of its expansion as a thermonuclear fireball—the bomb to end all bombs—in which the original stuff of creation (which he called *ylem,* a dense gas of protons, neutrons, electrons, and gamma radiation) was transmuted by a chain of nuclear reactions into the variety of elements that make up the world today. In the forties, Gamow and a group of collaborators wrote a series of papers spelling out the details of thermonuclear Genesis. Unfortunately their scheme didn't work. Some atomic nuclei were so unstable that they fell apart before they could fuse again into something heavier, thus breaking the element-building chain.

Gamow's team disbanded in the late forties, its work ignored and disdained. Some of them left science. Gamow himself retired and wrote popular books on science and cosmology that influenced the next generation.

Ultimately and perhaps ironically, it was the very revulsion for the idea of the beginning—thermonuclear or otherwise—of one scientist in particular that led to the completion of the great work on the life, death, and rebirth of stars, which Baade and Sandage and Schwarzschild and all the others had begun. The person who did it was Fred Hoyle.

An English astronomer and physicist with a froglike face behind thick glasses and a blunt manner, Hoyle absolutely detested the notion that the universe had a beginning. He compared it once to a party girl jumping out of a birthday cake; it just wasn't dignified or elegant. During a BBC broadcast Hoyle referred to Gamow's theory derisively as "the big bang" and was amused when it became standard terminology.

Hoyle had been born and raised in Yorkshire, in the north of England, the son of a struggling textile merchant; he attended Cambridge on a full scholarship, nearly penniless and mocked for his working-class accent and political conservatism by his fellow students. The experience left him embittered and suspicious of all establishments—social, scientific, and otherwise—a professional outsider. He was bombastic and he liked to pick intellectual fights. "I never waste time seeking the solution to a problem along conventional lines," he told *Discover* magazine, "because if the solution were to be found in that way, somebody would have done it already."

In 1946 Hoyle and two other upstarts, Thomas Gold and Hermann Bondi, had invented an alternative to the expanding universe that they called the steady state. In it the universe had neither beginning nor end. The idea was that as the galaxies spread out, new matter would pop into existence in the empty space left behind and coagulate to form new galaxies. On the large scale, then, the universe would always look pretty much the same.

When it was published in 1948, the steady state theory captured the hearts of scientists for whom the majesty of the universe depended on its being eternal and infinite. Hoyle's theory was considered more elegant than the big bang, and that was a quality much revered by physicists. Even Hubble regarded it as a healthy development. Having two rival theories to choose between sharpened the task of cosmology. In 1948, Hoyle visited Hubble in Pasadena and was cordially received. According to Greenstein, the physicist Charles Lauritson told Hoyle at a party, "I don't believe a word of what you say, but have a drink."

In 1953, Hoyle returned to Pasadena, trailing bad blood from Cambridge University, where his brash, argumentative style had not gone down well with the clubby dons. In England the cosmological debate had turned into a savage feud between Hoyle and a fellow professor, Martin Ryle, a tweedy liberal radio astronomer, easily given to tears.

Sandage had learned most of what he knew about the steady state theory from reading about it in the Sunday supplements in the newspapers. Initially he was not impressed with Hoyle and his gang, one of whom, Bondi, was notorious for a set of Murphy's Law–type pronouncements he called theorems.

"Bondi," Sandage recalled, chuckling dryly, "said that you shouldn't do any experiment now because two years from now you can do it much better. . . . [Bondi's] first theorem is that whenever there is a conflict between a well-established theory and observations, it's always the observations that are wrong. And when he announced the first theorem in England and it was read by the Mount Wilson astronomers, they just dismissed all the steady state boys. That was the beginning of the rejection. Bondi was associated with Gold and Hoyle, and they made such outrageous statements that they just couldn't be believed."

When the bad boy Hoyle came back to Pasadena, however, he came not so much as a maverick cosmologist but as a stellar physicist. It was on that basis that he ultimately gained Sandage's respect.

Hoyle realized that the steady state model shared with its big bang rival the problem of explaining the origin of the chemical elements like carbon, oxygen, gold, iron, nitrogen, uranium, and lead. Nature, scientists assume, always takes the easiest path. If matter was going to appear out of nothing—either at the beginning of time in enormous quantities and densities, or bit by bit, over eons, in the intergalactic voids—the odds were overwhelming that it would first appear in its simplest and most fundamental form: elementary particles like protons and electrons.

Both theories had to explain how and where the chemical elements had been built out of the more fundamental particles. How, both groups recognized, involved thermonuclear fusion, which required high densities and temperatures. Hoyle wanted to show that you didn't need a single big bang nuclear reactor at the beginning of time to account for the creation of the elements and proposed that all the nuclear transmutations actually took place in stars. In stars, thermonuclear furnaces that burned for billions of years, there might be time for nuclear reactions to take place that didn't have a chance during the few seconds in which the putative big bang was hot and dense enough to be thermonuclear.

For example, one of the gaps in Gamow's chain of reactions occurred when either a proton or neutron was added to helium, which contains two protons and two neutrons; the resultant nucleus of five particles fell apart before it could be used to make anything heavier. In 1952 a Cornell scientist named Edwin Salpeter had shown (using Sandage and Schwarzschild's computations) that in the cores of red giant stars, where freshly minted helium was plentiful, three nuclei might on rare occasions collide and stick together to form a carbon nucleus. The gap at atomic weight five would, therefore, have been bypassed.

On its own, Hoyle calculated, Salpeter's process was too slow to account for the amount of carbon in the universe today. But Hoyle, a crack nuclear physicist, realized that carbon production would be speeded up sufficiently if the carbon nucleus could be demonstrated to have certain properties. Nobody had thought to investigate this aspect of carbon, but the steady state theory depended on it.

So Hoyle showed up in 1953 on the doorstep of Caltech's Kellogg Radiation Laboratory. It was an old brick building. In the basement was a jungle gym of vacuum tubes and bulbous chambers that looked like old diving bells in which atomic nuclei could be fired at one another and the properties in nuclear reactions measured. Here the predisposition of the carbon nucleus to be made from three heliums could be examined. The head of Kellogg was Willy Fowler, a grizzled and stocky extrovert from southern Ohio who liked to pretend that he was just a hard-drinking country boy when in reality he was a powerful and shrewd physicist.

The experiment succeeded. Carbon had exactly the properties

necessary for it to be produced abundantly in stars. Fowler was convinced Hoyle was a genius and went back to England with him. Both Hoyle and Fowler returned a year later with a pair of English physicists in tow, Margaret and Geoffrey Burbidge. Geoffrey, portly with Falstaffian features and a fondness for waving a cigar in the face of his opponent while making a point talking a mile a minute, was a theoretical physicist; Margaret, petite with delicate features, was an astronomer.

Driven by Hoyle's vision, a four-way collaboration ensued on the quest to explain the origins of elements. Its members referred to their group as if it were a chemical formula, B^2FH—two Burbidges, a Fowler, and a Hoyle. Splitting their time between Cambridge and Pasadena, they slowly hammered out the history of the elements, a process that culminated in a long paper in 1957 in *Reviews of Modern Physics*. The B^2FH paper, however, was regarded less as a deathblow to big bang cosmology or as the clincher for the steady state theory (in fact, the words *steady state* do not even appear in the paper) than as a culmination of the ideas about stellar evolution, the work driven by fifty years of Mount Wilson observations. It laid out a new view of the cosmos, of the galaxy as a dynamic evolving organism, of stars that were not just autonomous balls of fire but an interacting community, breathing the interstellar wind in and out.

The story as now believed goes like this. The original stars were primordial hydrogen, the simplest element, and helium. An average star, like the sun, would spend a few billion years burning most of its hydrogen into helium and then fizzle out. In more massive stars, whose cores were squeezed under gravity's grip to extreme temperatures, the helium might ignite and burn into carbon and oxygen, while the carbon in turn could ignite and form neon, sodium, and magnesium. Depending on the mass of the star, neon and oxygen might subsequently burn, in a process that would go all the way up to iron, the most stable element, with each successive stage of burning lasting a shorter period of time. The star becomes layered, like an onion, with silicon, nickel, oxygen, sulfur, and neon. When the core finally poops out, unable to get hot enough to ignite the next round of fusion, it collapses; a shock wave rebounds and blows the outer layers of the star into space. In the most extreme case, that of a supernova, a galaxy's worth of energy erupts through the star, triggering one last frenzy of thermonuclear reactions, which make rare and heavy elements and throw them to the galactic winds.

The ashes from these various explosions, rich in heavy elements, drift and mingle with the clouds of gas and dust that clot the arms of the Milky Way and from which new stars are endlessly condensing. The whole process repeats itself, each succeeding generation of stars containing a slightly higher percentage of heavy elements, or metals (the astronomers' technical term for the elements heavier than helium). Stars and galaxies, in this scenario, were hydrogen's way of bootstrapping itself into oxygen and iron and nitrogen and carbon, into the constituents of life. The chemical evolution of the galaxy sputtered along in a series of nuclear explosions.

The thought of it made Sandage giddy. We were sitting in a restaurant by the beach in San Diego when he recalled that moment, his voice suddenly hitting the high register in the middle of a discussion of Chesapeake Bay seafood. "What's so amazing, it seems to me, is . . . you and I," he exclaimed. "Every one of our chemical elements was once inside a star. The same star. You and I are brothers. We came from the same supernova. Well, maybe many supernovas have gone into the mixing of a human being," he mused, then brightened.

"We were all together in the same nebula."

Sandage became friends with the Burbidges and, through them, Hoyle. He went to bat for Margaret against the Mount Wilson establishment when she was barred from using the telescopes. Women had not been allowed on Mount Wilson, by Hale's decree, since 1905. To observe Burbidge had to pretend to be her husband's assistant, and they had to stay in an ancient unheated cabin away from the Monastery. In the end, Mount Wilson was sexually integrated.

Once, when Sandage was away, the Burbidges and Hoyle stayed in his house, and Hoyle had a disastrous experience with Sandage's washing machine. For years afterward, Hoyle claimed half-jokingly that Sandage had sabotaged the machine as revenge for his heresies.

Hoyle's heresies and his personality plagued a brilliant career. His conflict with Ryle caused him to quit Cambridge in a rage when he was passed over to run its Institute of Astronomy. And his later espousal of such ideas as life originating in outer space and raining to earth from comets probably cost him a share of the Nobel Prize. In 1983 Fowler shared the Prize with Subrahmanyam Chandrasekhar who was being honored for work on stellar evolution.

Sandage reflected, "Hoyle, I think, was and is more interested in all the worlds that could be instead of the world that is. Anything that is logically accepted within the world of physics even though you have to change the laws a little bit is of interest for him to speculate on, whereas Hubble was an absolute empiricist, asking, 'What is it really like?' "

The cosmological debate continued nonetheless. It was okay with Sandage. Steady state, he thought, was a silly theory, but it was the universe that would ultimately decide.

3

THE WAR OF THE
WORLD MODELS

The Catholic Church had been quick to take sides in the big bang–steady state debate. In 1951, opening a conference at the Vatican, Pope Pius XII pronounced that the big bang was compatible with official Catholic doctrine. That was nice, but scientifically it didn't mean anything. All decisions were premature. Sandage knew that the so-called expansion of the universe, all the great debates so seductive to newspaper readers and religious mystics, rested on a slim reed of evidence gathered in the 1920s and 1930s. The decision about what kind of universe we live in, he knew, would rest on data he was now beginning slowly, systematically, to accumulate. Around the middle of the 1950s it began to take him over. By the end of the decade he would be the new Hubble, the man who knew more about the universe than any other human.

Sandage told the story of the discovery of the expanding universe as if it were an old family tale, the sort that gets told and retold every time the relatives gathered for holidays. In a way it was a family tale; the company of cosmologists in those days was small enough to fit around a dining table. To Sandage the history of cosmology was full of little moral lessons.

The first was that, as Pasteur said, "Chance favors the prepared mind." Back when he was working for Shapley as a night assistant, trying to learn astronomy, Humason had taken some plates of the Andromeda Nebula, M31, and brought them to Shapley. On the other side of the plate from the emulsion, Humason had marked in ink a few pricks of light that he said might be variable stars. Shapley, who believed that the Milky Way was the entire universe and that the spiral nebulae were local whirlpools of gas, wiped the marks off as he explained why they couldn't possibly be stars. "Shapley, having argued that the nebulae were not separate galaxies," said Sandage, "wasn't prepared to see the evidence that they were."

A few years later, Hubble identified specks of light similar to the ones Humason had found and that Shapley had erased in Andromeda and a handful of other nebulae. They were, in fact, a kind of star known as Cepheid variables. Cepheids (after Delta Cephei, the first one discovered) had the convenient property that their light output rose and fell in a regular sawtooth pattern whose period was inversely proportional to the star's average absolute magnitude—that is to say, the more luminous the Cepheid, the slower it pulsated. All Hubble had to do was measure the period and apparent magnitude of the Cepheids to calculate their distance. When he did that in 1924 the answer came out to 900,000 light-years. The true scale and identity of M31 as a full-fledged galaxy emerged. What was a smear of light a few degrees in diameter on a plate was in reality a luminous pinwheel 100,000 light-years across.

Being a gentleman of the Mount Wilson school, he sent a courtesy note to Shapley at Harvard before he announced his discovery. "And you know," said Sandage dryly, "about that animosity."

The expanding universe followed five years later.

Having established *what* galaxies were, Hubble naturally moved on to ask how they were distributed in space (a question his scientific descendants would be grappling with half a century later) and if, and how, they were moving. Once again he found himself the beneficiary of an earlier failure of nerve or imagination.

The first astronomer with the opportunity to discern the flight of the galaxies had been Vesto Slipher. He had been hired by Percival Lowell—a Boston aristocrat and fancier of Martian life—mainly to look for canals on Mars. Slipher also did some real astrophysics, however, and by 1914 he had collected spectra[1] of about thirteen of the mysterious

---

1. As a scientific tool, the invention, or discovery, of spectroscopy in the nineteenth century rivaled the telescope itself. Just as music from an orchestra is a blend of many sounds—trebles and basses and harmonies—so the light from a star or any light source is rarely "pure," but is usually a blend of light of different wavelengths. In spectroscopy, a prism or diffraction grating is used to "unblend" the light and spread it out into its different wavelength components, the way the sun's light is spread into a rainbow by raindrops, so that the physicist or astronomer can discern which wavelengths are present and which are absent.

What we call visible light consists of electromagnetic waves ranging from wavelengths

spiral nebulae. He was looking for evidence that the nebulae were rotating, as they should be if they were whirlpools condensing into stars—as one theory had it. He found something stranger.

When light from a star like the sun is dispersed into a rainbow spectrum, dark lines appear at certain of its wavelengths—always the same wavelengths—like notches cut out of the star's light. These were the wavelengths at which particular atoms in the star's gaseous body absorbed light. The signature of calcium, for example, appeared at the blue end of the spectrum as two dark lines that the pioneer spectroscopist Fraunhofer had labeled the H and K lines. In Slipher's spectra of the nebulae the whole pattern of lines was displaced from its laboratory wavelengths, like a mask that had shifted, slid down the spectrum. By itself that was no big deal. Light is a wave, and physicists had long known that light from a moving source changes its wavelength as perceived by a stationary observer, the way a car engine sounds higher when the car is approaching and lower when it's going away. The phenomenon is called a Doppler shift; astronomers had already used it to analyze the motions of stars and even the motions of different layers of material inside stars.

Objects coming toward us have their spectral lines shifted systematically to blue, shorter, wavelengths; objects going away show red shifts. The bigger the shift, the larger the relative velocity of approach or recession.

Stars in the galaxy dance in and out with respect to the sun; some have blueshifts, some redshifts. One might expect a similar random pattern from the nebulae. That was not what Slipher's data showed.

With the exception of the Andromeda spiral, all the nebular spec-

---

of around 4000 angstroms (0.00004 centimeters) at the blue end of the spectrum to around 7000 angstroms (0.00007 centimeters) at the red end.

Joseph Fraunhofer pointed a spectroscope at the sun and he found that the spectrum was broken by a myriad of sharp dark lines, like the cracks between your teeth, as if a notch had been cut out of the sun's light at these wavelengths. When spectroscopists trained their new instrument on burning or glowing gases in the laboratory, they also found lines in the spectrum—a different and unique pattern that constituted a "fingerprint" for each gaseous element they investigated. Calcium, for example, announced its presence by a close pair of lines at the blue end of the spectrum. Hydrogen produced a distinctive, harmonically spaced set of lines; going from blue to red (or short wavelengths to long) the wavelengths jumped from line to line by a constant factor of four to three. That was a pattern that every astronomer memorized, because hydrogen was the most abundant element in stars and the universe. It was by recognizing such familiar patterns of lines in the spectra of stars and the sun that astronomers learned that the rest of the (visible) universe is made of the same stuff as is the earth.

The explanation for spectral line patterns came in the 1910s and 1920s with the advent of quantum mechanics. According to that theory, atoms can only emit or absorb energy in discrete bits called quanta, determined by their internal atomic structures. This means atoms can emit or absorb radiation only at the frequencies or wavelengths that correspond to those quantum energies. If the atoms are situated between us and a light source, as are the atoms in the outer layers of a star, they will absorb radiation and leave dark bands in the spectrum. If the atoms themselves are being excited, as in a flame or neon light and certain kinds of gaseous (nonspiral) nebulae, they will emit radiation and their spectra will have bright lines in them.

tra were shifted slightly toward the red, to longer wavelengths. If this was a result of the familiar Doppler shift, it meant almost all the nebulae had pretty high speeds and were racing away from the earth, from the Milky Way.

A puzzled Slipher reported on his findings to the American Astronomical Society in 1914. He kept on collecting nebular spectra and velocities for the next ten years, but attempts to understand the overall pattern foundered on astronomers' ignorance of the true nature and scale of the nebulae.

In the twenties, following on his work on M31, Hubble was able to divine distances to a few dozen galaxies—or groups of galaxies—which until then had been smears of light on the dome of the night. He used Cepheid stars in the closest galaxies as a starting point and stepped outward. The most distant in his sample was a giant cloud of more than a thousand galaxies in the constellation Virgo whose distance Hubble estimated as 5 million light-years.

It was when he innocently compared the distances of the nebulae to Slipher's unpublished spectra (and a few of his own) that Hubble fell on what would be the prime fact of the twentieth century, the most amazing scientific discovery of all time—the first one that pointed beyond science altogether—the cipher that would haunt him and Sandage and future generations of cosmologists and drive them to the sky clawing for patterns in the darkness between the galaxies, seeking the signatures of particles, forces, dimensions, and energies only a heartbeat from eternity, demanding from the blank crumpled night some explanation of the origin and fate of time itself. All this from a cloud of dots on a graph, but for Sandage it was like an amnesiac confronting the marks on the closet door where he had been measured when he was a boy, reminding him that he had been young once, and small, that he had been born and perhaps therefore would die.

The pattern that Hubble divined was unexpectedly, disturbingly, simple. The farther away a galaxy was, the farther the absorption lines in its spectrum had been shifted to the red. If these redshifts were simply due to the Doppler effect—the most conventional explanation—it meant that the farther away a galaxy was, the faster it was flying away from our own. The most distant galaxies in his sample —in the heart of the Virgo cluster, which he put at 5 million light-years—were also receding the fastest, at 1200 kilometers per second.

Hubble looked at that cloud of dots and drew a straight line through them, thereby inventing (or discovering) what would forever be known as the Hubble law and the Hubble expansion: A galaxy's distance was proportional to its "redshift velocity."

Every time in later years that Sandage saw that graph, with its messy cloud of dots, reproduced in textbooks and journals, he marveled. "Science is full of false clues," he said. "There are more false clues than correct clues. Hubble had the incredible ability to go through the maze of false clues and his plates were not very good. He was not a good observer, but he unerringly went to the truth each time. Most people who say they're objective always reach the wrong answer. You have to know what to *neglect* and what not to neglect. Experimental science is not as pure as most people are led to believe."

What did the Hubble law mean? If it were right, at the very least Hubble would have discovered a powerful method for mapping the locations and distributions of galaxies throughout the universe: Just take a spectrum, measure the amount of redshift, and multiply by a simple numerical constant to get the cosmic distance. More important, if the Hubble law prevailed universally, then he had uncovered a central fact of nature. Gone forever would be the vision of a static unchanging universe that had prevailed since antiquity. In its place would be a dynamic evolving universe; nature would resemble nothing so much as a gigantic explosion. The sanity of astronomers depended on finding out whether Hubble was right.

Hubble and Humason, his observing lieutenant, commenced a kind of cosmic leapfrog. Find some gauge of the distances of galaxies, measure the redshift displacement of their spectral lines, and check to see if distance was still proportional to velocity. Then go farther still.

But the deeper they went into the sky, the more desperate and ad hoc became the distance criteria and the longer it took to record the requisite spectrum of the galaxy. Sometimes it took all night for the dispersed light from a single galaxy to seep into the photographic emulsion while Humason hung on the telescope like a human counterweight to keep it tracking smoothly.

"Hubble would send Humason out to get the redshift. Hubble got the distances," recounted Sandage. "And they marched forward. From 1929 to 1931—two years—they went from speculation all the way out to 20,000 kilometers a second. Hubble knew what it was almost immediately just because the angular sizes of the galaxies were changing in an inverse proportion to redshift depth. He knew he had something great. He knew it was the most crucial discovery ever made in science."

By 1935 Humason had measured a cluster of galaxies in the constellation Ursa Major that were traveling at a speed of 26,000 kilometers per second—one-seventh of the speed of light. The Hubble law still held at that distance; the universe was still expanding, if that was indeed what it was doing.

The slope of the line of Hubble's graph—the ratio between redshift velocity and distance—was a number that became known even before

Hubble's death as the Hubble constant, H_0. Its value in 1929, in the quaint measurement systems of cosmology, was 530 kilometers per second per megaparsec[2]—which is a mouthful, but just means that for every million parsecs (roughly 3.26 million light-years) farther out in space a galaxy was, it was going 530 kilometers per second faster. Hubble estimated the uncertainty in his value as about 15 percent.

Einstein, in fact, had already predicted that the galaxies were just like raisins in a rising cake, being pushed apart by the mysterious explosion of space and time themselves. But Hubble had done his work in ignorance of Einstein, and he had been reluctant to draw such a grand conclusion from his own flinty data. He cautioned, in fact, that the redshifts might not be classic Doppler shifts at all, but some new physics, and perhaps should be thought of as "apparent velocities." There were perhaps 100 billion galaxies in the sky, and so far the "expansion" of the universe was based on less than 200 of them.

However hesitant he was to make pronouncements about the implications of the redshift measurements, they had, he realized, given him an empirical yardstick to lay out the local cosmic surroundings. This was the universe that Sandage grew up into: The sun lay near the outer edge of the Milky Way galaxy, a flattened pinwheel of 100 billion stars. Attended by a pair of foggy concentrations of stars called the Magellanic clouds, the Milky Way anchored one end of a cigar-shaped collection of a dozen and a half galaxies traveling together through space called the local group. The other main members were the Milky Way's twin, M31, the Andromeda Nebula, and a smaller spiral called M33.

Similar groupings of a dozen or so galaxies—a big spiral and the crumbs left over from its formation—were scattered through nearby space. These were all dwarfed by the great cluster of galaxies in the constellation Virgo, where more than 1000—maybe 5000—galaxies swarm. It was the most distant landmark for which Hubble had been able to intuit an independent distance.

Beyond Virgo the only surveyor's stick was the so-called redshift velocity, and even those were hard to come by because the galaxies were so faint. Only the most brilliant in each cluster were accessible. Out there were more great clusters in Hercules, Coma, Ursa Major, Perseus, and Centaurus. On photographs of these clusters, the greatest agglomerations of matter in the universe, island universes swarmed like flies on a window:

2. For historical reasons, astronomers measure distances in parsecs. One parsec is the distance at which a star's position in the sky would appear to shift by an angle of one second of arc when it is viewed six months apart, from the opposite ends of the earth's orbit around the sun. It is equal to about 3.26 light-years or around 20 trillion miles. Such parallax measurements are the most reliable indications of a star's distance and form the basis in one way or another for the entire cosmic distance scale, but they are possible only for the nearest stars, astronomical positions in general being accurate only to roughly one-tenth of an arc second.

tiny pinwheels, blobs, slivers of spirals seen edge on, spiders with their legs tangled.

In the deepest Mount Wilson exposures, galaxies outnumbered stars and looked like soft burrs stepping lightly, gray level by gray level, into the night. From his counts of numbers and magnitudes, Hubble concluded that they were distributed uniformly; astronomers had penetrated the sky far enough, he said, to achieve a reasonable sample of the universe as a whole.

Hubble had found that galaxies came in a limited number of forms. Most galaxies were either ellipticals—reddish, round, and featureless—or spirals, like the Milky Way and M31. He had further subdivided spirals into normal ones and those in which the arms came off the end of a central bar. Spirals, 80 percent of the total, were found mostly loose or in small clumps like the local group; ellipticals lived in the large clusters like Virgo.

A diagram of these galaxy types looked like a tuning fork. Down the handle of the fork ran the ellipticals, the roundest, E0, at the end, the most elongated, E7, in the middle, where the two spiral tines branch off from something called an S0, a flat disk of a spiral with no discernible arms and a big nucleus. The spirals, barred and unbarred, ran from *a* types, with tightly coiled arms and bulgy centers, to *c* types, with loose spangled arms and anemic cores. It was the dream, or maybe just the conceit, of the tiny core of galaxy observers like Hubble that this tuning fork diagram would play the same role and be understood in the same way for galaxies as the H-R diagram had been for stars. But nobody knew enough about galaxies to know for certain that that would be the case. Did they evolve? If they did, which way? What made them look the way they did?

Hubble had ended his book *The Realm of the Nebulae* on a poetic and prophetic note. "Thus the explorations of space end on a note of uncertainty. And necessarily so. We are, by definition, in the very center of the observable region. We know our immediate neighborhood rather intimately. With increasing distance, our knowledge fades, and fades rapidly. Eventually, we reach the dim boundary—the utmost limits of our telescopes. There we measure shadows, and we search among ghostly errors of measurement for landmarks that are scarcely more substantial."

Sandage repeated those words with a chill, and a premonition of fatigue in his voice, like a sailor hearing some distant dangerous siren call.

One day in 1954 the siren call finally came in the form of Milton Humason, who had dropped by Sandage's office with a request. In a sense, he wanted Sandage to take over, to become Hubble.

In the twenty years since *The Realm of the Nebulae* had been published, Hubble, Humason, and Nick Mayall, an astronomer at the University of California's Lick Observatory near San Jose, had been engaged in a gargantuan effort to reach farther out into the universe, and reach the great clouds of galaxies in a more systematic way than Hubble had originally done. From the various catalogs compiled by Messier and his descendants, they had made a list of 850 galaxies visible from the Northern Hemisphere and had painstakingly measured their magnitudes and recorded their spectra to determine their redshift velocities. Mayall had done the brighter and, therefore, easier galaxies with Lick's 36-inch-diameter reflector. Humason did the fainter ones with the giant mirrors of Mount Wilson and Palomar. It was the most extensive survey yet of the universe. Hubble was supposed to have done the analysis of the cosmological meaning of all these data.

Now Humason and Mayall were ready to publish their data, and Humason asked Sandage to pinch-hit for Hubble. Sandage felt like an imposter and a little guilty at the prospect of coming in after twenty years of work and getting his name instead of Hubble's on the major cosmological work of the decade, and he had a little stage fright. On the other hand, "It was tremendously exciting to be asked to be involved in what was the most fundamental connection between the expanding universe and the observational data."

Sandage's specific job was to analyze the relationship between the apparent magnitudes of galaxies and their redshift velocities. Suppose for a moment that all galaxies in the universe had the same intrinsic luminosity—as if they were all light bulbs of the same but unknown wattage—so-called standard candles. In that case, the relative brightnesses of galaxies would be inversely proportional to the squares of their relative distances. Stated simply, a galaxy that appeared one-fourth as bright as another one on a photographic plate was twice as far away and, if the Hubble law held, should be moving outward twice as fast. Another that appeared one-hundredth as bright should be receding at ten times the velocity. The universe, Sandage knew, was not built to such exacting tolerances, but with large numbers of galaxies the Hubble law should show its hand as a statistical trend. And it did. On average, he found, that as the apparent magnitudes of galaxies rose (and then got fainter),[3] their redshift velocities rose proportionately, just as the Hubble law said it should. The data were noisy, like Hubble's original graph, but he could draw a straight line through his points. Humason and Mayall's data reached out to velocities of 100,000 kilometers per second—one-third of the speed of light—and Hubble's law still held, the universe still seemed to be exploding in every direction at the same rate.

3. Remember that fainter objects have higher astronomical magnitudes.

Moreover, there was a subset of galaxies that did appear to have been manufactured with General Electric precision. The survey included eighteen "rich" clusters, each containing thousands of galaxies. Such clusters contained large numbers of elliptical galaxies: reddish round or cigar-shaped conglomerations of stars resembling globular clusters but much larger and brighter. In every large cluster there was one elliptical—the "first-ranked" elliptical—that outshone the others; often it was at dead center of the great galaxy cloud. The first-ranked ellipticals appeared to be the most massive and luminous of all galaxies known. Suspecting that they represented some sort of natural limit on the size of galaxies, Hubble and Humason had speculated back in 1931 that, whenever they appeared, these galaxies might be intrinsically the same: nature's standard light bulb.

Sandage plotted the data for the eighteen first-ranked ellipticals separately and found that their magnitudes were exactly proportional to their velocities. It was a perfect fit with the Hubble law. They all fell on a straight line. They were all peas in a pod, as he liked to say.

To be sure, this was a circular argument. Sandage was using the Hubble expansion to prove that the giant ellipticals were uniform and was also using the giant ellipticals' uniformity to prove the expansion of the universe. It was consistent, but if the redshift velocities had some other source, the whole scheme would fall apart. The only way to prove absolutely that the ellipticals were uniform was to measure their distances (and thus their absolute magnitudes) independently of their redshifts.

However, Sandage realized with a sinking heart, the distance scale set up by Hubble was in abysmal shape. In the leap outward through the realm of the nebulae Hubble had resorted to more and more desperate gauges of distance. First it was Cepheid variable stars, but they were too dim to be seen in galaxies beyond the local group. Farther out, Hubble had relied on the brightest stars in each galaxy as standards. Finally he had to use whole galaxies.

Baade had made the first correction to Hubble's distances in 1952. Now Sandage found more problems. Hubble had confused gas clouds for red stars, and the stars by which he calibrated these measurements of stars and clouds and galaxies had been themselves mismeasured. It just showed what a genius Hubble was, Sandage said, that he had been able to discover anything at all.

In an appendix to the Humason-Mayall-Sandage paper, Sandage reported in 1956 that the distances of the galaxies were actually three times larger than Hubble's original estimate. In effect, the known universe was three times bigger. He concluded that the Hubble constant was therefore more like 180, measured in the units of kilometers per second per megaparsec. Making the universe three times bigger also made it three times older, to about 5.5 billion years. The reason was simply that at their measured redshift velocities it would now take the galaxies three times as long to spread apart to the new corrected threefold separation. Sandage had started a trend that he would continue for the rest of his life.

"You see," he explained once, his voice deliberate and almost apologetic, "Hubble started off at 530, and if you have God saying 530, it takes a long time to walk down Sinai. We were a tenth of the way down when 1956 came and Humason and Mayall got their redshifts out and in Appendix C we had to do something. That was what we did. It was the first teeny step of getting the calibrator and distance."

When the report was published in 1956, it cemented Sandage's role as the successor to Hubble. "Birth of Universe Traced to Blast" read the headline in the *New York Times*. Mount Wilson promoted him from "associate astronomer" to "astronomer."

A year later he was awarded the Helen Warner Prize, given for outstanding work by a young astronomer, by the American Astronomical Society. He had to give a prize lecture. "It was titled 'Current Problems in the Extragalactic Distance Scale,' " he mused. "We had not gone very far down the road to getting new data, which had begun in 1950 but hardly any of which had been reduced. In fact none of them were reduced. I attempted with the knowledge then available to correct every step in Hubble's chain. I said in that paper that by applying all these corrections the Hubble constant was 75. Now whether I said I think it could be 50, I don't know.

"It was clear that the campaign had to be taken directly to the sky with a vengeance."

Sandage was invited to give a series of lectures at Harvard in 1957. While he was there, freed temporarily, for the first time in years from monthly commutes to the mountaintop, he met Radcliffe graduate, Mary Connelly, who was teaching at Mount Holyoke. Connelly, like Sandage, was a midwesterner. She had gone to the University of Indiana as an undergraduate and then attended Radcliffe. They were married in 1959 and bought a house in Altadena, in the foothills climbing up to Mount Wilson east of Pasadena; that was the house with the washing machine that snared Hoyle.

Connelly's astronomical career, except for a few chores helping her husband with his math or editing a compendium, fell prey to the usual domestic pressures. Sandage once explained it like this: "She found it was very hard to stay in astronomy with me, because of the internal pressures I put on her. I tend to make quite strong demands about getting things accomplished, and it was just best that she do something else."

In time there was a pair of sons, David and John. Sandage once listed his hobbies as music and gardening, although later he became very interested in cooking. In fact he had little time for any of it, except listening to the opera on Saturday afternoons. After all he was gone half the month. Their happiest time as a family, he thought, was when they all went to Australia together for fourteen months and lived at Mount Stromlo Observatory while he observed southern galaxies.

Hitting his stride as an astronomer, Sandage was now grabbing down observing projects in the late fifties with both hands, long extensive ambitious jobs, any one of which would make an astronomer's reputation, but which taken together constituted nothing less than an attempt to rebuild astronomy. If he was going to solve the universe, if he was going to be Hubble, everything had to be perfect. Somebody had to calibrate the Cepheid variable stars. Someone had to calibrate the main sequence. Someone had to really test those assumptions about elliptical galaxies. Someone. It had to be him. He had a monopoly on the world's largest telescope. He took on a variety of partners and ran them into the ground.

As Baade and Humason slowly phased down their observing activities, their plates, as Hubble's had before them, accumulated in Sandage's office rather than in the Mount Wilson plate vault—a basement room under the stairs in which file cabinets were stacked to the ceiling. He kept those plates in a special file cabinet to which he had the only key. "When Baade and Humason retired, the entire nebular department consisted of me," said Sandage. "From then on, it was a matter of keeping up the good fight."

Sandage knew that it would be a long haul and that he was unprepared. It was one thing to follow Hubble's instructions—it was another thing to *be* Hubble and take charge of the cosmological quest. In the late fifties he had set out to relearn cosmology and general relativity, to understand and master the equations he had grazed in graduate school. He dreaded it. "Now, you have to remember that one of my teachers in graduate school was H. P. Robertson," Sandage explained. "H. P. Robertson was the greatest theoretical cosmologist of his time. He came from Princeton to Caltech in the last two years of my graduate work, so I took a class in mathematical methods of physics from him. He was there as the driver and Hubble was there as the driver, and I felt responsible to try to understand some of this stuff."

The theory of cosmology and the expanding universe had developed independently and largely in ignorance of the data and observations of Hubble, the observers in the West being obsessed with getting the most out of their telescopes and the theorists in Europe being concerned mainly with mathematical elegance—a division that has characterized the cosmological quest to the present day.

Einstein had been the first one down this dusty road; in 1917, a year after he invented general relativity, he had tried to apply it to the universe. According to general relativity, matter or energy (which were equivalent) warped space and time like a heavy sleeper sagging a mattress. The universe, Einstein realized, was the ultimate sagging mattress. He proposed that the weight of the whole cosmos could wrap space-time back around on itself, rather like the surface of a balloon. This resolved the ancient and philosophically embarrassing riddle of what was beyond the end of space, what was outside the universe. In Einstein's scheme there was no end, no outside. Shoot an arrow or a light beam infinitely far in any direction and it would eventually come back and hit you in the butt.

But there was a problem with the curved-back universe. Such a configuration was unstable, it would either fly apart or collapse. Einstein didn't know about galaxies. He thought, and was reassured as much by the best astronomers of the time, that the universe was a static cloud of stars. To explain why his curved universe didn't collapse like a struck tent, therefore, he fudged his equations with a term he called the cosmological constant, which produced a long-range repulsive force to counteract cosmic gravity. It made the equations ugly and he never really liked it. That was in 1917, twelve years before Hubble showed that the universe was full of galaxies rushing away from each other.

When Einstein heard about Hubble's discovery, he discarded the cosmological constant, calling it the worst blunder of his career. At the crucial moment Einstein had lost faith in the beauty of his own beautiful theory. (What he should have suspected, of course, was the "evidence" that the universe was static. His contemporary, the inscrutable Eddington, said that no experiment should be believed until it has been confirmed by theory.) Had he stuck to his guns, Einstein would have made one of the greatest predictions in the history of science, that the universe is dynamic, that is, expanding or contracting.

Meanwhile, in the early twenties, a wan Petrograd (now Leningrad) theorist with a drooping moustache, Alexander Friedmann, excited by general relativity, had done Einstein's work for him. First of all, he pointed out that Einstein had made a mathematical error and that the cosmological constant did not even work in stilling the universe. Without the cosmological constant, Friedmann demonstrated, Einstein's equations described a dynamic universe, an organism that, through the forces of gravity swelled and shrunk like a heartbeat. There were three possibilities for the shape that that universe could take: space on the cosmic scale could be curved convex, concave, or flat. Each of these geometric possibilities had its own predetermined history that came with it. Geometry, in this case, was destiny.

The convex universe was like Einstein's original spherical space-time, closed on itself. This closed universe began at zero time with zero radius. It expanded to a maximum radius of curvature and then contracted again back down to zero. The time it took for this cosmic cycle depended on the mass of the universe. Friedmann guessed a mass of a billion trillion (10^{21}) suns for the universe and got 10 billion years for its life span—not a bad number for the day. It was possible, Friedmann suggested, that such a universe could rebound again and again.

The second possibility was what mathematicians call an "open" universe, a concave curvature somewhat like the inside lip of a trumpet. Arrows or light beams shot out into this space would never come back. It was difficult to draw pictures of the space in this cosmos. Space had no limits. The radius of curvature started at zero and just kept growing forever. Galaxies in this universe would never come back together, but would disappear from one another into an ever-expanding night.

In between the two, balanced like a knife edge between immortality and recontraction, was a universe in which the overall large-scale geometry was flat, Euclidean. Space would still be locally curved around massive objects like the sun or clusters of galaxies. Mathematically it was the simplest of what were admittedly simple toy-model universes to start with; that gave it a special cachet among theorists. It was attractive enough that Einstein and Willem de Sitter, a Dutchman who had also taken a turn with the cosmological equations, formally adopted it in 1932. The flat universe would slow to a stop, but only after an infinite amount of time. In this scheme it was also possible for some cosmologists (the ones for whom a recollapse of the universe was a cheery thought—like a cosmic reunion) to believe that after another infinite length of time the universe would come back together, and thus to think of a flat universe as a potentially closed universe. It would expand forever but not a day longer. (This was a notion that would confuse them and the public for the next fifty years.)

Einstein quibbled with Friedmann's math and eventually admitted that, while it was technically correct, he was not sure that this notion of the universe beginning at zero radius had any physical meaning. Friedmann died of typhoid at the age of thirty-seven, and his work remained lost until a Belgian cleric, the Abbé Georges Lemaître, reinvented it independently in 1931. Lemaître had heard of Hubble's work and was the first to suggest that the expansion of the universe was real and caused by an explosion. The mystical Eddington played a central role in elucidating and popularizing this cosmological upheaval. During the next two decades theorists like Sandage's old teacher Robertson had reformulated and simplified the mathematics of cosmological theory and astronomers learned a new way of regarding the sky.

What does it mean to say the universe is expanding? Few statements sound so glib and cause so much trouble when people try to visualize what is really going on. Most laymen think of the galaxies as exploding from some point and flying outward through space. According to Einstein's general relativity, however, it is really space that is exploding, carrying the galaxies along like twigs on a current. A popular analogy is a raisin cake baking in the oven. As the yeast expands the dough, the cake gets larger, carrying the raisins farther apart from each other.

Although we see individual galaxies receding at different speeds, space expands at the same speed everywhere in all directions at once. In fact it is this uniform expansion that gives the illusion that everything is racing away from *us*. Suppose the universe doubles in size in an hour. In that time span a galaxy originally 1 mile distant will become 2 miles distant and thus seem to be moving away at 1 mile per hour. But a galaxy 10 miles away at the start of the hour will be 20 miles away at the end—it will seem to be going ten times faster than the other one.

These cosmic speeds are all relative. At the other end of the universe or of the raisin cake, a galaxy or a raisin would see itself at rest and its own neighbors moving modestly while *we* would appear to be dashing at high speed.

In the new expanding universe theories, the notion of redshift, the shifting of light toward longer wavelengths in distant galaxies, took on a broader and more powerful meaning: It became the primary marker of time and distance. Technically, the redshift, a quantity astronomers denoted z, was just the fraction by which the wavelengths of spectral features of a star or galaxy had been increased over their normal "laboratory" wavelengths. A redshift of .05, for example, meant the wavelengths had increased by 5 percent; a line that normally appeared at 5000 Angstroms was shifted to 5250. For small values—much less than 1.0—the redshift is proportional to the velocity of the receding galaxy.

Cosmologically, there was a simpler way of thinking about the redshift. As the universe expanded, light waves within the universe were stretched along with it. The redshift that Sandage measured in a distant galaxy was simply the amount by which the universe had expanded between the time the light left the galaxy and the time that it impinged on a photographic emulsion on Palomar. A redshift of 1.0, for example, meant that the universe had doubled in size during that time; according to some models, he was seeing light that had left a galaxy when the universe was half its present age, halfway back to the putative creation. The actual distance or look-back time to the galaxy depended on things like the Hubble constant, which astronomers didn't and don't know very well.

*   *   *

An astronomer who wanted to understand the history and geometry of the universe had only two kinds of data he could gather from the galaxies: redshifts and magnitudes. Sandage set himself the job of determining which of the Friedmann universes, if any, we inhabited by finding out how each of those world models would manifest itself in redshifts and magnitudes on the sky, through the majestic lens of the 200-inch telescope.

He sweated and stumbled through the mathematics. His wife helped him with his calculations, just as Einstein's first wife allegedly helped him with the equations of relativity. "So I had to learn what I was trying to do," Sandage explained. "I tried to learn some theory, and lo and behold, it wasn't quite as hard as it seemed. In fact, nobody had made the calculations. The calculations seemed easier and easier the further I got into it. But it took me five years, very hard work, from 1956 to 1961. But at the end of that period I had some insight, some intuition."

The galaxies in the expanding universe, as Sandage folksily envisioned it, were just like a bunch of rocks thrown in the air. If they were thrown fast enough, like a moon rocket, they would escape and never fall back. If they were too slow, if the universe's collective gravity was too strong, they would fall back down.

In their final formulation by a German mathematician named Mattig, the Friedmann equations were deceptively simple, given their portentous subject matter. At any point in cosmic history you only needed to know the value of two cosmological parameters to characterize the universe and know its type and destiny. One was the Hubble constant, H_0, which told you how fast the universe was expanding and gave a scale to the cosmos. The other number defined its shape—open, closed, or flat, gracefully curved or pathologically bent. It was called the deceleration parameter and designated q_0 (q-nought). Technically q-nought was a measure of how fast the expansion of the universe was slowing down, how the gravitational drag of the contents of the universe was slowing the flights of the galaxies. The universe was born, either to be free, or to yo-yo back to the cradle.

The magic value of q_0 was 0.5. A higher value meant that the universe would eventually end—Sandage's "rocks," the galaxies, did not have enough energy to get away. A lower value indicated a so-called open universe that would never stop expanding. Exactly 0.5 gave you the flat simple universe, one that coasted to a stop at infinity; in this case the galaxies had been thrown outward, improbable as it may seem, at precisely escape velocity.

In his papers, which appeared like a trumpet blast in the *Astrophysical Journal* (*ApJ*) at the beginning of the sixties, Sandage described cosmology as the search for two numbers. Could the universe be as simple as the Friedmann models? "I think it's a lovely poem," he said. "It's a closed problem, a problem that has solutions in a closed way: given H_0, q_0, you know everything about the geometry of the universe." His voice retreated into its dry tone. "That's marvelous, if true."

To Sandage the most beautiful and amazing models were of the oscillating universe. With an oscillating universe the world both ended and never ended. The cosmos was a cycle, but the cycles were embedded in eternity. The universe forever arose out of its own ashes.

"There were two or three papers on the oscillating models," Sandage said fondly. "Not that they made any sense, but they were just so beautiful, expanding and contracting."

The models were deliciously specific. Take a closed universe with q-nought equal to 1.0. The span from big bang or creation event or initial singularity or first light, if you will, to the final burning crash of recollapse—space-time Armageddon, last light—would be 82 billion years. That was a long time, but not forever. By 1961, Sandage's estimates of the age of the universe had risen to be about 11 billion years, which left about 30 billion more years of expansion, or cosmic progress, before the whole ensemble, taking its marching orders from the curve laid on it at birth, would begin to regress and fall back into chaotic fury.

Just for fun Sandage tried to calculate what a future astronomer living in those times, say 50 billion years from now, when the balloon of the universe was deflating, could see. "Those models are fantastically interesting," he gushed, "because in the light travel time you see the universe at different times. If there's an oscillation, you can see out so far that you see back in time before the expansion stopped. So you can see some redshifted galaxies at great distances whereas the nearby ones are blueshifted. From the most distant galaxies, that information that the universe is contracting has not yet had time to reach you." He exhaled. "Fascinating. So just as an exercise to keep myself off the streets, I did all those neat calculations as an amusement. And they were all published."

His calculations were in a sense his initiation rites, his official acknowledgment of having taken up the burden of determining what kind of universe we live in. The observational program that Hubble had laid out divided cosmology naturally into two almost-independent parts: one to measure H_0, the Hubble constant, which tells how big and old the universe is, and the other to measure q_0, which meant determining the curvature of space to find out if it was a sphere, the inside of a tulip, flat, or something else. If you could see far enough, Friedmann and Einstein promised, you would see the fantastic molding of the cosmos, you could see the mind of its architect made manifest in the sky.

One way to measure the curvature of space was to simply count the numbers of galaxies at ever increasing distances. How the counts rose would tell you how the volume rose and whether the geometry of the universe was warped. Hubble had tried to discern the cosmic geometry from such "nebular counts" back in 1936 and reported that it appeared to be positively curved, that is, closed. He and Sandage had begun to repeat the measurements when Hubble had his first heart attack. The problem with counting galaxies was that the farther out one looked, the more that faint galaxies fell below the threshold of detectability and went uncounted. "So the whole thing in principle is fine," Sandage griped, "but in practice it decays."

There was another way, in principle more tractable, to detect the deceleration of the expansion of the universe directly, and that was where Sandage decided to place down his chips. It was called the Hubble diagram, the same test Sandage had performed with the Humason and Mayall data to show that the universe was expanding evenly.

The Hubble diagram, as mentioned earlier, was a variation of the Hubble law, which predicts that a galaxy's redshift velocity is proportional to its distance. Suppose again that there were standard candles, bright enough to be seen at great depths in the universe, so that their relative distances could be gauged accurately by their relative brightnesses. On the Hubble diagram the redshifts of some purported standard candles were plotted against their apparent magnitudes.

According to the Hubble law these should lie on a straight line, but only for a while. The Friedmann equations predicted that over very large distances, at truly cosmic depths in the universe, the Hubble law would break down, just a little. And what would have been a straight line should curve.

The reason was gravity. Like rocks tossed in the air, the galaxies should be slowing down as they rushed outward, braked by the combined gravity of everything in the universe. The universe should have been expanding faster in the past than it is today. The light from very distant galaxies, Sandage reasoned, has been traveling to us for a long time—the farther out we look, the further back in time we look. Therefore, we should see these galaxies as they were during an earlier epoch when the universe was expanding faster; they should be traveling slightly faster than the Hubble law would predict. Exactly how much faster would reveal how fast the universe was decelerating and thus how it was curved.

The beauty of the Hubble diagram was that Sandage didn't have to know the absolute distances or absolute magnitudes of his standard candles. He just had to be able to judge *relative* distances from apparent magnitudes. To do that he had to know that he had a true standard candle, that something *here* and now and something way over *there* and then were intrinsically the same.

"And so," he said, "the search for q-nought is the search for twins."

Sandage concluded that the Hubble diagram was the battleground on which the war of the world models would be fought. The Humason, Mayall, and Sandage redshift survey had given him a standard candle, one he could pursue far into the intergalactic shadows. Nature was being very kind, indeed, if it had made the biggest and brightest galaxies also the most uniform.

On the Hubble diagram the ellipticals fell on a straight line, but that was only for now, for the local cosmic moment, for a mere eighteen clusters in the neighborhood. If Sandage could extend that simple graph far into space, follow the galaxies around the curve of the cosmos, the line of the Hubble diagram should bend. He would be seeing back to a time when the universe was expanding more vigorously than it is today and when the galaxies were receding faster.

On the extended Hubble diagrams that he drew for all those *ApJ* papers, then, the theoretical universe curled away from that boring monotonic straight line—one way for the closed universe, another way for an open one, a more radical deviation entirely for the steady state universe, which could also be fitted into the two-number straitjacket. On the plane of the Hubble diagram, a multitude of manifolds flowered and parted company, laying tracks for the galaxies to discriminate between.

Sandage just had to drive far enough out into the universe to extend the diagram to the point where the tracks betrayed themselves and diverged, cleanly, if they ever did from the shadows of the ghostly errors, and committed themselves to one universe. He was a digger with a treasure map and a shovel not knowing how far underground—a blade's worth, a mile—either the gold or the water table that would swamp all further digging was. But on some evening in his study or in the paneled Santa Barbara Street offices, maybe by the reddened cloistered light in the 200-inch dome, plying a sheet of graph paper and a pencil, he might be the first man to know, and know he knew, the fate of everything.

The imperative was simple—follow the galaxies outward, measure their brightnesses and redshifts, plot them on the diagram, and wait for it to curve. One more thing to do during his fourteen-hour nights.

"It's like this," Sandage enthused, talking fast, eyes burning, his hands flipping imaginary plates in and out of the chemical baths, "you go to the telescope and you get a spectrogram of a galaxy. And in the old days you plop it in the developer and then the hypo and then out, and you saw those bloody K [calcium] lines march right down the spectrum. It's reality in the hypo; it's absolute reality in the hypo. The first time I ever took a direct photograph of a galaxy—I'd seen the reproductions in the books—but the first time I got it in the hypo and turned the darkroom lights on and there was M101 or M51 just like it was in the books. I cannot explain it.

"That was the contact with reality, like the plates in the hypo, like the spectrum when you see the redshift. As Eddington said, it's a few photographic grains on a plate, and out of that you deduce all this stuff about creation. Yet it works every single time. That's more real, those few grains on the photographic plate than the wave functions or the false vacuum or the gauge theories out of which the theoretical constructs come. It's an event, a pointer reading in the laboratory. Still a little far from reality, I mean it's an indication of reality but it isn't reality itself."

Watching the universe grow. It was like seeing the marks on the closet door—the reminder that he had once been a kid and small. A reminder of the Creation Event.

Sandage's public pronouncements on the problem amounted to cheery postcards remarking on the splendid view and eagerly anticipating the arrival at his destination. In 1958 he told the *New York Times* that he had seen to a depth of 2 billion light-years (by the Hubble constant of the day) and there was already some evidence of deceleration. He reported that he was temporarily blocked from seeing any farther because of night glow from intense sunspot activity, but when the sun's complexion cleared and the sky calmed down and maybe they could see to 4 billion light-years, the answer might fall into their laps.

In fact, Sandage was treading water in the late fifties for a number of reasons. The redshifts of the cluster ellipticals that had been identified and measured, by Humason and Mayall, had been measured quite accurately. Most of the uncertainty and confusion in the Hubble diagram came from the measurements of magnitude, which were done by comparing the sizes of black fuzzy blobs on photographic plates by eye. That method was about to go out of fashion. Bowen had hired a Caltech physicist, William Baum, to build photoelectric instruments for Palomar and Mount Wilson that would measure the magnitudes of stars and galaxies without the intervention of human judgment. A Hubble diagram redone with photoelectric magnitudes, Sandage thought, would be cleaner and more revealing. He didn't think it was worth mounting a serious campaign to extend the diagram until the new technology was ready and proven.

The other question was where to extend it. The universe, at the depth to which the 200-inch could penetrate, was almost completely uncharted and unexplored. Could he see far enough? How could he pick out the magic ellipticals that were his standard candle against the gray shadows of the sky at the extreme distances where the universe would show its hand?

While he was waiting for the first problem to be solved, radio astronomers came in and solved the second. Within a few years, guided by whispery electromagnetic emanations from the sky, too faint by the time they reached earth even to power a Christmas tree light, Sandage and his friends and a sudden growing band of competitors would be able to see halfway across the universe.

That the heavens were crackling and hissing with radio noise was an accidental discovery made by a phone company physicist named Karl Jansky in the thirties. It was not pursued with any vigor until after the war, when physicists and engineers flush with wartime radar technology turned themselves loose on the sky. The English led. Martin Ryle, Hoyle's bitter rival, and his students strung the fields of Cambridge with cable antenna to catch the radio signals from different part of the sky as they passed over, to measure their strength and their positions. He began to publish a series of catalogs of radio sources, and by the late fifties the numbers ran into the hundreds. The statistics, Ryle thought, suggested that most of the radio sources were far out in space beyond the Milky Way galaxy; the fainter ones vastly outnumbered the strong ones, which implied that there were more and more radio sources the farther out one went. These data were the basis for the argument he used against Hoyle's steady state theory: They implied the universe was not the same at all times and places.

Ryle's data were not good enough, however, to allow optical astronomers to identify what the radio-emitting objects in the sky actually were. Radio and light are both electromagnetic waves, but radio waves are a thousand to a million times longer. To resolve the sources of radio waves as accurately as the 200-inch resolved the sky at optical wavelengths would take a radio dish miles in diameter. To Ryle's equipment even a pinpoint of radio noise, like that from a star, came through as a hazy, indistinct blur. Telling astronomers that there was a radio source, say, in the northern half of Cygnus, was like telling the Secret Service there was an assassin in New York City: good luck.

A competition rapidly developed between radio astronomers, especially in England and Australia (where they could see the southern sky) to narrow the positions of radio sources and thus find out what they were.

Acting on a tip from the Australian group, in 1952 Walter Baade had traced a source known as Cygnus A to a strange, agonized-looking galaxy with what appeared to be two nuclei in the middle of a distant cluster of galaxies. For a while Baade thought he had discovered colliding galaxies and even won a bottle of Hudson's Bay Procurable whiskey over it from Rudolph Minkowski, the other Palomar astronomer who specialized in hunting radio sources.

More data proved the colliding galaxy theory wrong. Cygnus A was in fact a single galaxy, a giant elliptical with some sort of dark dust lane appearing to divide it. One by one the radio sources were all traced back to bright elliptical galaxies in the centers of clusters. Virgo A turned out to be M87, a huge elliptical near the center of the Virgo cluster that had an enigmatic little spike of light sticking out of it. Centaurus A was tracked to a galaxy known as NGC 5128, a ball of light that also had a dark slash of dust across its belly.[4]

In fact, most of the radio galaxies proved to be none other than Sandage's standard candles, but crackling, it seemed, with a secret violence that belied their starry symmetry, crying out electrically in the long night.

The nature of that violence was a mystery. When the distances to these galaxies were factored in (on the basis of their redshifts and the Hubble constant of the time), the static that barely stirred the electrons in Ryle's grassy antennae was revealed as a shattering roar that strained the boundaries of known physics. For a 1958 meeting in Paris, Geoffrey Burbidge calculated that to produce the energies seen in a typical radio galaxy two million suns would have had to be converted completely to energy. If nature knew how to do that, physicists did not; thermonuclear fusion was not efficient enough. The theorists went home from Paris shaking their heads.

Sandage recognized that these radio galaxies might be an easy way to extend the Hubble diagram of magnitudes and redshifts out far enough to get an answer about the universe. If they really were the same as his standard candle galaxies, the radio ellipticals would be easy to locate even at incredible distances. They were raising their hands, like kids in a classroom, vying for attention: *Here, over here, measure me. No no, over here, measure me.* Sandage became very interested in the radio sources.

According to the delicate division of cosmological labor in Pasadena, however, radio sources were not Sandage's turf. After Baade retired back to Germany in 1958 that beat belonged to Minkowski and a new Caltech astronomer from Holland named Maarten Schmidt. But all Sandage had to do was follow along in the footsteps of the hunters, once they had found a radio galaxy and tagged it with a redshift, and measure its apparent magnitude to add it to the Hubble diagram.

Late in 1959 Minkowski got a tip on the position of a source called 3C 295—"3C" meant it was from the third Cambridge compilation of radio sources, while the "295" referred to its position in the catalog sequence. His first step in identifying a radio source was taking a

4. The 5128th item in the *New General Catalogue,* a list of nebulae and star clusters begun by J.L.G. Dreyer, building on the pioneering work of the English astronomers William and John Herschel.

photograph of that location in the sky so he could see what he was dealing with. The second step was to take a spectrum of the most suspicious object and likely source, to see what it was made of or to measure its redshift if it was a galaxy. Sandage was on his way to the telescope when Minkowski got his tip. Minkowski asked him to take a picture.

Sandage was always happy to oblige. He had a whole routine for occasions like this, which involved taking two photographs. First he made a long exposure through the Ross corrector lens, which enlarged the telescope's field of view. While that would show the radio object itself, presumably a faint fuzzy blip, the lens distorted the star patterns around it. When he came to the telescope Minkowski would not be able to see the galaxy—if that was what it was—with his eye through the eyepiece; to take its spectrum he would have to offset the telescope blindly from bright stars he *could* see. So Sandage took a second photograph without the corrector. Halfway through the exposure he turned off the telescope's drive so that the stars would make trails on the plate. From the trails, which run east-west, Minkowski would know exactly how the field was oriented in the sky, and he could navigate to his invisible mystery object.

On his next trip to Palomar, Minkowski performed the necessary offsets, steering the slit of the spectrograph to what looked like a dark hole between the stars, and made a 4-hour exposure. He was rewarded by a faint spectrum, just enough to see that the object looked like a galaxy. He needed a much longer exposure to measure the redshift. It happened that Minkowski was retiring. His next run, in the spring of 1960, was to be his last. He resolved to nail 3C 295 as a retirement present for himself.

There was normally a certain amount of drama any time Minkowski went to the telescope, not just because of the unworldly objects he was chasing, but because of Minkowski himself. To accommodate his girth, the seat in the prime focus cage had been removed and a wide bench seat from an old tractor put in its place. He had a reputation for being able to break almost any piece of equipment, no matter how sturdily foolproof it was made. Once he had spent four whole nights guiding the telescope on one particular very faint object to get a spectrum and then had run into the darkroom and dropped the plate into the wrong chemical, erasing the exposure.

Minkowski was assigned four nights for his last observing run. The first two nights were cloudy. On the third he exposed all night long on 3C 295. The next night, his last, he continued the exposure. Around midnight Minkowski ended his marathon exposure and came down from the telescope. In the darkroom, on the still-wet plate, Minkowski measured the redshift from 3C 295 as .46, by far the largest redshift ever recorded. On his last night at Palomar he had set the cosmic distance record as well as the speed record. 3C 295 was retreating at 46 percent of the speed of light.

Sandage happened to be on the mountain that evening with the night assistant Robert Sears. "We sat in the library while he developed the plate," Sandage recounted, "and we could tell from the sound of his footsteps coming down the corridor that it was a success, because he was just bounding down the corridor, and he opened the library door with a bottle of bourbon in his hand and three glasses."

Sandage was thrilled. Sandage was blown away. Later he calculated that, according to what model of the universe he used, the "lookback time" to the galaxy 3C 295 was between a third and a half the age of the universe. Sandage measured the galaxy's magnitude—the photoelectric devices were up and running—and added it to his Hubble diagram. 3C 295 made a lonely point far out in the upper right-hand corner. From it he made a preliminary estimate of q_0, the deceleration parameter, and got a value of 1.0—seemingly indicative of a closed universe. But Sandage knew that the diagram was full of uncertainties and assumptions yet to be tested about elliptical galaxies. When he tried to apply these corrections, the case between an open and closed universe became too close to call, although it did look bad for Hoyle's steady state model. In the observatory's annual report he wrote tersely, "the problem remains unsolved."

Sandage was discouraged. They still had to see much farther if the Hubble diagram was ever going to bear unequivocal fruit. But how could anyone match or exceed Minkowski's feat? He was only halfway home, and already Hubble's ghostly shadows of the sky were closing in on him.

4

BONFIRES ON THE
SHORES OF TIME

In the summer of 1960, not long after Minkowski's spectacular retirement feat, a young Caltech radio astronomer came to Sandage with a proposition. He had an unpublished list of radio sources that he wanted Sandage to help him identify. Sandage, who picked up projects like this as a magnet picks up iron filings, accepted in an almost blasé manner. But this was no ordinary list—it should have trembled and glowed in his hands—and in accepting yet another little task, Sandage was in fact opening a door to the ends of the universe. Neither Sandage nor astronomy would ever be the same.

The young radio astronomer's name was Tom Matthews. He had come to Caltech in 1956 after getting his Ph.D. at Harvard and had helped build a pair of 90-foot-diameter radio dishes in the Owens Valley, east of the Sierra Madres, which he was using to hone the positions of the Cambridge radio sources. Matthews was also an enterprising member of the radio astronomy grapevine. At a recent workshop a friend had given him the unpublished list of measurements, which he had made at Jodrell Bank in England, of the diameters of radio sources.

Radio diameters—or how smeared out in the sky the radio signal appeared—were potentially important. Typically most of the emission from a radio galaxy came from a pair of lobes that flanked (and dwarfed) it like the ends of a dumbbell. The farther away, the smaller the dumbbell structure should appear. That meant that small sources might be very far away.

Matthews noticed right away that Minkowski's object, 3C 295, was one of the smallest radio sources on the Cambridge list and, within the limits of the radio telescope, was virtually indistinguishable from a point. Its double structure, if it had one, was too small to be resolved. And it had turned out to be the most distant galaxy in the known universe. Matthews culled his list for other radio sources that were bright and very small, thinking they might turn out to be equally spectacular and faraway galaxies.

He came up with ten. None of them had been identified optically because their positions weren't known accurately enough yet. Matthews next looked at plates of the National Geographic Society–Palomar Sky Survey—which comprised a photographic atlas of the heavens—and didn't find anything that looked like obvious radio galaxies. So he took his list to Sandage. All Sandage had to do was photograph a few promising patches of sky with the 200-inch telescope for him. "There was absolutely no science involved," Sandage joked.

In the late summer, during his August-September run, Sandage pointed the 200-inch in the direction of the coordinates Matthews had given him for a source called 3C 48, in the constellation Triangulum. In the listings it had no measurable radio extent at all. Matthews took the exposed plate and plotted the location of the radio source relative to the stars. The coordinates fell right on a tiny sixteenth-magnitude dot that, under the magnifier, appeared to have a little wisp of nebulosity attached to it. A single star had never been a radio source before. He gave the plate back to Sandage a few days later with the comment, "Gee, a radio star."

In October, Sandage went back to Palomar and slewed the big mirror back to the "star" in Triangulum. The first thing he did was the standard astronomical diagnosis. What kind of star was it? He used the photometer to check out its colors. This involved measuring the magnitude, or intensity of the star's light, through filters that isolated wide chunks of the spectrum. The star gave back strange color readings; it was radiating abnormal amounts of energy in the blue and ultraviolet bands. Next he took a detailed spectrum and found the lines were unreadable. "It was like nothing that had ever been seen at that time," he recalled. "I think I took a total of some five or six spectra, and came back and measured the positions of the lines and it made no sense at all." The wavelengths of the spectral lines matched the fingerprints of no known elements.

Sandage took his spectra to Ira Bowen, Mount Wilson and Palomar's former director and a stellar whiz. Baffled, Bowen suggested showing it to Greenstein, who also couldn't make anything out of it. But Greenstein considered himself something of a crack decoder of pathological stars. The strange spectrum of 3C 48 was a challenge to his pride. He kept wrestling and fiddling with it. On his own trips to Palomar he obtained more spectra. Somewhere along the line, he considered the possibility that it was a normal spectrum grossly redshifted—more than most galaxies—and then dismissed it. Stars didn't exhibit redshifts of that magnitude. Greenstein worked up some elaborate model in which the star was the naked burning hot core of a recent supernova.

Meanwhile Sandage was monitoring 3C 48 and discovered that its brightness varied by as much as half from one fortnight to the next. That clinched it, as far as Sandage was concerned, that 3C 48 had to be some kind of weird star. Stars varied, galaxies never did. A galaxy was the massed light of billions of stars spread over 100,000 light-years; for it to waver or brighten appreciably in a short period of time, billions of stars had to quench or flare in unison. The variability suggested that 3C 48 could not be more than a couple of light-weeks in diameter, a few times the size of the outer solar system.

Sandage flew to New York, where the American Astronomical Society was having its annual winter meeting, burst to the podium unscheduled, and announced the discovery of the first "radio star." In its account of the meeting, *Sky & Telescope* magazine reported, "Since the distance of 3C 48 is unknown, there is a remote possibility that it may be a very distant galaxy of stars; but there is general agreement among the astronomers concerned that it is a relatively nearby star with most peculiar properties."

More of these peculiar radio stars turned up over the next couple of years as Sandage and Matthews and other astronomers made their way down Matthews's list. As if to emphasize the tentative nature of Sandage's diagnosis, they were tagged "quasi-stellar radio sources," which a NASA physicist shortened to "quasars," to general disgruntlement; the name stuck anyway. Later, for reasons that will become obvious, the name changed to "quasi-stellar objects." Quasars were points of light, like stars, everyone agreed, but nobody knew what they *really* were or even if they were made out of ordinary matter.

Sandage and Matthews continued to accumulate maddeningly incoherent data on 3C 48 for three years. Sandage, who was an expert on stars and their evolution on the H-R diagram, kept measuring 3C 48's magnitudes in different color bands, trying to understand what kind of star it could be. In all that time they didn't publish anything, "because I didn't understand it," said Sandage. "And I'm so conservative that I won't publish anything unless I think I understand something about it. That's my downfall in a certain sense."

While Sandage dithered, the key that would unlock the mystery of the quasars—and overturn observational cosmology—fell into the hands of Schmidt, who had succeeded Minkowski as the hunter of radio sources. The key was a source called 3C 273. Matthews knew its position well enough for Sandage to have taken a photograph of it in 1962, but not well enough to identify any of the objects that appeared on the resulting plate as the radio culprit. Sandage had noticed a bluish 13th-magnitude star with a little spike of nebulosity sticking out from it at the center of the field, but he didn't think much of it.

In the fall of 1962 the moon passed in front of 3C 273, and by timing the cutoff of radio signals a trio of radio astronomers in Australia were able to pinpoint the location of the source. They sent the coordinates to Schmidt, who found that they fell exactly on the little blue star.

A 13th-magnitude star can be seen with a small backyard telescope; its spectrum can be recorded in a matter of minutes, not hours or nights, which is what Schmidt proceeded to do. He got the usual bizarre set of spectral lines and was about to give up on it when he recognized a familiar pattern—hydrogen lines he had seen many times in stars, but in this reading they were over on the wrong end of the spectrum, in the red. With an old circular slide rule, he quickly calculated the redshift: .16—the lines were redshifted 16 percent. That meant that the "star" was flying away at some 47,000 kilometers per second and was more than 1.5 billion light-years away. It was more distant than all but a handful of known galaxies, and if Schmidt's calculations were correct, it was the most luminous object in the universe.

While this was sinking in, as luck would have it, Greenstein walked by Schmidt's office. Schmidt told him about 3C 273's redshift— he had cracked the spectrum. And in that instant Greenstein remembered his old spectral calculations and realized that he had known all along the solution to 3C 48. It was a redshift after all. The lines in 3C 48 were redshifted *37 percent;* next to Minkowski's galaxy, it was the most distant object in the universe.

They looked at each other, checked each other's little chips of spectral film again, and then began to celebrate. Word spread rapidly around Robinson Hall and then Pasadena that something truly astounding had been found. Schmidt went home and in his imperfect English told his wife that something "terrible" had happened. Then he realized he had meant to say "wonderful."

Maybe *terrible,* with its biblical connotations of power and grace, was the appropriate word. Schmidt didn't sleep much that night.

There were objects in the sky masquerading to the eye as ordinary stars that were in reality incredibly bright and so far away as to make the distant galaxies seem like cozy neighbors to the Milky Way. How far could you see such an object—from what depths of time and distance? What were they and how many of them were there? The doors to infinity were blasting open, ancient light from mysterious beacons running down the corridors of time.

Sandage got the news by telephone that the mystery that he had discovered had been solved by someone else. He was naturally incredulous. "Now, I didn't believe that the identification was with redshifted lines at first, because it was so foreign. Everything I had ever learned about redshifts was connected to galaxies," he explained later. Within a day the evidence had become unequivocal. The atmosphere in Pasadena was electric. Cosmological revolution was in the air.

Schmidt dashed off a paper to *Nature* titled "3C 273: A Star-like Object with Large Redshift." Greenstein, listing Matthews as a coauthor, rushed off another paper on 3C 48. They were printed in the March 16 issue along with a paper by another Caltech astronomer, Beverly Oke, detailing photoelectric inspections of 3C 273, and one on the Australian radio observations.

According to general relativity, there were only two possible explanations for so large a redshift, Schmidt explained in his note of eight paragraphs. The first was gravity in a very dense massive star; in that case, some of the spectral features of 3C 273 that required very rarefied conditions would be impossible to explain. That left the other possibility: "The stellar object is the nuclear region of a galaxy with a cosmological redshift of 0.158, corresponding to an apparent velocity of 47,400 km/ sec. The distance would be around 500 megaparsecs, and the diameter of the nuclear region would have to be less than 1 kiloparsec [3000 light-years]. *This nuclear region would be about 100 times brighter optically than the luminous galaxies which have been identified with radio sources thus far.*"

Sandage, the discoverer of quasars, was left on the sidelines watching his colleagues stampede into print with barely a nod to his own work. The failure—Sandage's crushing failure—to solve the outstanding observational puzzle of his career and to discover a whole new class of cosmological objects had been compounded by a massive invasion of his turf. Outwardly he was correct and calm; inwardly the volcano was seething.

Greenstein got the message in a fairly direct way. A picture of Greenstein had hung for some time on Sandage's office door. The next time Greenstein went over to Santa Barbara Street he saw his picture hanging in shreds on Sandage's door. The door was closed.

"At Mount Wilson and Palomar the tradition was that you didn't intrude into somebody else's field unless you thought they were idiots," Greenstein explained, reminding me of the venerable gentleman's code in Pasadena. The quasars, 3C 48 in particular, had been Sandage's stars.

Greenstein realized, to his chagrin, that in pursuing the spectrum of 3C 48 on his own and then leaping into print with the solution to it without at least including Sandage, he had violated that unwritten code. Greenstein had rushed in too quickly to solve Sandage's star for him.

"During this time Sandage had been working on the colors of 3C 48, but he could charge," Greenstein paused and continued softly, "a terrible charge, that I, a metapapa, had helped destroy his work."

For Greenstein the quasar imbroglio represented a double pain. Not only had he sandbagged his greatest student, but he, too, had failed to recognize the answer to the puzzling spectrum of 3C 48 even after he had solved it. It reminded him of why he was disinclined to cosmology in the first place. There was something more than science at stake. "All astronomers are trying to find the beginnings," growled Greenstein. "These are big stakes. Quasars hurt too much. I left them."

Matthews, too, admitted that he felt like he had been trampled. It rankled him a little bit that Sandage had been hailed as the discoverer of quasars and now Schmidt was hailed as their explicator. He wound up collaborating with Schmidt for a while and eventually left quasar research.

Even before Schmidt's unmasking of the true nature of quasars, the energies of ordinary radio galaxies were bumping against the limits of physics, as Geoffrey Burbidge had pointed out in Paris in 1958. The ingenious Fred Hoyle had suggested in 1961 that radio galaxies were powered by the gravitational collapses of million-solar-mass stars, pointing out that, according to Einstein's general relativity, such a collapse would be a hundred times more efficient than thermonuclear fusion at turning matter into energy. General relativity, alas, failed to predict what the outcome of such a collapse would be for the matter that was left, except to say that it might, well, disappear from the universe. Now quasars seemed to demand that astronomers take such a radical idea seriously. What could produce the energies of 100 galaxies in a volume not much larger than the solar system? Sandage played an important role in providing a seeming answer.

In the summer of 1963, he and Roger Lynds, a graduate student, happened to investigate a relatively nearby galaxy known as M82, an irregular-looking blob in Ursa Major. M82's core sputtered weakly with radio noise (as does the core of the Milky Way). They photographed the galaxy through a filter that isolated the pink light given off by interstellar hydrogen, thus producing a picture that emphasized not so much the stars of M82 as the gas in the star lanes. The plate that emerged from the developing tray shook Sandage. The image looked like the comic-book shards of a grenade explosion: jagged streamers of gas flowing in every direction out of M82. The galaxy looked as if it were coming apart at the seams. It was apparently exploding.

In December, Lynds showed his and Sandage's beautiful and disturbing photographs to a meeting of cosmologists, physicists, and astronomers called to Dallas to consider the frightful power and promise of quasar astronomy. The pictures wound up on the cover of the conference proceedings. In a separate paper, Sandage and the Burbidges ran down evidence, ranging from the modest core of the Milky Way to the lordly quasars, that explosions in the nuclei of galaxies were perhaps a common occurrence in the universe, the new violent universe.[1]

The conference was named the Dallas International Symposium on General Relativity and Relativistic Astrophysics. Sponsored partly by big oil, it proved to be the forerunner of a biennial event usually referred to as the Texas Symposium. It is a sort of traveling congress of cosmology and relativity theory that has been held as far from its Lone Star origins as Jerusalem. The first meeting was hosted by a small outfit called the Southwest Center for Advanced Studies, which under the leadership of a German expatriate named Ivor Robinson, had become one of the outposts for scientists interested in general relativity and gravity. In a pithy after-dinner address, Gold described it as an historic meeting, which would be remembered for "the display of strong men wrestling with even stronger facts."

"Everyone is pleased," Gold reported. "The relativists who feel they are being appreciated, who are suddenly experts in a field they hardly knew existed; the astrophysicists for having enlarged their domain, their empire, by the annexation of another subject—general relativity. It is all very pleasing, so let us hope that it is right. What a shame it would be if we had to go and dismiss all the relativists again."

Schmidt was invited to give the keynote address describing his quasar discoveries, but he was too shy and Greenstein had to give it. The meeting marked a fork in the path of cosmology. From Dallas some astrophysicists steered their attention inward to concentrate on trying to fathom the physics responsible for the inferno at the centers of galaxies— a journey that would lead to the black hole. The others were more interested in using quasars as beacons from deep time and space to find out what they had to say about the universe.

Sandage and Schmidt were among the latter group, which swelled rapidly. There was indeed a gold rush in the sky. Margaret Burbidge became a skilled quasar hunter, as did astronomers at other observatories,

1. Ironically, on investigation, astronomers eventually concluded that M82 was not exploding after all. The "explosion" was a complicated optical illusion caused by drifting intergalactic dust. The idea that quasars and radio galaxies were related to explosive events in galactic nuclei survived, however.

all over the world. Quasars were often bright enough to be seen with small telescopes; the problem was finding them in a sky already full of billions of dots. The paucity of well-determined radio positions was still a bottleneck. The lists grew slowly. Within a year there were nine quasars. In time all of Matthews's compact radio sources would turn out to be quasars. Astronomers leapfrogged each other setting redshift records. It turned out that 3C 273 was the closest—or most *recent*—quasar.

Schmidt devoted himself to searching for quasars and mulling their statistics. Again and again he went back to Palomar and launched new surveys to find new quasars. How were they distributed in time and space? What were they telling him about the evolution of galaxies and the universe? The further back in time astronomers looked, the more quasars they found. The redshifts of quasars reached gaudy heights: 1.0, 2.0, 2.5. All the quasars were long ago, faraway.

In the expanding universe, these redshifts corresponded to look-back times of 80 or 90 percent of the age of the universe.[2] Beyond a redshift of about 2.5 the numbers of quasars seemed to fall. Quasars with redshifts of over 300 percent were rare enough for newspaper headlines. For almost ten years the record stood at 3.53.

Schmidt began to suspect that astronomers were seeing the edge of the universe—not in space, but in time. If the universe had erupted once in a fiery big bang, then there could not have been galaxies forever. If he looked far enough out, he would be seeing back to the time before galaxies even existed, before they condensed out of the big bang fog and lit up the expanding space-time with stars. The only problem was, at that point there would be nothing to see.

The quasar counts told Schmidt that galaxies had come into being in the universe at the look-back times that corresponded to redshifts of around 2.5 to 3—when the universe was maybe a quarter of its present age. A few precocious galaxies had struggled to get their fires going earlier, the most distant quasars marked the boundary of the galactic era of cosmic history. Schmidt yearned for the day when he stood on that shoreline, when he had looked so deep that he could see no more quasars.

One night Schmidt stood in the dome at Palomar while the telescope was turned to train on the then most distant known quasar. He was tall, slim as a starving graduate student, with glasses and wavy gray hair. It was a cool fall evening, but he was lightly dressed for the observing ahead, in a bright red shirt and a sport coat. He said he preferred a little

2. To translate redshifts into ages and distances, according to the Friedmann expanding universe equations, requires knowledge of the values of the Hubble constant and the deceleration parameter—both of which remain controversial. The most consistent recent estimates suggest a universe of some 15 billion years old, which would mean that the most distant quasars happened some 12 to 13 billion years ago.

discomfort; it made observing more romantic. "Before the sixties, in general senior astronomers didn't let on that they didn't know something in their field," he said in a cozy Dutch accent. "That was changed a lot in the sixties when things happened that people who were so certain of themselves had no inkling of, wild things they'd never dreamed of." He paused. "Senior people now are not certain of anything."

He tapped a pair of dots on a television monitor. "What is the difference between these two?" he asked wonderingly. "One is probably a star, a few thousand light-years away. The other is a quasar. At maybe 12 billion light-years, it's the champ. In fields like this, quasars are anonymous, but they are the most distant objects in the universe. That is still one of the biggest surprises of my life." Then he rode, humming a little Bach to himself, up to the cage.

Sandage's wounded pride at having missed out on solving quasars recovered when he realized that quasars might help him solve the universe. With their enormous luminosities and hence, their visibilities over vast reaches of space and time, quasars might be the tools he needed to determine finally the shape and fate of the universe. If, by chance, they were standard candles and he could use them to extend the Hubble diagram of magnitudes and redshifts out to the edge of the theoretically observable universe, then surely the fateful deceleration parameter, q_0, the number that spelled the future of the expansion of the universe—whether the galaxies would fly away forever or come back crunching home together— would at last pop cleanly up out of the ghostly shadows of error.

In the early sixties, Sandage had already had some success in extending and improving the Hubble diagram using radio galaxies like Minkowski's object. He felt that he was closing in on an answer. The answer was that the universe was closed: Some day space and time would fall back on itself.

While Schmidt pursued the redshifts of radio sources, Sandage came along behind and measured their magnitudes with the new photoelectric photometer. By 1964, a year after the quasar revolution, he had accumulated enough radio galaxies to make a new Hubble diagram. The radio galaxies, he found out, fell on the same straight line as the first-ranked elliptical galaxies in clusters—the ones he had already determined were standard candles. That told him that radio galaxies *were* the same as the bright ellipticals; he could use them as standard candles in the universe diagram.

Perhaps the quasars were also standard ellipticals. Just for fun, he appended to that Hubble diagram the redshifts and the magnitudes of the nine quasars then known. They, too, fell on the same straight line that now continued on the Hubble diagram all the way out to a redshift of 2.0. It didn't mean anything with such sparse data, but the "formal" value Sandage calculated for the deceleration parameter, q_0, was 1.6, which if it held up, would indicate a solidly closed universe.

Sandage told the newspapers that he needed more data, maybe about thirty more quasars, and then it would be decision time. Already the data were ruling out the steady state universe, which required q_0 as measured on the Hubble diagram to be -1.0. The universe was beginning to show its hand.

There was a great caveat in all of this. The success of the Hubble diagram depended on the assumption that two standard candle elliptical galaxies at the opposite ends of space-time, billions of light-years apart, were twins. But if all galaxies had formed at about the same epoch in cosmic time—a reasonable assumption—then he was seeing the more distant galaxies in his sample as they had been when they were younger, because he was looking back in time. The light that he recorded at the telescope had left the most distant galaxies when they and the universe had been, say, 5 billion years old, while the light from nearby ones had left them when they and the universe were more like 10 or 15 billion years old. Before the fate of the universe could be read off it, the Hubble diagram had to be corrected for the possibility that the characteristics, particularly the brightness, of elliptical galaxies, might change over those billions of years.

Unable to grow a galaxy in a laboratory, Sandage tried to estimate from his knowledge of stellar evolution and studies of ancient star clusters in the Milky Way how elliptical galaxies would evolve. The common wisdom was that elliptical galaxies lack the gas and dust that are the raw material for new stars; like globular clusters, all their stars dated from the beginning. As their most luminous stars burned out, Sandage reasoned, elliptical galaxies should get dimmer and redder with age. When he corrected his Hubble diagram calculations for this putative dimming, the deceleration parameter came out lower—the more dimming, the lower it was pushed—and the case between the open and closed universe became too close to call. "However, the result is very sensitive to the stellar content assumed," wrote Sandage in a Carnegie report. "Until the stellar content is known from observational work, the problem remains unsolved."

Every new batch of redshifts, every new quasar, was the occasion for a paper and another extension of the Hubble diagram. Again and again, Sandage came to the same "formal" conclusion of a closed universe, the 82-billion-year heartbeat.

Again and again this was followed by the obligatory paragraph in which Sandage hedged his bet and explained how much (or how little) the standard-bearing ellipticals would have to have evolved to knock q_0 down to the value for an open universe, sometimes with the dismal observation that the universe may never make itself perfectly clear on this point. By and large the newspapers trumpeted the first part of the statement and buried the caveat.

In fact, despite his public pronouncements, Sandage gave other astronomers the impression that he thought the evolutionary correction was solved in principle and that the Hubble diagram would give the answer to the fate of the universe—a faith that not all astronomers shared. Eventually, Sandage figured, someone would do it right. Geez, wasn't it better to try and solve problems, even if he was doomed to be wrong, he used to say, than just retire to the nearest bar?

Hungry for more and better data, Sandage tried to think of how he could gather more quasars for the Hubble diagram. In 1964 he had a brainstorm that would lead him, he thought, to the brink of cosmic truth. Waiting for radio positions so precise that they singled out one star was too slow. He had seen enough quasars by then that he knew what they looked like—the large ultraviolet excess made them extremely bluish—well enough to pick one out of a crowd.

Sandage arranged for Ryle to send him unpublished positions of 3C radio sources. Sandage would then go to the 100-inch telescope on Mount Wilson and make a photograph centered on these coordinates—three photographs actually, through ultraviolet, blue, and yellow light filters. The field of view of the 100-inch was wide enough to encompass the uncertainty in the radio data. By comparing the intensities of the dots on the different plates, Sandage should find one near the center of each field that was heavy on ultraviolet. That would be the quasar.

Clever. Except that when he and his assistant, a research student from France named Phillippe Veron, examined the first few fields, they found instead of one blue star, four or five weird blue objects in each field. At first Sandage figured that he had made a mistake. He went back to the telescope and measured each of the blue objects with the photometer. They were blue; they all had that old ultraviolet excess.

With a little effort Sandage was able to identify the true radio source in each case. He and Ryle published a short paper announcing the identification of four new quasars, including 3C 9, which set a new redshift record. But Sandage continued to worry about all the other blue stars on the plates. He called them interlopers.

It was not the first time Sandage had encountered mysterious blue stars out where they didn't belong. He remembered that in 1952, at a meeting in Sweden on galactic structure, there had been discussions about anomalous blue stars at high galactic latitudes—the part of the sky that is far from the plane of the Milky Way. Their nature was baffling. Astronomers Guillermo Haro and Wilhelm Luyten had cataloged thousands of these blue objects, and nobody had been able to explain them. It was the sort of mystery that a good stellar astronomer like Sandage should be able to solve.

Now Sandage wondered if these interlopers were actually the Haro-Luyten blue objects and, furthermore, what they had to do with quasars, which they seemed to resemble. In December, when the Texas Symposium met for the second time, Sandage mentioned his suspicions during a talk.

Philip Morrison, a Cornell astrophysicist, raised his hand and asked, "How do you know that these blue objects are not quasars in their own right?"

"I don't," he answered.

When the meeting ended, he went racing back up to Palomar. Picking three of the blue objects at random, he measured their redshifts. One of them turned out to be the second most distant object in the universe. Bingo. By golly, as Sandage said afterward, they were quasars. Quasars with their radios turned off.

Sandage began to get very excited. There were thousands of stars on the Haro-Luyten list. Could they all be quasars? It would be the greatest explosion of observational limits in the history of astronomy. Could nature have really been so abundant? How could he find out? It would take the rest of his life to verify the whole list.

Rather than measure all those redshifts, Sandage resorted to statistics. He analyzed the numbers and magnitudes of the blue objects, and he found a break in the distribution. He concluded that the brighter stars on the list were, in fact, stars in the Milky Way. The fainter half of the sample bore the statistical earmarks of being far away. Sandage concluded they were quasars that for some reason did not broadcast radio waves. Actually, most quasars didn't make radio waves. According to his work, radio-quiet quasars outnumbered the noisy brand by 500 to 1. Quasars were like dust. Nature had been very generous indeed.

While Sandage was getting excited, in the spring of 1965, thunderous news came out of New Jersey. A pair of Bell Labs radio astronomers had discovered a uniform microwave hiss that seemed to permeate the sky. Princeton theorists had suggested that the signal was the lingering redshifted and weakened radio remnant of the big bang fireball itself. Sandage gushed in one of those Sunday wrap-up features in which he was a regular actor, "We are on the verge of knowing what happened when our universe was created—of rewriting the book of Genesis. That is clearly more important than landing a man on the moon."

The air of discovery was infectious. Back in Pasadena Sandage was bouncing off the walls as he worked on his paper about the blue objects. The other astronomers at Mount Wilson had never seen him so high. It was going to be the greatest paper in the history of astronomy, he crowed. He had discovered a thousand quasars, a thousand objects on the other side of the universe. It was like owning a thousand unicorns. "The excitement," recalled Sandage, "was otherworldly."

With 1000 quasars he now had a statistical blunderbuss with which to do cosmology, and he proceeded to brandish it. Hubble had shown that you could determine the shape of space-time just by counting galaxies at different faintness levels, assuming that faintness was an indication of distance—if you could see far enough and count accurately enough. That was how he had first tried to measure the geometry of the universe in 1936. Assuming that the Haro-Luyten stars were all what he dubbed blue quasi-stellar galaxies, Sandage analyzed the numbers on the Haro-Luyten list cosmologically. The blue galaxy counts increased with faintness in a way that suggested positive curvature, a big q_0, a closed universe.

As an added fillip, he took the three redshifts of the blue galaxies and plopped them on a Hubble diagram with the nine known quasars. That, too, indicated a closed universe. The answer, subject to the usual hedges, was getting very close.

Sandage rushed his paper, modestly titled "Blue Quasistellar Galaxies: A Major New Constituent of the Universe," to the *ApJ*. The editor, Subrahmanyan Chandrasekhar of the University of Chicago, a mathematically gifted and versatile theoretical astrophysicist, rushed it into print. The paper appeared in the May 15 issue, with the notation that it had been received May 15.

Shortly thereafter, while the paper was in press, Sandage unveiled his discovery to the Pasadena astronomy community in an enthusiastic lecture at Caltech. In the audience was Zwicky, who had been saying for years in his private convoluted language that some of the blue objects were what he called "compact galaxies," and in fact some of them had already been identified with large redshifts. It was not always easy to keep track of what Zwicky was doing. Sandage either didn't know, or had forgotten; in any event he had failed to try to find out. Zwicky was incensed. After Sandage stopped speaking, having wrapped up the universe, having made a gift of a thousand quasars to cosmology, he looked out expectantly, Zwicky slowly stood up and pulled himself to his full height. "So," he asked, sneering, "vot's new?"

The room was silent.

Sandage and his wife had been planning to go to England for the summer. They left shortly before the paper actually appeared (the May 15 issue never came out on May 15), leaving behind a press release that Caltech issued the next day. It quoted Sandage as saying, "The clues indicate that our universe is a finite, closed system originating in a 'big bang,' that the expanding universe is slowing down, and that it probably pulsates perhaps once every 82 billion years." Since the universe was then thought to be about 13 billion years old, that left 69 billion years to the apocalypse.

The press seized on the story, especially the formal prediction of the end of the universe. The only evidence that the press release or the newspapers gave in support of the end of the world was the redshifts of nine quasars and three quasi-stellar blue galaxies. In a page-one article in the *New York Times* in May 1965 Walter Sullivan wrote,

> Not only do these discoveries have great philosophical and scientific implications; it is hard to see how they can fail to influence the creative currents of our time.
>
> The realization, during the Renaissance, that the earth is not central to the cosmos deeply affected the creative geniuses of that time. If Dr. Sandage's preliminary interpretation of the data is correct, and it is established that the universe is ultimately bound to collapse upon itself, this too will have enormous implications.

So Hubble's boy had come through, had made his rendezvous with destiny.

While he was in England, however, Sandage heard distant reports that he was in trouble. In his excitement Sandage had overlooked something crucial. He had forgotten that the stars called white dwarfs, the cooling cinders of burned-out suns, also tended to be faint and blue, and thus mimicked quasar colors. The Haro-Luyten list was polluted with them. Moreover, there was a technical problem with the magnitude measurements. He had rushed in too soon. Now his greatest paper was being shot down in flames at the American Astronomical Society meeting.

A bitter reaction had set in. Part of it had to do with the understandable envy of other astronomers for the Palomar crowd and the perception that they were arrogant and enjoyed special privileges. The idea that Sandage's paper had been received and published on the same day, without outside scientific review, reinforced the stereotype. "How could that be?" asked Sandage twenty years later, his voice snarling as he imagined the indignation of other astronomers. "It was not ever sent out to referee, and that's where the enormous pressure came from the astronomical community. 'How can Sandage circumvent the whole process, write a blatant paper which turned out to be essentially right, but in its principal detail wrong?' Where it said all of the blue objects—if I'd said 'most'—all the equations are right, but every blue object is not a quasar." His voice fell off.

"And in the meantime I'd been involved in extending the Hubble diagram, to apply the classic test for deceleration. I'd been doing that for years, the answer there was q_0 equals 1. You didn't worry about evolution, so there's no question at the time I wanted q_0 to be greater than a half, neglecting evolution. And somehow in the radio-quiet quasar counts I could justify that in my mind."

A few weeks after the meeting he ran into another American astronomer, Owen Gingerich from Harvard, standing in line at the Tower of London, and asked him if it was safe to go back to America. Gingerich replied that three of his four engines were shot out, but maybe he could limp home on the fourth.

"So that was a pretty grim time, to have shown yourself too rapidly and too much to the press," Sandage said. He wrote to Horace Babcock, the Mount Wilson director, and offered to resign. Babcock wrote back and said that that wouldn't be necessary. As far as he was concerned, Sandage had a job until "the cows come home." "There being no cows on Santa Barbara Street, I assumed that was for a long time," Sandage chuckled.

"So that was my lesson in humility and pride that I've never forgotten."

Chastened, Sandage spent two years redoing the blue galaxy work, correcting the magnitudes. Overall the cosmological conclusions were little changed when he was done, although he expressed them with a tad more caution. The same image of the universe appeared in the newspapers. In 1969, a *Washington Post* headline reported "Astronomers See Universe Exploding in 69 Billion Years." The collapsing, doomed universe became part of the background music of the times.

Radio-quiet quasars, Sandage calculated when he was done, grossly outnumbered the radio kind. He had indeed discovered a new constituent of the universe, but he gradually came to the conclusion that they were useless to him for the Hubble diagram. Quasars were not standard candles; no two of them were identical, and even individual quasars varied wildly from year to year or week to week. Sandage abandoned them.

Only Zwicky never forgot or forgave. Zwicky was convinced that Hubble and his sycophants at Mount Wilson were always stealing ideas from him and failing to give him credit. Sandage's paper drove him to new heights of invectiveness. A few years later, in an introduction to a catalog of compact galaxies, Zwicky termed Sandage's paper "one of the most astounding feats of plagiarism." He noted the lack of referee (and the arrogance it implied) by suggesting that any competent astronomer in the field would have known about Zwicky's prior work.

"Again," he wrote, "disregarding all previous statistical studies made on the distribution of faint blue stars and stellar objects in breadth and depth of space Sandage advanced his own analysis, drawing from it some of the wonderful and fearful conclusions about the large scale structure and the evolution and of the universe that were completely erroneous. Among these perhaps the most ridiculous is that 'he gave the first determination of the rate of change of the expansion of the universe' as entered by Sandage himself, for the benefit of the general public, in his column in *Who's Who in the World of Science.*"

The chief night assistant on the 200-inch, Gary Tuton, was no stranger to practical jokes and liked to twit the serious-minded Sandage. For years afterward, on nights that Sandage was scheduled to use the telescope he would take Zwicky's book down from the shelf and leave it open in the dome's lounge and reading room. The next morning he would always find the book stuck back on the shelf.

Sandage's psyche simply walled off the whole incident of the blue galaxies and the collapse of the universe. When I asked him about it, my curiosity provoked by a few yellowed newspaper clippings in the library, he frowned. He said he didn't remember why he said the universe was ending, if, indeed, he had said that at all. That was another Sandage, an historical person he was vaguely related to, that he could criticize with typical Sandage gusto. "Some reporter asked about closure," he said dryly, "but that was not the issue. The issue was that quasars exist that don't emit radio waves, and, therefore, the whole plethora of quasars was bigger than we had thought. But the object was not to close the universe, I think, now. I think."

He protested that he had been misunderstood in any event; the closure of the universe had been based on quasar counts, not redshifts. "If I get the *Carnegie Yearbook* and see that it was tied to quasars then I was crazy at the time. If I was responsible for those sentences I was crazy then," he growled.

For Sandage the quasar period marked an end of innocence to the astronomers in Pasadena. They had tasted publicity and liked it. He resolved to keep his head down in the future. Thus began a long slow withdrawal from the columns of the *New York Times*.

"Those are the days when you'd come down from Palomar and everyone would expect you to come down with a pot of gold," he said in a low voice. "Sometimes, almost always, it worked: new quasars, the biggest redshift, variability, are they galaxies or are they nearby? It was exhilarating but it was also very cutthroat. It essentially split the department and split Caltech and Mount Wilson.

"Everyone was out to impress the media. The most important person in the middle was the press officer of Caltech." His hands chopped the air above a dining table, his countenance sagged. "I don't think science should be done that way. In some ways the desire for press releases in those early days drove the science. In retrospect, it isn't so bad because out of it all has come a generalized knowledge of understanding of what was found, and an understanding that nothing was that worthwhile."

He leaned forward and scowled. "But if you tell me the media said nine quasars indicate the universe is collapsing"—his face closed in disgust—"well, that's just a bunch of hooey."

One of the first victims of Sandage's withdrawal was Arp, his old observing partner. Arp had watched the quasar derby darkly, from the sidelines. He had been frozen out of the traffic in secret radio coordinates, even though his research specialty was unusual galaxies.

Arp felt that the midwesterners who ran Pasadena suspected him of being unstable, perhaps because of his New York artistic background, or his penchant for collecting photographs of weird galaxies. He had bent over backward to pay his dues, spending a year on Mount Wilson measuring Cepheids in M31 for Baade, instead of going as he wished to South Africa to study them in their natural laboratory in the Magellanic clouds. Still his artistic sensibilities went against the grain. While Sandage was standardizing the universe, Arp was collecting novelty and revolution. His big work was an atlas of strange galaxies. Most of them looked like tortured amoebas, but there were expressions of delicate symmetry: ring galaxies with fingers of star clouds pointing through them.

So Arp was poised for sour grapes. One night, sitting out the clouds in the Palomar library, he remembered a chance remark by a colleague that there seemed to be a lot of radio sources near his peculiar galaxies. Arp started checking the catalogs and found out that most of those sources were quasars. He wondered if there could be a connection between these wild disturbed galaxies and the quasars near by them. The miraculous powers of quasars, after all, derived from the assumption that their enormous spectral redshifts represented enormous distances in an expanding universe. If they were closer, there wouldn't be so much energy to explain. Suppose, rather, that the quasars were just local objects, ejected, say, from those disturbed galactic cores, and their redshifts were the effect not of distance but of some new unknown physics? Surely it was not too absurd to suppose that there could be laws of physics that science didn't know yet.

Arp gave himself over to combing the sky for evidence that redshifts were not ironclad indicators of cosmological distance, mindful that he was striking not just at quasars but at the foundation of cosmology. He turned out to be a genius at finding mystery. Every funny galaxy he inspected turned out to have a quasar tucked under an arm or at the end of a tendril of gas, or lines of them nearby. He photographed luminous bridges of gas that appeared to link galaxies whose redshifts projected them to be billions of light-years apart. The most famous was an object called Markarian 205, a high-redshift quasar that photographic analysis seemed to place in front of a rather nearby galaxy.

Most astronomers dismissed Arp's diagrams and his computer-processed photographs as coincidences, but they thought the challenge to the mainstream was healthy, especially because quasars and the discovery of the leftover big bang radiation had eliminated the steady state theory. Some of them periodically urged him to do a proper, statistically controlled survey instead of just roaming the skies looking for random trouble. "I don't have time," he always said in his nasal New York accent.

Except for some moral encouragement from the Burbidges and Hoyle, Arp got little support. When he persisted he began to have problems getting telescope time. He was not invited, for example, to share his results at subsequent meetings at the Texas Symposium. He felt martyred. His biggest and most implacable foe was his old friend Sandage.

Sandage felt that Arp was trying to create mystery where there was no need for it. In the meantime, the greater, deeper mystery, the creation and expansion of the universe, was getting trashed. Arp, Sandage thought, was trying to destroy cosmology. Arp complained that he couldn't talk to certain people about cosmology any more. Some people took it all too seriously.

There was never any conscious decision, but one day Sandage found that he just wasn't speaking to Arp any more. "It was like breaking with a brother," he said. Which was difficult, because they were both knights of the 200-inch and Mount Wilson. They avoided each other in the corridor their offices shared on Santa Barbara Street. From the walls the gentlemen founders of cosmology looked down, stiff and formal. Their doors stayed shut; they didn't look each other in the eye when they did meet, passing like galaxies in the night.

5

GOD'S TURNSTILE

In 1963, while the rest of the world went crazy about quasars, a young physicist, a twenty-one-year-old graduate student named Stephen Hawking, lolled listlessly in his rooms in Cambridge, England. Wagner blared from the record player. Science fiction books lay about. He had a flop of long brown hair, steel-rim glasses, and clear blue eyes with a glint of mischievousness, or perhaps impudence. Two traits stuck out in Hawking's personality: He was a smart aleck and he was stubborn. But for the last year he had watched and felt his limbs slither out of his control. His speech had slurred like a drunk's, and his feet refused to stay reliably under him. On top of that, for the first time in his life his casual brilliance in physics had deserted him. He was just spinning his wheels, making no progress; 1963 was not a good year for Stephen Hawking. It was the year they told him he was about to die.

Up until 1963 life had been rather a jaunt. He had been born, as he liked to point out later in his life, on the 300th anniversary of the death of Galileo—January 8, 1942—the oldest of four children, and raised in the London suburb of St. Albans. As his father was a researcher specializing in tropical diseases, a scientific inclination came easily to him. He recalled, "I always wanted to know how everything worked. I would take things apart. Often they didn't go back together again. I wasn't terribly good with my hands."

Somewhat to his father's chagrin, at first, Hawking chose not to follow in his footsteps. "I felt that the biological subjects were too inexact, too descriptive," he explained. "They often seemed to consist of doing detailed drawings. I was never any good at drawing. Nowadays it is much more exact than it was then, particularly molecular biology, but that didn't exist at the time when I was growing up.

"It was really my interest to study the basic laws. I want to understand. I want to understand how and why we're here, and what is the mechanism of the universe."

When he was fifteen, his curiosity led him and a few friends into an exploration of parapsychology. They read up on the famous experiments performed at Duke University in the fifties that purported to show some statistical evidence in favor of telepathy and telekinesis, and they conducted their own dice-throwing experiments. Then he heard a lecture by a scientist who had analyzed the Duke experiments. "He found that whenever they got good results the experimental techniques were faulty. Whenever the experimental techniques were really good they did not get results. So that convinced me that it was all a fraud."

Hawking applied, in the English parlance, to "read" physics and mathematics at University College at Oxford, his father's alma mater. By now he was interested in astronomy as a career. "I went to the Oxford University observatory and looked around. They had no telescopes. All they had for observations was a spectrohelioscope." A miserable summer spent measuring double stars at the Royal Greenwich Observatory with a 28-inch refractor left Hawking "not impressed" with observational astronomy. He was no budding Sandage.

As a student, Hawking was bright but lazy. He impressed his professors and fellow students as the sort who would rather point out the mistakes in the textbook than do the problems at the end of the chapter. He grew long-haired and famous for not studying, being smart enough to do the work without cracking the books. He rarely took notes. He fell asleep in classes and made a show of tossing his own papers contemptuously in the wastebasket. The lecturers, he contended later, weren't that good anyway; the system was made to be abused. Hawking was a free spirit, popular with his classmates. He had a predilection for making bad jokes. He coxed on the crew. His grades were good but not great.

Hawking had his heart set on going to Cambridge, after graduation, and studying for his Ph.D. with Fred Hoyle, the champion of the steady state and a national hero, but first there was the little matter of an honors exam that the outgoing students were required to take. Hawking didn't score high enough for "first honors," which meant among other things that he wouldn't get into his first choice of graduate school.

He was allowed a special oral exam, but in the meantime he applied for a job with the civil service tending monuments.

He needn't have bothered; the oral examiners were impressed by his intelligence. Afterward they asked him what he was going to do next. "If I get a first I'm going to Cambridge," Hawking said. "If a second, I will stay here. So I expect you will give me a first." And he got one.

Strolling down the cobblestone streets of Cambridge, past its soaring chapels, its shops full of striped ties, and its walled residential colleges, each with its sacred square of lawn reserved for the feet of that college's fellows only, one is likely to feel transported to some heaven of intellect, youth, and culture existing in a bubble outside time. The River Cam, barely 30 feet wide, winds behind the colleges between grassy flowered banks arched by numerous footbridges. On a soft June evening I watched and listened to a choir sing madrigals from a raft as the sun went down and evening twilight began while a thousand people, many in tails and top hats, gathered on the green banks. In response to dinner bells, students and faculty alike whizzed down the streets on bicycles, the traditional black gowns that students were required to wear having shrunk to small black capes fluttering like vestigial wings over more traditional collegiate clothing—Levis and sweatshirts.

For a young, ambitious English scientist, Cambridge had been the place to be for at least 300 years, ever since Isaac Newton, confined by the plague, sat in his rooms at Trinity College and worked out the laws of gravity. In the early sixties Cambridge was becoming a crossroads and a center for cosmology—the other kind of cosmology, which didn't require a telescope. Every summer Fowler and the Burbidges, along with their students, returned to Cambridge to hike and work with Hoyle on the evolution of stars and the nucleosynthesis of the elements.

Hawking's hopes of studying under Hoyle were futile, as it turned out, because Hoyle had no students, which was just as well, since his own research was so heretical. Instead, Hawking found himself under the tutelage of Dennis Sciama, a theorist of Hoyle's generation.

Sciama, deeply tanned with a shock of wavy white hair, is an excitable man whose voice rises to a high pitch when he is enthused. He had gone to Cambridge and studied physics against the wishes of his father, who felt betrayed that young Dennis didn't want to take over the family clothing business. Sciama, too, had at first failed to make it into Cambridge graduate school and wound up doing radar work for the army, where his scientific work earned him enough respect to gain admittance to the university. Sciama had a philosophical bent—he had taken a course from Ludwig Wittgenstein—and it inclined him toward the steady state theory of the universe. He became good friends with Bondi and Gold. It was his nature to be passionate, Sciama admitted, and he would get very worked up about the steady state. "I think I liked the steady state theory because it's the only one in which life will always be possible somewhere," he said. "I demanded strong evidence to disprove such a fine ideal." It seemed to him at the time that it was the theory preferred by more imaginative and creative people.

In 1961 Sciama had set up a group at Cambridge to do research in general relativity and cosmology. At the time, he was the only one in England teaching relativity. "John Wheeler at Princeton was doing a bit of what I was doing at Cambridge, which was to train a generation of general relativists," Sciama explained with an evident paternal pride. "In England Cambridge dominates mathematically the way no place in America does. Now I'm at Oxford." He laughed. "So you have to believe what I'm saying. I was strategically placed to get good students. Hawking was one of the best."

Like everyone else, Sciama recognized right off that Hawking had a smart, if undisciplined, mind. Whether that would make him a brilliant scientist was an entirely different matter—you couldn't tell about people, Sciama always said, until they started to do research.

During Hawking's first year of graduate school a tendency that had first manifested itself at Oxford—stumbling and slurred speech— grew worse. In 1963, his condition was diagnosed as amyotrophic lateral sclerosis (ALS). In Britain, ALS is called motor neuron disease; in the United States, it is commonly known as Lou Gehrig's disease. ALS is a progressive deterioration of the nerves that control the voluntary muscles. It usually leads to immobility and then death within two to five years. It's only the voluntary motor functions that are lost; the brain is unaffected, trapped in an insolent, decaying body. The cause of ALS is unknown, and so is the cure. Hawking's future was laid out for him: a cane, a wheelchair, a bed, increasing dependence, and decreasing mobility and communication. Eventually even breathing would be affected; pneumonia is often what proves fatal in the end.

Soon after the diagnosis, Hawking's father paid a secret visit to Cambridge and asked if Sciama would expedite his degree. Sciama was made of sterner stuff than that, however, and he refused.

Hawking went into a tailspin of inactivity. He retreated to his room. "When it first came on," he said, "the disease progressed very rapidly. And then I was very depressed, because I thought I would be dead in a few years. There didn't seem to be any point in carrying on."

The deterioration and depression went on for a couple of years. Hawking floundered scientifically as well, for which Sciama blamed himself. Part of the reason for Hawking's gloom, Sciama thought, was that he hadn't been given an interesting enough problem to work on yet. It was the thesis adviser's job, after all, to find a project that would engage and stimulate his student. Nevertheless, Sciama said, Hawking never lost his ability to tell bad jokes.

As it turned out, not every part of Hawking's life went flat. Shortly after his "condition" was diagnosed he met a woman and began seeing her. Jane Wilde was actually a former classmate of Hawking's from St. Albans who was now studying medieval poetry at the University of London. Jane was red haired, lively, and as stubborn as Hawking would prove to be. "He already had the beginnings of the condition when I met him," she recollected in a precise little slip of a voice. "So I've never known a fit, able-bodied Stephen. It's just something that's progressed very gradually. In personality he was very, very determined, very ambitious." While Stephen was introverted, intense, and analytical, Jane was a classicist, extroverted, and interested in art and music. She opened him up. They resolved to live intensely.

They were wed in 1965. Jane betrayed no qualms about marrying a man who might not have a future. "I decided that was what I wanted to do and that was what I was going to do, so I did," she said.

Marriage was the turning point in Hawking's life. Things suddenly didn't seem so bad; he had a family to support. He couldn't just think about himself and his predicament any more. "My marriage made me determined to live," he said. "It gave me a reason for continuing on, striving. Without the help that Jane has given I would not have been able to carry on nor have the will to do so."

In the meantime his nosedive had flattened out; the deterioration had slowed and Hawking was still alive. Sciama told Hawking that because he was apparently going to live he'd better finish his thesis.

Sciama had a new idea for Hawking. He sent him down to London to meet an old mathematician friend named Roger Penrose, who was giving lectures on one of the most bizarre and disturbing predictions of general relativity: that there could be regions in the universe called singularities, where the laws of physics broke down, where matter and energy and even space and time themselves were destroyed—or created. Hawking went to hear Penrose, and he knew he had found his lifework.

In mathematics a singularity is what happens when you divide by zero—the answer is infinite, undefined. Something like that, Einstein's equations seemed to say, would happen if you piled enough matter or energy in one place: The equations would blow up and give meaningless answers; the universe would give way. "This is a great crisis for physics," Hawking explained, "because nobody can say what will come out of a singularity."

That Einstein's theory should contain such an apocalyptic prediction was something that physicists had been slow to catch on to and even reluctant to accept. General relativity was the very model of an elegant and successful theory of physics, a triumph of principle and mathematical beauty over the messy world of data. It was based on the principle of equivalence, in which Einstein surmised that a person out in empty space being pulled upward in a windowless elevator would not be able to tell the difference between gravity and acceleration, and that, therefore, there was no difference between them. Gravity, Einstein concluded, was not a force, it was just an illusion caused by the geometry of space-time. What determined that geometry was simply the amount of mass and energy present.

Empty space-time, according to general relativity, was normally as flat as an empty mattress, but matter and energy warped space-time the way a heavy sleeper sags a mattress. Under their influence, so-called straight lines would bend and even light beams would curve. The more densely the mass-energy was concentrated, the more "curved" the space around it became. When a baseball fell to the earth, in this view of things, it was not being "pulled" down by some force, it was simply following the track of least resistance through curved space-time. In effect, space and matter manipulated each other. In the words of Sciama's Princeton counterpart Wheeler, "Matter bends space, and space gives matter its marching orders."

One consequence of this, we have seen, was the expanding universe. Another was the dreaded singularity.

Too much matter would bend space too far. Space would sag and warp like chewing gum stretched beyond its limits. The density of mass and energy would become infinite. In Wheeler's words again, "smoke would pour out of the computer." The result would be an absurdity of nature and law, a point where nothing was predictable. A singularity could be a cosmic dead end, where particles and energy simply went out of existence, a free-fire zone where anything was permitted and possible. The marching orders that space give matter would become the insane ravings of a Captain Queeg.

Einstein himself reluctantly admitted that singularities were *mathematically* possible in general relativity, but thought they were nonsense as far as the real world was concerned. After all, a real physical object could never be squeezed down to a point, could it? That was one reason he was uncomfortable with Friedmann's equations suggesting that the entire universe had expanded from a point.

In 1939 two physicists—J. Robert Oppenheimer, later head of the Manhattan Project, and his graduate student Hartland Snyder—did a thought experiment, a *Gedanken*, in the best Einsteinian tradition. They asked themselves what would happen when a massive star finally burned itself out and collapsed of its own suddenly dead weight. A lightweight star like the sun, they already knew, would compress itself to the size of the earth and the density of iron before its weight was balanced by the forces of electrostatic repulsion between atoms. A slightly heavier star, it had been hypothesized by Zwicky and Baade in earlier, friendlier times, would fall even further, until it had scrunched itself into a ball 10 miles in diameter; its very atoms would collapse under the enormous weight. The star would end up as a giant atomic nucleus composed entirely of neutrons held up, in effect, by the pressure of neutrons pushing against one another.

Adding even more weight to this so-called neutron star, Oppenheimer and Snyder innocently figured, would compress it even further and increase the already unfathomable pressure inside the star, but here they ran into a curious thing, the diabolical catch-22 of general relativity, the clause in God's contract that doomed dead stars and perhaps the universe. Because pressure, too, was a form of energy, by the laws of relativity it was equivalent to mass, and thus helped to bend space-time as surely as did a stack of rocks. As the star crunched smaller, this pressure would actually add even more weight to the star, increasing its woes and speeding its collapse. So the very forces that were trying to save the star eventually would turn traitor and doom it. The harder the star tried to fight its shrinkage, the faster it would shrink. The result would be a runaway collapse. Nothing could stop the star from shrinking so small and becoming so dense that it would wrap space completely around it and disappear from view as if swallowed by a magician's cloak. The star— or whatever it became inside that dark zone of infall—would be cut off forever from the rest of the universe. To a distant observer, only its gravitational field would be left behind, ghostlike, surrounding empty crushed space.

None of this was taken too seriously by most physicists, until the advent of quasars. Ordinary thermonuclear fusion could not account for their stupendous energies, which had inspired Hoyle, ever the master of unconventional explanations, to suggest that superstars millions of times more massive than the sun might power radio galaxies, not thermonuclearly, but gravitationally. Gravity is the weakest force in the universe, but it is also a universal force, the ultimate particle accelerator. Hoyle pointed out that if you just let such an assemblage of matter as enormous as that of a star collapse under its own weight, up to 90 percent of its mass would be converted to energy, enough to fuel a quasar. But the end result might be the dreaded singularity.

Hoyle's suggestion—made before Schmidt's unmasking of the quasar—was the spur for the Dallas meeting in December of 1963 at which Sandage showed his exploding galaxy photographs.

That first Texas Symposium was a knot in which the destinies of many cosmologists and astronomers would become entwined. Penrose and Sciama were there. Hawking was home in England feeling sorry for himself, unaware that his lifework was about to be spelled out from a Texas stage.

The spelling out was done by Princeton physicist John Wheeler, who had been worrying about singularities and gravitational collapse for ten years. Wheeler didn't look much like a prophet. He was a short, solid, but unassuming man, fifty-one years old, easily overlooked in a crowd. He dressed like a banker, and he comported himself with the unfailing courtliness of a diplomat; I could never get him to go through a door before me. What was left of his thinning hair was combed straight back over his crown. He had a large head, and his face seemed oversize compared to his body; it was a heavy, solemn face, weighted as if by the cares of the universe, but one that turned cherubic when he smiled. In fact, Wheeler was a wild man, a professor with a taste for metaphor and firecrackers. He kept a chest full of fireworks on his office desk and he used them often. He had been known to shoot off Roman candles in the hallways. "What good is a discovery if you can't celebrate it with a firecracker?" he asked me with a sly grin. He had helped build the atomic and hydrogen bombs and he had a fallout shelter in his backyard. The son of peripatetic librarians, Wheeler had lost part of a finger to a blasting cap when he was a kid in Ohio. He became a peripatetic sort of physicist; he liked to compare himself to Daniel Boone. "Whenever anyone moved within a mile of him, he felt crowded and moved on." Wheeler bounded onto the Dallas stage that afternoon halfway through the conference with a mission; he was about to send physics into a decade-long encounter with death. He felt he had been preparing for it his whole life.

Wheeler was a disciple with two heroes. One was Niels Bohr, the legendary Danish theorist who was one of the founders of quantum theory. Wheeler spent an impressionable year at Bohr's Copenhagen institute and became his lifelong collaborator. "The people Bohr had there furnished a kind of loyal band of collaborators that transcended national frontiers. You can talk about people like Buddha and Jesus and Moses, Confucius, and so on," said Wheeler fondly in a soft voice. "But the thing that really convinces me that such people existed were the kind of conversations one would have with Bohr."

The other was Einstein, who was in Princeton at the Institute for Advanced Study when Wheeler became a young professor at the university. "It took a special summoning up of initiative to go and see him," Wheeler recalled, regretting his timidity and the fact that he never pressed Einstein on the issue of singularities. "On the atomic bomb, the biggest secret was that it could be made. With Einstein, by the same token, the biggest thing that he did for everybody was already out, that you could hope to understand the universe. All comparisons are odius, but as much as I admire Einstein, I think Bohr was the greater human being. I would regard Einstein as declaring the goal that I would feel we should work toward, and Bohr as providing the method and style: continued discussion, keep the ball rolling."

Wheeler was already middle-aged, with a successful career as a nuclear physicist behind him, when he took up general relativity and gave himself the job of finding what he called "the glittering central mechanism" of the universe. He once described the quest like this: "Paper in white the floor of the room, and rule it off in one-foot squares. Down on one's hands and knees, write in the first square a set of equations conceived as able to govern the physics of the universe. Think more overnight. Next day put a better set of equations into square two. At the end of these labors, one has worked oneself out into the doorway. Stand up, look back on all those equations, some perhaps more hopeful than others, raise one's finger commandingly, and give the order 'Fly.' Not one of those equations will put on wings, take off, or fly. Yet the universe 'flies.' "

What made physics and the universe fly? Like many early general relativists Wheeler had the hope that all the phenomena of physics, not just gravity, would ultimately be explicable in terms of space-time geometry. In 1956, he proposed that electric charge could be explained by little wormholes in space-time—the positive charge being one end of the tunnel, the negative charge being the other end. Ultimately, he says, he decided that geometry was not crazy enough to explain particle physics.

Wheeler figured there was no better place to look for the secret of the universe than where it ended, in the apocalypse, metaphysical and real, of gravitational collapse. His calculations convinced him that matter, and many of its essential properties—properties that were the proud underpinnings of powerful laws of physics, such as the distinction between matter and antimatter—simply disappeared at singularities.

To Wheeler, the collapse of a star was just nature's rehearsal for the grand cataclysm: the "big crunch" in which the universe would end if, as Sandage kept hinting, the universe turned out to be closed. One day billions of years from now the cosmic expansion would drag to a halt and then the universe would fall back in on itself. At some definite date in the future, predictable from the Friedmann equations and Sandage's measurements, its density would rise without limit; smoke would finally pour from the metaphorical computer. All the galaxies, all of existence, would disappear into a singularity like a forgotten thought. With them would go space and time, and even physics itself, which is a game played on a field of space and time—the equations all had x's and t's in them. How can that be? Wheeler asked himself. Physics must go on, he argued, because by definition physics is that which does go on its eternal way despite all the shadowy changes in the surface appearance of reality. It was a paradox he relished repeating: "Physics ends, but physics must go on."

Bohr had drilled into him a rather dialectical notion of science. Science only advances by renouncing its past, he said, and there is no hope of progress without paradox. Wheeler thought that gravitational collapse, the notion of physics predicting its own death, had to be the greatest paradox and thus the greatest hope in the history of science. To survive the crisis of death by gravity, Wheeler thought, physics would have to renounce space and time themselves. General relativity in the end would have to surrender to quantum theory, the equally paradoxical laws that governed the world of subatomic particle physics. He spoke wistfully of "the fiery marriage" to come between general relativity and quantum theory.

"Things have got to come to an end, and this end has to be in some sense not an end," he explained. "Think of time as symbolized by a long, woven, shallow basket. At the end it turns around and comes back, and the weaving is no different at one place from another and another. And yet we're not upset that the basket doesn't go on forever and forever. In fact we're glad it doesn't. How is space-time woven? What's it put together out of and how? That's exactly the issue."

Wheeler walked around with a bound notebook in which he wrote lecture notes, calculations, odd jottings, and observations—everything that related to his professional life, including pep talks to himself, with a black fountain pen. The previous volumes lined one shelf of his office. He kept a box of note cards on his desk (next to the fireworks chest) full of aphorisms, his own and others', and quotations—the raw material for speeches and popular articles. One, of his own invention, read, "In the middle ages a man was not a man who could not wield a sword. Today the word is the sword. A man is not a man who cannot project a thought with power."

Another: "We will first understand how simple the universe is when we understand how strange it is."

Wheeler saw his role in the battle for the ultimate theory half as a kamikaze old fogy who could afford to make a fool of himself and half as a sort of phrasemaker, encourager behind the lines. He used the word like a Samurai or a Madison Avenue whiz on his graduate students and his colleagues. When Kip Thorne, a graduate of Caltech, arrived in 1962 to begin his Ph.D. work at Princeton, Wheeler dragged him into his office, closed the door, and gave him a three-hour private lecture on gravitational apocalypse. "He blew my mind," said Thorne grinning, now a bearded, long-haired Caltech professor who favors pendants and Mexican shirts. "I emerged a convert."

Wheeler kept looking for better and better ways to dramatize the plight of physics and the universe. It would be at a conference in New York in 1967 that he made his public relations master stroke, when he coined the term "black hole" to describe the results of gravitational apocalypse. "Sometimes the patient doesn't believe the doctor that he's sick, until the doctor gives it a name," explained Wheeler. The name touched a nerve.

"For years," Wheeler said, "death has been something I think about every day. To me, a picture without a frame is not a picture. Life without death would lose its value. It gives a finite scope for our activities, it gives a sense that we cannot live for ourselves alone. We have to live for what comes after us. Attitudes toward death, I think, are indicative of attitudes toward life." That Sandage's investigations would ultimately reveal a closed universe, Wheeler had no doubt. The universe and even time itself would have to die. Space-time had a disease, and he was the doctor.

Three hundred physicists and astronomers watched as Wheeler, fist clenched, strutted his dark pronouncements in Dallas. For whom does the bell toll? he asked rhetorically. And then answered: all of us. "It is sometimes argued that there is no reason to trouble about the kind of physics which goes on in the collapsing matter during the final stages of the dynamics," he concluded. "The object becomes redder and redder as seen from outside. It keeps its mass-energy—because it cannot radiate it—and therefore preserves also its gravitational attraction. Nothing that goes on inside in the final stages is relevant, it is occasionally said, because there is no way for word of these events to reach the outer observer. Therefore regard questions about these events as meaningless, and forgo any attempt to analyze them. However, to an observer located on the falling system the collapse takes only a limited and very modest amount of proper time. . . . In this event, one can point to the Einstein-Friedmann picture of the expanding and contracting universe, according to which we are located on just such a cloud of galaxies, and have no choice but to live with its eventual collapse. Far from being a bystander safely outside of all collapsing systems, one from this point of view is right within the most interesting expanding-and-collapsing system of all, the universe. To try to analyze *its* physics would not seem to be a meaningless occupation."

To the astronomers in the audience, more used to arguing about galactic redshifts or the colors of globular cluster stars, this kind of talk was almost metaphysics. Wheeler recalled the Dallas meeting as a standoff. Even Oppenheimer, he says, didn't "resonate" to the idea; by then he had become a particle physicist. "The astronomers were inclined to wait and see," he said laughing mirthlessly. "There was clearly a different breed of cat performing up there on the stage." Nevertheless his charge hung over physics.

It was not until a year later that Wheeler first met Hawking, who was to be in a moral sense his greatest student, who would inherit and inhabit most firmly the aura of the black hole. Initially, though, it was Penrose who pursued what Wheeler called delicately "the issue of the final state" most brilliantly and elegantly.

Penrose was the scion of one of those famous British intellectual families like the Huxleys, whose achievements go back generations. His father was a well-known eugenicist. Penrose's own specialty was geometry. He had a talent for design; among other things he collaborated with the German artist M. C. Escher on his famous paradoxical drawings. Penrose is handsome with wavy, mussed brown hair and a slightly rumpled, distracted look; his shirttails are often out by the end of the day. His workmanlike demeanor belies an almost unworldly elegance of mathematical technique. It was Sciama who had convinced Penrose that he should apply those talents to general relativity. Penrose approached general relativity as if it were a geometry problem in four dimensions (which, of course, it really was), applying powerful theorems and elegant proofs. His papers and calculations looked like little hieroglyphs, and in the end they drove the nails into Wheeler's coffin.

Many scientists had hoped or believed that singularities would turn out to be mathematical idealizations, like the frictionless slides or perpetual pendula in freshman problem sets, unrealizable in normal life. A real star would probably not collapse in a spherically symmetric manner, especially if it were spinning, which it almost assuredly would be. Perhaps the matter would slosh and swirl around the center and miss crashing together to form the final singularity. No such luck, Penrose said in his elegant theorem language, and in 1965 he proved that the runaway collapse of a star would pinch down to a full-fledged singularity. You couldn't have a black hole without having a singularity in it. Sciama called it the most important contribution to general relativity since the theory was formulated. Penrose proved, in effect, that wherever enough matter and energy were gathered space-time could have an end.

Hawking was a quick student and he absorbed Penrose's methods. In the last chapter of his thesis he applied those techniques to the case of an open universe (one that was going to expand forever), which could be thought of as the collapse of a star in reverse. "That was the best part," Sciama said fondly. If general relativity were right, Hawking concluded, then sometime in the history of the universe at least one singularity had to exist. In the case of the expanding universe, there was surely one in the past, some 10 or 20 billion years ago. If he ran the expanding universe backward, like a film in reverse, the density of matter and radiation grew and grew without limit until finally the very light in the sky became like a lead blanket, dragging all of creation down to a singularity. Later Hawking and Penrose in a joint paper extended the argument to cover all models of expanding universe. Hawking, who had a gift for understatement, allowed aloofly that this singularity might be regarded in some sense as the birth of the universe.

Hawking cut a lean figure around town in the late sixties, walking with a donnish air, stumbling down to sit in the first row of somebody's important seminar and asking penetrating questions. After his Ph.D., he stayed on as a research fellow at Cambridge, commuting between the department of applied mathematics and theoretical physics, an industrial-looking building in the heart of the city, and Hoyle's low-slung modern-istic Institute of Theoretical Astronomy (the word *theoretical* was dropped in a merger with Ryle's group in 1972, when Hoyle quit) on its outskirts in a little three-wheeled cart. His body continued its slow decline. By 1969 he had abandoned the cane for a wheelchair. His speech was becoming difficult to understand. Hawking's stubbornness kicked in. He declined help whenever possible, preferring whatever independence he could muster. Guests to his Cambridge townhouse watched him take fifteen minutes to climb a flight of stairs alone. Nevertheless, life went on: in 1967, Jane had given birth to a son, Robert.

Jane showed me a black-and-white photograph from those days. It shows them on the way to the wedding of a friend, he in a suit, she in a minidress, with a jaunty but fragile windblown air about them. He looks slim and stylishly collegiate leaning on his cane, his hair fashionably shaggy, owlish in dark-rimmed glasses, and a shade preoccupied or be-mused. They resembled any of thousands of young academic couples that one might have known, rushing into the light of 1967.

During the late sixties the English theorists hit their stride. Playing off one another like a pair of determined squash opponents following each brilliant ricochet with another, Penrose and Sciama's group produced a number of theorems about singularities, the structure of space-time, and the fate of matter caught in its dark path. A famous theorem by Penrose decreed that the collapsing matter or anything that fell into a black hole would either hit a singularity and be crushed out of existence, or if the black hole were spinning—having formed from a rotating star—would go through a wormhole like the bull's-eye on a target and fountain out at another place and time, perhaps even in another universe, as a white hole.

These theorems and problems were posed in a kind of geometrical shorthand developed by Penrose and Brandon Carter, a tall, debonair, bearded Frenchman who was another of Sciama's students. In these Penrose diagrams, as they came to be called, the entire past and future of the universe was mathematically transformed into a triangle. Its vertices and two of its sides represented various kinds of infinity (the infinite past, infinite distance) and the third side was a time axis. Singularities appeared as jagged lines either perpendicular or parallel to the time line. In the former case, which represented a nonrotating black hole or the universe as a whole (Hawking showed one year that the universe wasn't rotating), the singularity was at a particular time, and you could not avoid it anymore than you could avoid your fortieth birthday.

In the second case, the singularity was at a certain place, and you could in principle swerve to dodge it. But you couldn't escape the black hole in which it was embedded, so you found yourself, having swerved, in a new triangle, back-to-back with the old one, representing a new universe. The English space-time jockeys soon learned to stitch Penrose diagrams together into long chains like some kind of cosmic wallpaper and chart the paths of lucky astronauts into black holes and out of so-called white holes from universe to universe.

Mathematically, a "white hole" was the opposite of a black hole—a fountain of energy connected to its cousin by a so-called wormhole through space-time. As it happened, however, both wormholes and white holes, on further reflection, turned out to be physically improbable. All the mass-energy spewing from a white hole would crinkle space right around and turn it black, so there was no escaping a black hole through its back door. Only the circus unpredictability and the nihilistic hunger of the singularity remained for the black hole astronaut.

Luckily, it seemed that you need not have this metaphysically shattering encounter unless you volunteered for the plunge into a black hole. In all the simple calculations that people like Hawking and Penrose could do, the only places singularities formed were in the centers of black holes; they were always surrounded by event horizons—an imaginary surface that marked the point of no return. Inside the event horizon the velocity needed to escape the gravitational grip of the black hole was greater than the speed of light, nature's absolute speed limit. It was actually worse than that. Inside the event horizon space and time exchanged roles; just as you cannot look *out* of the universe, you couldn't look out of the black hole—all roads led to the singularity. The only way out was backward in time, a hard ticket to book.

The event horizon that imprisoned the astronaut also kept word of the breakdown of physics from getting to the outside universe. Whatever emerged from a singularity would never escape from the black hole that enclosed it! Nature seemed to shield itself from such unpredictability.

Penrose wondered, but could never prove, if there was a "cosmic censorship" principle that forbade the appearance of a naked singularity. Hawking had mixed feelings about it. On the one hand, physicists had to believe the universe makes sense; you had to play your cards as if you could win. On the other hand, we are all products of the absurd, the one unclothed singularity in which the universe had begun and its twin in which we might meet our maker. Ever solicitous of disaster victims, he did not think that the black hole astronaut would take much comfort from the cosmic censor as tidal forces stretched him like a noodle and flattened him like a pancake and the singularity pulled him in like the eye of God. If he failed one of us, worried Hawking, the cosmic censor could fail us all.

Another theorem that consumed Hawking's attention described a black hole's disturbing ability to cover its hungry tracks by destroying information about what it had swallowed. The equations that described black holes turned out to be embarrassingly simple, *too simple*. A black hole could have mass, charge, and angular momentum, and that was it. But by the grown-up sixties, particle physics had discovered a whole laundry list of properties like strangeness, baryon number, lepton number, and isotopic spin that distinguished the myriad subatomic particles popping up in the detector chambers of particle accelerators. Somehow in a black hole gravity destroyed all these distinctions. A black hole cared not whether it ate blue stars or interstellar dust or neutrinos or pure radiation or matter or antimatter—the outcome was the same. What sort of clue was that to whatever created and destroyed the universe? Hawking suspected that the singularity was somehow to blame.

Wheeler phrased it in a koanlike manner: "A black hole has no hair." Carter proposed the no hair statement as another theorem. Hawking, Carter, Werner Israel, and David Robinson proved it in steps. Despite its enigmatic nature, the no hair theorem revolutionized black hole physics, because by including fewer variables, it simplified the choices nature could make and the equations physicists had to solve.

Much of this work was finally and elegantly encapsulated in a slim volume Hawking had cowritten with fellow Sciama student George Ellis, a South African who returned to his native country to carry on a two-pronged battle against both the universe and apartheid. The book is called *The Large-Scale Structure of Spacetime*; it was published in 1971 and is regarded now as a classic. I held it in my hands once and turned the pages, admiring what seemed as beautiful and as incomprehensible as Egyptian hieroglyphs.

*   *   *

"Why study black holes?" Hawking asked rhetorically. "Why do people climb Mount Everest? Because it is there. That plus the fact that I had some techniques that could be applied to them. I was interested initially in the big bang, the beginning of the universe, then I turned my attention to the end of space-time."

Hawking was the ideal black hole cosmonaut. The same illness that had chained his body had loosened his mind. He liked to joke that it was lucky he had chosen such a sedentary occupation. He counted himself as the luckiest of men. "I think I am happier now than I was before it started," he said. "Before the illness set on I was very bored with life. I drank a fair amount, I guess, didn't do any work. It was really a rather pointless existence. When one's expectations are reduced to zero, one really appreciates everything that one does have."

He had more and more trouble writing but he had been able to visualize Penrose diagrams in his head. When he moved to black holes, the work was more traditionally mathematical, and he had to memorize equations, something he said he wasn't good at. He tended to think in pictures. A lot of problems he might have pursued were ruled out because their computations were too lengthy. What was left for Hawking were the most abstract and fundamental questions; he was free to roam conceptual labyrinths for which equations had not yet been written.

"Sometimes I make a conjecture and then try to prove it," he explained when I asked him how his mind worked. "Many times, in trying to prove it, I find a counterexample, then I have to change my conjecture. Sometimes it's something that other people have made attempts on. I find that many papers are obscure and I simply don't understand them. So I have to try and translate them into my own way of thinking. Many times I will have an idea and start working on a paper and then I will realize halfway through it that there's a lot more to it.

"I work very much on intuition, thinking that, well, a certain idea ought to be right. Then I try to prove it. Sometimes I find I'm wrong. Sometimes I find that the original idea was wrong, but that leads to new ideas. I find it a great help to discuss my ideas with other people. Even if they don't contribute anything, just having to explain it to someone else helps me sort it out for myself."

He retained his sense of humor; he fought a running battle with *The Physical Review* over his predilection for jokes. "Once I wrote in a paper, 'Suppose you have a race of little gnomes stationed . . .' *The Physical Review* changed it to *observers*," he said, eyes flashing.

As writing and even speaking became increasingly difficult, Hawking learned to hone his thoughts to a Zenlike brevity and clarity. "Five minutes with Stephen Hawking is worth an hour with most other physicists," Bernard Carr, one of his graduate students, told me. Even so, the circle of friends that could understand his paralyzed speech slowly narrowed. His graduate students translated for him in interviews and conversations. When he gave talks, he projected his text on a screen overhead.

Hawking claimed not to miss freedom of physical movement. "It is occasionally inconvenient, for example, not being able to handle papers, or not to be able to just pick up a book and turn to a page. But then, on the other hand, I've found that it's often easier to work things out in my head than to look them up, which is probably better," he said, echoing a common habit of physicists. "I would like to be able to play more physical games with children, like croquet or football. That's really my main regret. People are very helpful, and I have things so well arranged that it really doesn't bother me."

Every other challenge that his own black hole, sucking at him gently but unceasingly, offered, he answered. He and Jane traveled and entertained as much as possible. Hawking became synonymous with black holes, and black holes became synonymous with everything that was hip and morbid in astrophysics.

In 1971 Hawking invented the concept of mini–black holes, which made the subject of black holes practically tabloid material. Only the most massive stars could collapse all the way to a singularity under their own dead weight; smaller objects would need a prodigious squeeze to overcome the initial resistance. Hawking proposed that the enormous pressures of the big bang could have produced that squeeze and jammed random little clots of matter much smaller than a star to black hole densities. The galaxy today could be sprinkled with trillions of miniholes, each of them harboring its singularity like a razor blade in a Halloween apple. He and Carr computed that the smallest black hole around today might have a mass of a billion tons—about the same as an asteroid or an iceberg—and the diameter of a proton. Just a pinprick in the fabric of space-time, but if you stepped on it, it would take off your foot.

Mini–black holes were a hit with astronomers. There was virtually no astrophysical puzzle that couldn't be solved by the judicious postulation of a strategically placed black hole. If experiments to detect neutrinos from nuclear reactions in the center of the good old sun were coming up catastrophically short, maybe there was a black hole in the center of the sun. What caused the strange explosion that flattened trees over fifty miles of Siberia in 1906? A mini–black hole crashing through the earth (and presumably out through the North Atlantic). Lowell Wood, a Lawrence Livermore Laboratory scientist later of Star Wars fame, proposed that if we could find a little black hole in the solar system and drag it back to the vicinity of earth, its tight little gravitational field would make the perfect confinement vessel for a thermonuclear reactor.

"The black hole," one astronomer commented, "is the scapegoat of the seventies." Wherever you looked in the sky, or maybe under your own feet, it seemed as if the invisible cosmic maw was open and waiting. Like everyone else during a decade of war, paranoia, and assassination, astronomers had become necrophiliacs. They dwelt in dark mysterious places looking for the Word, and the Word was *annihilation*.

It was hard not to think of Hawking as his own metaphor. Wheeler could have invented him. Those were the days, in the late sixties and early seventies, when we believed Wheeler and Sandage, that the universe was closed and that one day the galaxies would all come home. God would eat us.

Wheeler and Hawking could invent the *idea* of black holes, but not the holes themselves. Only nature could do that. The evidence that black holes actually existed in the universe originated in the ideas of a third school of general relativity and cosmology. It started up in Moscow in the early sixties under the leadership of Yakov Boris Zeldovich, a fiery little pepper pot of a man, who like Wheeler had a history of bomb work. While the Cambridge school concentrated on proving theorems and the Princeton school on predicting the future, Zeldovich was more interested in seat-of-the-pants physics.

Zeldovich, too, fell under the sway of the black hole, but he concerned himself with the practical side of things. What would a black hole look like in the real world? How would matter fall into a black hole? It wouldn't go straight down the gullet, Zeldovich realized. More likely a gas cloud in the vicinity of a black hole would be spinning; it would collapse along the axis of spin into a pancake called an accretion disk and swirl around the hole like water around a drain. This disk, God's turnstile, would be heated by compression and friction to millions of degrees, a bruising incandescent wheel sparking X rays and gamma rays as it rubbed its skirts on the edge of oblivion. In 1965, Zeldovich and O. H. Gusneyov suggested that the place to find these demon carousels was in a double star system, in which one of the pair, having become a black hole, can feed on gas drawn from its unfortunate companion. Look for stars orbiting nothing and which are spitting X rays, they advised.

In 1970, a sort of flying Geiger counter was launched into orbit by NASA off the coast of Kenya; it was christened *Uhuru,* Swahili for "freedom." Its purpose was to map celestial X rays, which can only be observed from orbit because the atmosphere blocks them. It was the brainchild of a bunch of young physicists at the American Science and Engineering Corporation in Cambridge, Massachusetts, led by an ambitious and paternalistic Italian named Riccardo Giacconi. By the time the *Uhuru* data were finally analyzed, Giacconi and his brood were ensconced at the Harvard-Smithsonian Center for Astrophysics.

The final *Uhuru* catalog of X-ray sources had hundreds of entries that had to be identified, explicated, understood. One of them was in the constellation Cygnus, called Cygnus X-1, where the optical astronomers found a bright blue supergiant star that was whirling around an unseen object every 5.6 days. The best estimates—notoriously conservative—of the invisible companion's mass were about 10 solar masses, far too heavy to be anything but a black hole.

Finding alternate explanations for Cygnus X-1 became a cottage industry. "In a way a black hole is the most conventional explanation for Cygnus X-1," Hawking said. "If it isn't a black hole it really has to be something even more exotic."

In a hedging manner, Hawking weighed in on the side of the skeptics. Scientists are notorious bettors; he bet Kip Thorne, Wheeler's student and a good friend, that Cygnus X-1 was *not* a black hole. The stakes were a subscription to *Penthouse* for Thorne and a subscription to *Private Eye*—"the only magazine in this country that dares to publish scandals"—for Hawking. "It would really be easier for me to win than for Kip to win. There's a number of observations that could disprove that it's a black hole—for example, if they found that it was emitting absolutely regular pulses [like a pulsar]. By the no hair theorem that would be ruled out for black holes."

But in his bones, Hawking knew better; he knew he would be paying up some day, one of these days. Six thousand light-years away, that is to say, here in our own quadrant of the galaxy, there was a gate to eternity. After a decade of being only a bad dream, black holes were in the world.

6

THE KING OF BLACK HOLES

If the black hole theorists needed any encouragement to continue their morbid-sounding investigations into the nature of objects that were only barely in the universe, the discovery of Cygnus X-1, followed by other likely black-hole X-ray sources, supplied it. Ultimately, no one would look farther down the well of gravity than Hawking, and what he saw down there surprised even him.

One evening in November 1970, two weeks after his second child, Lucy, was born, while he was getting ready for bed, Hawking had an idea that would eventually change the direction of black hole research and in fact shake the foundations of general relativity. But for a long time it seemed like more of a nuisance than a breakthrough. As Hawking likes to tell it, his disability made getting ready for bed a slow process. "I turn problems over in my mind at odd moments. Getting to bed is not something that requires great concentration.

"I was thinking about the problem of colliding black holes," Hawking says. In the framework he had worked out, a black hole was like a dark bubble in space. The skin of the bubble was the event horizon, the place where the escape velocity was the speed of light, a membrane that could only be crossed in one direction—inward. The diameter of the event-horizon bubble was proportional to the mass inside. For a stellar-mass black hole it was a few miles; a mini–black hole was trillions of times smaller.

When two bubbles collided there were a lot of complicated things that he could imagine happening but could never calculate. They could deform each other, merge, splinter, shake, rattle and roll, emitting gravitational waves. Hawking remembered, lying there, that in such an event, or any interaction to which a black hole was party, the black hole could only grow larger, and consequently the area of its event horizon—the skin of the bubble—had to increase. The reason was simple. Nothing could leave a black hole, so all it could do was eat. And every pound consumed extended its sway microscopically outward.

"I was excited, yes," Hawking recalled. "I didn't sleep very well that night. It was really rather obvious when you are familiar with global techniques. I already knew that the area always increased. What I realized that night was that it placed a constraint on the behavior of black holes. The increase in area placed an upper bound on the energy that could be radiated in a collision."

What this meant was that Hawking didn't need to compute all the messy details of a black hole collision; all he had to do was wait until the dust cleared, when a predictably bigger hole would be sitting there. The next day he called up Penrose who had noticed the same tendency of holes to grow but, because he had evolved a slightly different definition of the boundary of the hole, did not know what to do with that fact.

Hawking did. It reminded him of thermodynamics.

Thermodynamics. It was as if he had popped the hood on a Ferrari and found an antique steam engine chugging away inside. Thermodynamics is the study of heat, gases, temperature, pressure, and the efficiency of engines. It was one of the oldest fields of physics, and in some ways one of the most elegant in a classic sense. What did it have to do with black holes?

Well, a key thermodynamic concept was the notion of entropy. Entropy was a measure of the amount of wasted heat or disorder in a system. The second law of thermodynamics—one of the most famous and far-reaching commandments of physics—stated that the entropy in a closed system—for example, the beating cylinder of an automobile engine or the universe—always stayed the same or increased, but never decreased. What it meant was that you could never have an engine that was 100 percent efficient, a perpetual motion machine—a little energy was always wasted. Order inexorably decayed to disorder unless energy was expended to clean things back up, in which case still more heat was left behind. The melting of an ice cube, the eroding of the Rockies, and the gradual disarray of a spare closet were all triumphs of entropy.

Roughly stated, the second law[1] said that things always get worse. Right away you were tapped into the spirit of black holes.

It was entropy that Hawking was thinking of when he thought about the inexorable increase in black hole sizes. "The area behaved an awfully lot like entropy," he said. Both increased almost mystically with time. Both enabled a physicist to know the outcome of a complicated event without having to compute all the details in between.

Black holes had a mathematical resemblance to thermodynamics. Was there a connection? Hawking didn't think so, but it was a nice analogy. Still, he didn't take it too seriously until 1972, when he and Carter and James Bardeen, a theorist now at the University of Washington, found more thermodynamic analogies. They began to use these analogies in their theoretical work, all the while believing that they were nothing but analogies, useful computational shortcuts.

Meanwhile, across the Atlantic in Princeton, one of Wheeler's students was taking Hawking's analogy much further and more seriously than Hawking himself did. This was where Hawking's bedtime insight started to become a nuisance. The student's name was Jacob Bekenstein. He claimed on the basis of a tour de force calculation in his 1972 Ph.D. thesis that the area of a black hole was not analogous to entropy, it *was* entropy.

Bekenstein's work was based on a twentieth-century spin on the idea of entropy: that it was negative information, that is to say, that disorder destroys meaning. Consider the Scrabble letters *k, i, a, g, h, w,* and *n*. There is only one way to arrange them so that they spell out *hawking,* and 5039 ways to arrange them so that they spell gibberish. The second law of thermodynamics said that if you arranged the letters to spell *hawking,* put them in a box and shook it, gibberish would probably result—the original message would be lost.

According to the no hair theorem that mass, charge, and angular momentum were the only qualities that black holes possessed, black holes also destroyed information—scrambled the Scrabble tiles—of the material that had fallen inside. There was no way of knowing whether it had been old shoes or antimatter, what color it had been, or how long ago it had vanished.

That enforced ignorance increased disorder and uncertainty, that is to say, entropy. Bekenstein, like Hawking, had a taste for bold argu-

1. There are really three laws that govern thermodynamics. The first law says you can't create energy; the second says you always squander some. There is a third law that says you can never cool anything to absolute zero—heat from the rest of the universe will leak in. Cynics and physics students put it like this:
 1. You can't win.
 2. You can't break even.
 3. You can't get out of the game.

ments. Information can be quantified mathematically into "bits," like the contents of a computer memory. In 1972 he calculated the number of bits it took to characterize the details that had been erased about matter lost inside a black hole and showed that they were proportional to the area of the event horizon. In effect, what Bekenstein proposed was that what a black hole really eats and is swelled by is information.

In Cambridge Bekenstein's breakthrough was greeted with derision. Hawking was outraged. He knew this was nonsense. Bekenstein, he felt, had pushed the analogy way too far.

The big flaw in taking the thermodynamic analogy literally, Hawking pointed out, was that if black holes really had entropy, then they also had to have a temperature. But a black hole couldn't have temperature—hold a thermometer up to it and the black hole would suck all the heat out of it. The thermometer would read absolute zero. Heat being a form of energy, the black hole could not radiate any warmth back to stir the mercury.

So the temperature of a black hole was absolute zero, and that was it. All the rest was bosh. "I was very down on Bekenstein," said Hawking.

A war of metaphors ensued between Cambridge and Princeton. When Bekenstein published a paper called "Black Hole Thermodynamics," in 1973, Hawking, Carter, and Bardeen responded with "The Four Laws of Black Hole Mechanics," with the prefix *thermo* pointedly absent.

Both papers, as it happened, were wrong, Sciama pointed out much later, but nobody noticed. He shrugged, "That just shows how difficult this work is." For a year or two the debate between Cambridge and Princeton was a standoff.

Most relativists sided with Hawking but Bekenstein, seduced by the potential richness and power of thermodynamics as applied to black holes, stood his ground. In 1980 he wrote in *Physics Today,* "In those days in 1973 when I was often told that I was headed the wrong way, I drew some comfort from Wheeler's opinion that 'black hole thermodynamics is crazy, perhaps crazy enough to work.' "

The resolution to the Hawking-Bekenstein argument came from an unexpected quarter. It came from quantum theory. And it changed everything. Some physicists become world beaters with a succession of breakthroughs while equally talented colleagues waste their talents on the wrong problems, they do the wrong experiment. The kind of intuition that is drawn to the right problem is called a good nose. Hawking had a good nose; it invariably led him to productive questions. In 1973 it led him back to miniholes and the overriding quest of late-twentieth-century physics. He announced his new project in a typical understatement, neither modest nor immodest.

"One ought to look at the quantum aspects of gravitation," he told Sciama one day. As if nobody had ever thought of doing it before. Sciama practically rubbed his hands and cackled. Hawking was going into the quicksand now. He'd met his match at last.

Quantum theory and general relativity were the yin and yang of twentieth-century physics, seemingly with nothing in common. Gravity ruled the large-scale structure of space-time, while the paradoxical laws of quantum mechanics described nature on the submicroscopic scale of atoms and elementary particles. Horrible things happened to people who attempted to reconcile these quarreling opposites. Their calculations were bedeviled with infinites and anomalies. The ordinary laws of physics, such staples as conservation of momentum or charge, went haywire. A researcher would spend months computing term after term of an endless series of mathematical expressions, wondering if the infinite terms would eventually cancel each other out, not understanding why when such miracle cancellations did occur. The would-be quantum gravity theorists floundered in a foam of mathematical despair.

Now Hawking proposed to join them. He was not in a mood to be deterred by the difficulties. He was probably by then a candidate for the title of stubbornest man in the universe; he had lived ten years with a killer disease, fathered a family, founded new fields of physics, become a doctor of the universe.

Hawking being Hawking, and clever as well as stubborn, he decided to come at the problem sideways. The way to quantum gravity was, perhaps, through mini–black holes. Suppose an electron went into orbit around a tiny black hole—far enough outside the event horizon so it wouldn't get sucked in—the way it would normally be the captive of an atomic nucleus. In this case, gravity, rather than electrical forces, would be the architect of the "black hole atom." Hawking became obsessed with figuring out the properties of a black hole atom. He started boning up on quantum theory, reading old textbooks.

His colleagues were not reassured: It seemed like a dubious voyage. At the time Hawking was spending half his time at the Institute of Astronomy. One of the people he talked to a lot out there was Martin Rees, who was two years his junior and had followed him through Sciama's tutelage. Rees was involved, among other things, in showing how the billowing accretion disks around giant black holes in the centers of galaxies could power quasars. He is short with wavy black hair, dancing eyes, a visage sharp as an eagle's talon, and a hunched back, the result of a childhood bout with scoliosis. Talking to Rees is like swimming with a shark; I always got the idea he was digesting my ideas faster than I could say them, nodding his head impatiently. "Ahh, yes, umm, yes, uh-huh." The effect was to make me talk faster, which made him nod faster, which made me talk even faster, like a teenager being egged on in a drag race, until the conversation veered off into incoherence.

Rees had been admiring Hawking for years but now he wondered if he was witnessing the end. His condition, it seemed to Rees, was getting worse. Hawking would ask him for a quantum theory textbook. Rees would set it up for him, open to the right page. Then Hawking would sit there for hours motionless, staring. Who knew where his mind was? Down in a realm so strange that curved space seemed sane by comparison.

General relativity had sprung practically full-blown from the brain of one man, the supreme triumph of pure thought over nature. Quantum physics had been a group effort; scientists across Europe in the first three decades of the century struggled to make sense of what came out of their laboratories when they probed into the heart of the atom.

Quantum theory said that there was an ineluctable fuzziness, a kind of chaos at the microheart of reality. It took its name from the initial simple but puzzling observation that atoms absorbed and emitted energy only in certain discrete amounts called *quanta* after the Latin *quantus* for "how much?" Observed on the smallest scale, nature exhibited a curious duality between wavelike and particlelike behavior. An electron, for example, could act like a wave or a particle depending on the circumstances. One of the most mystifying consequences of this duality was an epistemological nightmare called the uncertainty principle, which seemed to draw a final line in the dust against mankind's centuries-old quest to know the world in finer and finer detail. The uncertainty principle stated that knowledge came with a price. The more precisely you knew one trait of a particle like an electron—say, its position—the less precisely you could know something else, like its momentum. All the things worth knowing about the world, in fact, came in incompatible pairs: position and momentum, energy and time, wave and particle. Knowledge of one somehow destroyed the possibility of knowing the other.

According to Bohr, the Danish lover of paradox who became the philosopher king of quantum theory, this uncertainty was not just the result of inevitable experimental clumsiness—it was built into the fabric of reality. The electron had no position or momentum before it was measured. In some sense the electron itself did not exist before it made its mark on a laboratory apparatus. Bohr once commented that a person who wasn't outraged on first hearing about quantum theory didn't understand what had been said.

What existed instead of the electron was a metaphysical contrivance called the wave function, which encapsulated all the possibilities of the electron and was smeared out over all of space. Bohr interpreted the wave function as a measure of the likelihood that the electron was in some state or other. The wave function went around corners and through walls, which implied that, on rare enough occasions, so could material objects. Anything was possible. A baseball could pass through a plate glass window without harm to either, theoretically, and a flea could hop to the moon—theoretically. (Such magic was what made transistors work.)

At the moment of actual scrutiny of an electron or a baseball, Bohr concluded, the wave function magically "collapsed" to a specific answer to whatever question was being asked. But the scientist had to ask—otherwise nature didn't answer. On average the wave function collapsed to its most likely value, but there was no law saying it had to.

Quantum theory roughed up most of the comfortable and sensible notions of Victorian physics. One serious casualty was cause and effect. The physical laws that were so ironclad when collections of trillions of elementary particles—in the form of baseballs or ballistic missiles—were involved were only a good guess when it was two particles interacting. Newton's famous laws were only a statistical average in the subatomic domain. Anything could really happen. It was like the difference between trying to predict the progress of an anthill and that of an individual ant.

The rules that governed the subatomic world of blurry particles and nuclear forces were called quantum mechanics. According to quantum mechanics forces were transmitted by little quantum bundles of energy called bosons that were shot back and forth between elementary particles like bazooka shells. In the case of light, these bundles were called photons. Quantum mechanics was the language of particle physics; all the forces of nature were framed in this manner, except gravity.

"Why bother trying to 'quantize gravity'?" I asked Wheeler once.

"What is the alternative?" he shot back. "Which would you give up: quantum theory or gravitational theory? They've got to coexist. If you apply quantum theory to one field like electromagnetism, and not another, like gravity, and yet you allow interaction between the two like we know we have, then you arrive at inconsistency."

Wheeler argued that if you took quantum theory seriously, then even the geometry of space-time would have to pay its dues to the uncertainty principle. This uncertainty would only manifest itself on the smallest of scales, the so-called Planck scale, 10^{-33} centimeters, seventeen orders of magnitude smaller than a proton. In effect that length was the "quantum" of space. The result, said Wheeler, was that space-time was like an ocean viewed from far above, in an airplane; it looked smooth. The closer you looked, the rougher it got. From a little closer you could see waves, from closer still, ripples and swells and spray. Finally, under a microscope, there was nothing but foam, wormholes forming and dissolving, connecting and separating different points. Geometry, space, and time lost all meaning. Wheeler thought of these wormholes as incredibly tiny black holes; in effect, that meant that space-time was *made* of black holes, packed 10^{100} per cubic centimeter.

Hawking's initial foray into quantum gravity was more modest than Wheeler's and other particle physicists', a sneak approach. He first wanted to know what the effect was of an ordinary, classic, curved-space gravitational field on a quantum system. He called this the semiclassical approach. Until that day, most quantum mechanical calculations had been done as if gravity didn't exist—they were hard enough without it in normal flat space-time, just as basketball is strenuous enough in a regulation gym with two baskets and five men to a team. By envisioning an "atom" whose nucleus was a catastrophically powerful black hole, Hawking was getting ready to play basketball on the craters of the moon.

Little did he know that it would turn out that Bekenstein had been playing point guard.

In September of 1973 Hawking went to Moscow to confer with Zeldovich and the rest of the Russian relativists on the quantum mechanics of black hole atoms. Among them was Alexander Starobinsky, a skinny, intense young theorist with a stutter. Starobinsky ventured the opinion that rotating black holes would spray elementary particles. Hawking did not think that was completely crazy. It was known from Penrose's work, among others, that you could extract energy from the spin of a black hole just like any other dynamo; the energy would come out in particles and radiation just like it did from a particle accelerator.

But Hawking didn't like the way Starobinsky had reached this conclusion. When he got back to Cambridge, he resolved to redo the calculation for himself. Just to get things straight, he decided to warm up first, by calculating the rate of emission from a nonrotating quantum hole. He knew the answer should be no emission. Doing a trivial example for which you already know the answer is a common practice, a good way of drawing a bead on general principles when you are about to venture into uncharted territory.

Not that the general principles in this case were simple. It took Hawking two months just to frame the calculation in his mind. In November he traveled up to Oxford, where Sciama had moved to begin a new research group. He gave a talk describing the method for his upcoming calculations, but at that point he didn't know the answer.

Ten days later the equations he had elaborately set up finally unreeled themselves across the blackboard of his mind. It was lucky he had such privacy, because his results were embarrassing. His imaginary black hole was spewing matter and radiation like a mad volcano.

Hawking knew he'd made a mistake but he didn't know where. Stationary black holes swallowed things; they didn't spit them out. "I was very sorry when I got this infinite number of particles being produced because it destroyed my framework," he said. "In hindsight this idea of black hole atoms was not really a good one. I didn't want the particles coming out; I wasn't looking for them at all. I merely tripped over them. I was rather annoyed."

Christmas came. He spent lonely holidays replaying his calculations over and over again, doing everything he could think of to make the particles go away. At one point he locked himself in a bathroom to think.

In the meantime he was reluctant to tell anybody but his closest friends what was going on; he was afraid Bekenstein would hear about it. The radiation coming from the black hole had the same characteristics, the same spectrum, as thermal radiation from a hot body, so-called black-body radiation, like the heat your hand feels from a stove or a forehead. It meant that holes had temperatures, just as Bekenstein's work implied. "It was just the relationship that he had been sneering at," Sciama commented. Hawking didn't want to give Bekenstein any more ammunition, though, because he still didn't trust his findings.

In January Sciama walked into the Institute of Astronomy and encountered Rees, pale and trembling, hanging onto a door knob. "Have you heard?" he gasped. "Stephen's changed everything."

Sciama heard his story and convinced Hawking to go public with it at a quantum gravity meeting he was organizing the following month at Oxford. Hawking attended. "As the conference approached, it appeared more and more reasonable," Hawking recalled. "So by the time I gave my talk in February I had come around to believing it. But not many other people did." He called his paper "Black Hole Explosions?"—the question mark a testament to his own doubts.

Hawking wheeled to the front of the room and while a projector flashed his garbled words on a screen, described how his quantum mechanical calculations resulted in a flux of particles and radiation pouring from the hitherto ungiving surface of a black hole. Like all quantum effects, this one was only relevant for small holes. The temperature of a black hole was inversely proportional to its mass. A "normal" hole with the mass of a star would radiate at a temperature of only one ten-millionth of a degree; for all intents and purposes it would be a black sucker of doom. But tiny black holes, if they existed, would be hot. A tiny billion-ton primordial black hole formed in the big bang, a proton-size nick in space-time, would be white hot, 100 billion degrees K, spewing gamma rays.

As a black hole radiated, it would lose energy and mass, reasoned Hawking, thereby shrinking its event horizon. (Because it was radiating pure heat and disorder, the shrinking event horizon did not violate the entropy laws.) The smaller the black hole got, the hotter it got. The hotter it got, the more fiercely it radiated, and the faster it shrunk. The black hole was trapped in a runaway process. Sooner or later, Hawking concluded, it would simply explode in a Fourth of July shower of gamma rays and exotic elementary particles. A hole exploding. It sounded like a koan. When is a black hole not black? When it explodes.

When Hawking was through talking, the moderator, John Taylor, an English theorist of some note, dismissed it as nonsense.

So did most people for a while. It took Zeldovich, who in some ways was godfather to the exploding black hole, two years to accept Hawking's new work. In 1976 Thorne was in Moscow. On the last day of his visit he was called to Zeldovich's apartment. There, Zeldovich threw up his hands. "I give up," he cried. "I give up. I didn't believe it, but now I do."

Hawking was not surprised by the initial disbelief, for there was still a problem with his model. The calculations showed only that Hawking radiation—as it quickly came to be called—was coming from the black hole, but not *how* it was produced. Hawking spent the rest of the winter racking his brain for an explanation physicists could believe, or at least relate to. The radiation couldn't be leaking, like air from a tire, from matter inside the hole because there was no matter in the black hole; whatever went in was crushed out of existence at the singularity. A black hole was empty curved space. How could particles rain out of empty space?

The answer that he hit on was the uncertainty principle. One of the strange implications of the uncertainty principle, Hawking knew, was that there was no such thing as absolutely nothing; space could never be completely empty. The reason was because the amount of mass-energy in some volume—say, in a box—was always uncertain, no matter how carefully it was measured. The shorter the measuring time, the more uncertain. It was in quantum mechanics that this uncertainty—vacuum fluctuations was the technical term—manifested itself in the form of elementary particles dancing briefly into existence and then winking out again. These so-called virtual particles were created in complimentary pairs—a particle, say an electron, and its antiparticle opposite, a positron. They lived on borrowed energy, for an infinitesimal fraction of a second, and then met, canceling each other out of existence and paying back the energy debt to the universe. Crazy as it sounded, the physicist W. E. Lamb, Jr. had won a Nobel Prize for measuring the effect of these ghost particles on atomic transitions in hydrogen atoms. The vacuum was really an invisible fountain of creation.

Suppose, Hawking thought, this fountaining of particle pairs took place just on the edge of a black hole. One of the temporary particles could then drift over the edge into oblivion, escaping the possibility of remerging with its birthmate. The abandoned particle would be free to wander away from the danger zone and, like Pinocchio, become real. To a faraway observer it would appear to have popped out of the black hole. And its energy would indeed have come at the expense of the black hole. By swallowing one of the particles, the black hole would also be swallowing that particle's energy debt. When nature's books were balanced, that debt would be paid back out of the black hole's mass. In effect the black hole was swallowing negative energy. The black hole would appear to be radiating and losing mass at the rate predicted by Hawking's equations.

Hawking had an even more risqué explanation based on Richard Feynman's idea that an antiparticle is mathematically the same as a particle going backward in time. In this view, the particle that falls into the hole is actually an antiparticle going backward in time and coming out of the hole (which can only happen in the backward-time direction). Once out of the hole, it hits the strong gravitational field and turns into a regular particle going forward in time, like a car backing out of a driveway, hitting a garbage can someone had left out, reversing direction, and proceeding down the road. That allowed Hawking to connect his radiation directly with the singularity, a connection he seemed to relish.

Sciama, when I spoke to him, had a simpler explanation that he was very proud of. He pointed out that the event horizon itself would fluctuate because of the uncertainty principle. As it slid up and down the steep slope of the black hole field, it would radiate copiously. "When you think about it," he said, "you can just *see* that event horizon wiggling."

When I asked Hawking about it, he cautioned that none of these "heuristic" explanations should be taken too literally, that the real justification for the result was in the mathematics. "The most important thing about Hawking radiation," he said, "is that it shows that the black hole is not cut off from the rest of the universe. It has many suggestive similarities with the big bang. So far I haven't seen how to treat the big bang in the same way. You can regard the particles coming out of the black hole in some way as coming from the singularity at the center. And the singularity in the big bang might have behaved like such a singularity."

Did it have any relevance to the big crunch?

"You can ask what will happen to someone who jumps into a

black hole," he answered. "I certainly don't think he will survive it. On the other hand if we send someone off to jump into a black hole neither he nor his constituent atoms will come back, but his mass-energy will come back. Maybe that applies to the whole universe."

It was only in the effusiveness of the collective sigh of relief that greeted Hawking's deed, that it was possible to appreciate how heavily the psychological aura of black holes had weighed on the souls of physicists. Sciama said it was the most beautiful paper in the history of physics. An exploding black hole returning its mass-energy, gorged and stored over how many millennia, to the outside universe again was a candle of resurrection.

Scientifically, the main point was that gravity had met quantum theory. Hawking's treatment of miniholes was the first successful combination of the two theories. Both, in some strange way, paid homage to entropy. The final unification, when it did occur, would have to match his result. In the old classical and mystical realm of thermodynamics, of all things, Hawking had found the first common ground between gravity that bends the universe and the quantum chaos that lives inside it.

His discovery set off a search of the sky for exploding mini–black holes, using gamma-ray satellites and cosmic-ray detectors. None were found, but the search capabilities were modest. The instruments found that the level of gamma radiation in the sky was low enough to limit the number of black hole explosions to two per cubic light-year per century. There was a lot of room in the galaxy, but astrophysics was the main point. Because of its deep implications about gravity and entropy, Hawking's work was important even if tiny primordial black holes did not exist. Even if regular black holes did not exist.

Hawking became famous again. Or rather his fame and his legend ratcheted upward. By the end of the seventies he had won every prize a physicist can win except the ones they give in Stockholm. Every fall the rumors went around that it was Hawking's turn.

From 1974 to 1975 Hawking spent a year at Caltech, a year in which Jane pretended they were on vacation so that she could enjoy the beach and the mountains without worrying about what it would be like to actually raise a family in the mechanized chaos of Los Angeles. When he returned to England he could no longer drive the invalid car any more. He bought an electric wheelchair. As usual, turning adversity to his advantage, he learned to maneuver it dramatically and, being self-powered, it increased his mobility and independence. The Hawkings began to take in graduate students to help with the increasingly time-consuming task of taking care of Stephen. One of the first was Carr, who said it was like participating in history. At such comments Hawking snorted in return that it was hard for a student to be in awe of his professor after he has helped take him to the bathroom.

The drama of Hawking's life and his subject matter tended to obscure Hawking himself. He hated the attempts to put the mask of death on him. People made too much of his disability, he complained, when in fact it really just gave him an excuse to sit and think about the things he was interested in anyway. Not many people, including other scientists, could penetrate the psychological and perceptual screen of his illness. Hawking made a point of going further and longer than his colleagues. They got used to the whir of his wheelchair as he cruised the darkened aisles of scientific gatherings in Boston, London, Tel Aviv, San Francisco, and Moscow. At receptions and conferences where physicists talk casually, he was often alone with his companion, not mobbed as you might expect a brilliant thinker to be.

Well, as far as Hawking was concerned, it was their problem. He was in a predicament; he was not the predicament.

One of Hawking's star turns came at the 1976 Texas Symposium in Boston. The first time I ever saw him he was squeaking through the ornate lobby of Boston's Copley Plaza, crumpled fragilely in his wheelchair, a bulky square briefcase sticking out the back, his long fingers curled around a black control knob. Hawking looked as though he weighed about ninety pounds. His suit draped him. His feet cocked together on the footrest of the wheelchair like a child's. His face, smooth and unlined as a twenty year old's, was topped with a boyish haircut, but his age showed around the eyes, blue and wrinkled. He had a broad generous mouth, exaggerated by the emaciation of the disease.

Although I knew hardly anything about Hawking, he struck me instantly as a charismatic figure. On some level that I didn't understand I felt that I had always known about him.

There was no ramp to the stage of the ballroom, so when it came time for Hawking's lecture I joined half a dozen astrophysicists in the Iwo Jima chore of lifting him to the stage. The wheelchair ran on heavy car batteries, which made it a struggle. Hawking's head lolled less than confidently. On the stage he seemed even smaller, squashed beneath a giant screen that projected his words. Hawking's speech sounded like a series of murmurs and sighs, punctuated by an occasional drooping of the head when he was tired. He spoke for forty-five minutes.

Hawking had come full circle from the days when he resisted Bekenstein's notion that black holes had entropy. Now he seemed to see black holes as almost pure entropy, wreaking disorder and randomness on the universe, roaming like hungry sharks, eating information, and spreading unpredictability in their wakes. Because it came from the singularity, he said, the radiation from a black hole had an unpredictability that went beyond the already famous unpredictability of the uncertainty principle. In the latter case one could know either the velocity or the position of a particle. In the case of black hole radiation, he contended, you couldn't predict either one. This extra degree of randomness he called the "principle of ignorance." He concluded by alluding to a famous statement that Einstein had made once in arguing against quantum theory: "God doesn't play dice."

"God not only plays dice," Hawking announced, "he sometimes throws them where they can't be seen."

Afterward I talked briefly with Hawking and Carr. When I asked him if I could take a picture, he and Carr proceeded to spend half an hour trying to make Hawking's tie stand straight up in defiance of gravity.

Defiance was one of his traits all right. In the midst of a babble of gravitational despair, Hawking had cruised serene and hardheaded, holding in his mind curly little tensor equations and light-cone diagrams. It was hard not to think of him as a St. George who had gone into the mouth of the dragon and emerged reborn and victorious.

On one level, I thought, black holes were just the newest name for an old dream, a dream that even scientists share. The brain that looked at a photographic plate or the chalky scribbles on a blackboard and interpreted them had some components hundreds of millions of years old. The cortex read and reasoned; the lizard brain reacted. Like a good poet, Wheeler was in touch with the lizard brain, he knew how to evoke the fear and the loathing. Hawking—I wondered what Hawking was in touch with.

The history of science is full of stories of problems solved in dreams and other vats of the subconscious. Kekulé, for example, who discerned the cyclic structure of the benzine molecule, dreamed of a snake eating its tail. I wondered if in Hawking's own mind there was a connection between death and black holes. It was too astounding that of all the men who could have assembled the equations to look down the eye of a black hole and find there the glint of creation, transformation, it had been a man who was himself staring at personal oblivion who had done the job.

Why hadn't Wheeler done it? Or Bekenstein, who was so close, or Carter or Penrose or any of the Russians? The whole world had a psychological block about black holes. Only Hawking had been able to shatter it.

It seemed important to ask him these questions, so in 1978 I went to Cambridge for a week. I had in mind to play the role of Hawking's shrink; I wanted to know how his mind worked.

On the day before I was to take the train up to Cambridge from London, my birthday as it happened, I was in the Tate Museum staring at William Blake drawings of creation and damnation. They all looked like people being thrown into black holes to me. I was nervous because I had never attempted to do this kind of interviewing before, a normal nervousness exacerbated by the fact that Hawking was such an intimidating figure. You could talk to him, but he couldn't talk back; I felt I was about to go on stage. I realized that I couldn't remember what Hawking looked like. I could imagine a slim figure in a wheelchair, but he had no face, only darkness.

There was a Black Holes Are Out of Sight bumper sticker on the door of Hawking's office at the Department of Applied Mathematics and Theoretical Physics, a grimy, industrial-looking brick building off a narrow side street. The office inside was long and narrow with a shiny concrete floor and a stage for maneuvering a wheelchair, surrounded by an automatic page turner, a knob-activated computer, other mechanical aids, and dusty packed bookshelves. On every vertical surface papers were taped up and books were propped open so that Hawking could look at them. The office was also empty; Hawking wasn't there.

I had arrived at eleven, just in time for morning tea. Like Sciama before him, Hawking had gathered a crop of graduate students around himself. In age, dress, pallor, and evidence of nutritional deficiency, they resembled the road crew for a rock and roll band. I found them gathered around a white Formica table off a hallway, cracking jokes and scribbling equations on the tabletop. They were talking about politics. "I'm a right-wing socialist," Hawking announced to general laughter, adding, "I supported Carter and I supported Ford. But I never supported Nixon." The tabletop, I noticed, was veined with faint blue writing and mathematics, testimony to last year's hot subject. It turned out they never cleaned it. "When we want to save something we just Xerox the table," Hawking quipped.

Hawking's speech was no longer intelligible to any but a few of his regular confederates and his secretary. Most of my conversations with him were translated either by Don Page, a tall cheerful American postdoc from Caltech who characterized himself as a born-again Christian and who was living with the Hawkings and tending to Stephen, or by Gary Gibbons, a former graduate student with whom Hawking was collaborating on some of the far-out implications of Hawking radiation as well as sharing his office. Hawking stretched the syllables out in a low rumble. I often thought he had said a whole sentence when it turned out he had completed just a word. He maintained the scientific output of a normal man by dictating papers laboriously, a page a day, to his secretary or a coauthor.

Stripped of his mask of illness and the other messy details of his physical existence, Hawking was breezy, stubborn, unpretentious, impatient with mysticism or fuzzy thinking, barely tolerant of ambiguity, but happy to talk to you and answer questions as long as they were intelligent. The effort to speak often twisted his face in a grimace; his eyes lit up in anticipation of jokes. When he was tired his head would fall forward.

Fame, he volunteered, was a nuisance. Quacks as well as cranks sent him letters. There were invitations to snake venom camps and advice from mystics piled in the bookcase. His humor ran to corny undergraduate jokes or technohumor of the sort that you hear in the halls of MIT. Once in an elevator, I asked Hawking if he thought there would still be singularities in the final theory of gravity, whatever that might be. "There will always be singularities in the complex plane,"[2] he snapped, and drove out of the elevator, just brushing the tips of my toes.

I pursued the subject because the possibility that singularities might exist in the universe and be visible, experienceable, had always seemed miraculous. In a sense, the singularity was the closest a physicist could get to God. It was God. Would there be a singularity left behind, naked, after a black hole evaporated?

"That's what we all wonder," he said. "But my view is that a black hole completely evaporates, leaving behind just empty space. There would be these particles that had gone out, but there would be no singularity that just persists in time." He must have seen my disappointment. "That's not based on the mathematical treatment. We're still not sure how to treat the final stages of evaporation."

So the cosmic censor was still doing his job?

He frowned. "It's a question of what does cosmic censorship mean in the quantum domain? When a black hole produces particles, you do have, in a sense, a breakdown of the cosmic censor because of the particles getting out, but nevertheless you can quantify how cosmic censorship breaks down. You can say you know what the probabilities are. It doesn't break down in just an arbitrary way; in a sense it's the maximum randomness possible.

"We physicists have to believe that the universe makes sense, in order to go on working. In the back of all, perhaps most, physicists' minds there is the idea that there is some complete theory to describe the universe. We're finding out bits of this theory and trying to fit these bits together. Hopefully the universe is not merely capricious."

On my second morning in Cambridge I asked Hawking if he

2. A mathematical space of great importance in physics and mathematics in which numbers have both real and imaginary components. Imaginary numbers contain the factor i, which is the square root of -1. A complex number would take the form of $3 + 4i$, the first term being the real part and the second, the imaginary component.

thought coping with his illness had made him a better person and a better physicist. "Yes," he answered rather formally. "When one's expectations are reduced to zero, one really appreciates what one has."

Thus emboldened, I wondered out loud if he thought there was any psychological connection between his own predicament and the great black hole work. Was it possible that a man who had accepted death had become more open somehow to the possibility of life?

"You shouldn't exaggerate," he said gently.

I pressed on, reminding him that in the eyes of many popular New Age thinkers, scientists were just dredging up the same old archetypes and ideas that had been rattling in the human unconscious since the dawn of prehistory. One of the favorite myths of the black hole junkies was the so-called black Hindu goddess Kali, who eats her child as she is giving birth to it, the personification of the twin engines of creation and destruction that drive the universe. I began to read from a fan letter that quoted Joseph Campbell, the anthropologist and collector of mythologies.

As I read Hawking looked more and more disgusted. I thought he would surely run over my toes again.

Campbell had described Kali as "the terrible one of many names 'difficult of approach' *(durga)* whose stomach is a void and so can never be filled, whose womb is giving birth forever to all things." And he quotes the Hindu mystic Shri Rama Krishna, "When there were neither the creation nor the sun, moon, planets, and the earth, and darkness was enveloped in darkness, then the mother, the formless one maha-kali, the great power, was one with maha-kala, the absolute. . . . After the destruction of the universe she gathers together the seeds for the next creation."

Hawking could barely restrain himself. "It's fashionable rubbish," he snorted. "People go overboard on Eastern mysticism simply because it's something different that they haven't met before. But as a natural description of reality, it fails abysmally. It fails abysmally to produce results. I get quite a lot of letters from people with that sort of viewpoint. If you look through Eastern mysticism you can find things that look suggestive of modern physics or cosmology. I don't think they have any significance.

"There is probably a psychological connection. Calling these things black holes was a master stroke by Wheeler because it does make a connection, or conjure up a lot of human neuroses. If the Russian term *frozen star* had been generally adopted, then this part of Eastern mythology would not at all seem significant. They're named black holes because they related to human fears of being destroyed or gobbled up. So in that sense there is a connection. I don't have fears of being thrown into them. I understand them. I feel in a sense that I'm their master."

Hawking bridled at the suggestion that he had anything in common with the black hole astronaut. "I've always found I could communicate."

Hawking was not ready to be eaten.

The Hawkings lived just over the River Cam, a ten-minute wheelchair commute, in the ground floor of a large Victorian house owned by Cambridge University. I walked over one warm afternoon and sat drinking orange juice with Jane in the living room. Black holes seemed far away. Their daughter Lucy, with straw blond hair and sky blue eyes, wandered in and out. Late-afternoon light billowed through tall south windows behind a piano.

Jane, a freckled redhead with a chipper voice, displayed the sort of true-grit optimism that won the West in America. "We've been lucky," she said, "things have gone well for us. Always when things have seemed very difficult, something good has turned up. Things have adjusted themselves in a certain sort of way so they're almost better than they were before, given the progressively deteriorating condition.

"For instance, when we came back from California two years ago, we found that Stephen couldn't drive the three-wheeler invalid car that he's had on loan from the National Health Service, when in fact it really didn't matter. It was a blessing in disguise, because the roads are so dangerous out to the institute anyhow. It didn't matter because we could afford to buy the electric wheelchair, which, as you can see, he runs along in, and is really much more convenient for him because he doesn't have to be sure of having people to help him in and out as he does with the car. So he's completely independent in the electric wheelchair. There's always some compensating factor that makes deterioration acceptable. For another thing, of course, when we came back from California we couldn't live in our own house, which is on three floors. But the college was very good to us and offered us this ground floor which is absolutely magnificent, as you can see."

She spread her arms. "It's absolutely glorious, for us and the children. Wide wide doorways so Stephen can get in and out very easily. A beautiful garden, tended by college gardeners. So you see things always have a way of sorting themselves out.

"We go out a tremendous amount," she went on. "We've got lots of friends. We go to a lot of concerts. We're very lucky, there's a concert hall just two doors away from us. We go to the theater a lot. Stephen has had a campaign with our small local theater—it's actually a very lively theater. And they've installed a removable seat in back of the stalls. On the nights when we're going they take out the seat. He can drive in and park himself there in his chair."

Jane, a medieval linguist by education, did not pretend to follow her husband's work and she was resisting the installation of a computer in the household (although with a ten-year-old boy afoot, that was hopeless). "I think it takes great single-mindedness," she said about science, "and I think it means that you cannot really devote yourself to anything else at all. In a way, you see, there is a certain poetic justice in Stephen's condition. Because, given the initial tragedy of his situation, in fact it has meant that he's been able to devote himself completely and utterly to his work. There's simply nothing else that he can do."

Was that a problem? I wondered. "Well, for me there's nothing unusual about it because I knew there was nothing else he could do. I can imagine how frustrating it must be for some physicists' wives when they expect help from able-bodied husbands which is not forthcoming. I have no illusions on that score.

"I think Stephen has achieved an important ambition, to extend the bounds of human knowledge by just one step. Not many of us are in the position to do it. When we're looking at what he's done, I find it very satisfying, though I can't understand it. I feel he has been a success in the sense that he has used his talents to the fullest. He hasn't wasted any of his talents at all. In fact he's intensified them."

About then Hawking and Page came rolling up the driveway.

I went for a walk and wound up in the Kings College Chapel—a gray stone edifice with flying buttresses—and listened to the choir perform an evening service. Watching the day roll dark and slowly becoming mesmerized by the mournful voices haunting the vaulted ancient stonework, I began to feel lonely and mortal. Hawking's universe, it seemed to me, was a slippery and disturbing place.

Hawking had told me earlier in the week that he wanted now to really quantize gravity. That was his main ambition. In a recent controversial paper he and Gibbon had extended the notion of Hawking radiation beyond black holes to the expanding universe. Because gravity was the same as acceleration, the Canadian theorist William Unruh had shown that anyone being accelerated through empty space would see a flow of radiation and particles coming at them. Moreover, differently accelerated observers would count different numbers of particles and see a different temperature of radiation.

Hawking and Gibbon concluded, unhappily for the memory of Einstein, that reality itself began to diverge for differently accelerated observers, or in very strong gravitational fields. "Two different observers might even encounter different histories of the universe," they had written. Hawking had hinted in an interview with *New Scientist* that the laws of physics themselves may be somewhat observer dependent.

But what then was real?

"You shouldn't ask questions like that," he answered. "I don't think there's one unique real universe." He went on to explain that when you did quantum mechanics for the entire universe, as he was trying to do to quantize gravity, there were severe conceptual problems with Bohr's idea of the collapsing wave function. Who is it, for example, who observes the entire universe? And where does he stand? As a result Hawking had adopted a renegade interpretation proposed by Wheeler and one of his students, Hugh Everett. "The Everett-Wheeler approach," he explained, "says that the wave function for the universe contains many different branches, which all correspond to different measurements. Or different results of a given measurement. There are different branches of the universe, and each branch corresponds to a different possible measurement." So instead of collapsing, the wave function just kept splitting, like the branches of a huge elm.

Was this like parallel universes?

"They are parallel universes."

In his philosophy it seemed that every moment, then, was a creation event, an instance of prodigious invention, the spurring outward of a billion billion universes—universes in which the ball didn't go through Bill Buckner's legs in the sixth game of the 1986 World Series. There were a billion billion universes for each of the billion billion ways the ball could bounce as it advanced toward his frail frozen ankles.

But they were all cards from the same deck, those billion billion cosmos, the deck of the big bang, the ultimate creation event. We can only live in one universe, one moment, at a time, and that universe had a beginning, a birth of infinite density and insane laws (or lack thereof); it could have an end. If Hawking and general relativity were right, the entire deck of universes could be shot through with singularities, little imitations of the big crunch or the big bang, where space and time and the laws of physics as we normally understand them and take for granted broke down. Once in a magazine article I had quoted Hawking as saying this was a "great crisis for physics," but a typesetting error had transformed his statement to "a great *circus* for physics." I was tempted to let it stand. A circus seemed closer to the spirit of singularities than a crisis: anything was possible, even though it would probably kill you. In my mind they stood out in color (like a circus tent, vivid bright colors) against the gray predictability of science crunching its way like a giant adding machine calculator from inevitability to inevitability. All my life I had been looking for some way out of the gray doom of ordinary existence. To me, singularities were like hippies—killers in clown masks—running through the staid halls of the Federal Reserve Bank. They were at once liberation from the gray law and enforcers of the ultimate unknowable law of laws, tangible evidence of a mystery more powerful than anything we could think of, a truth that would fry your brain or blind you if you saw it— like the face of God, waiting there at the end of time.

They were magic. The idea that the laws of physics—gray sober relativity—should predict the existence of singularities was astounding. The singularity theorems, to me, were like evidence of a miracle, of a magic outside of physics itself. I wanted to know from Hawking if such miracles, such singular terrible transformations were real. If we couldn't see God, could we at least know God was there, even if sulking in a black hole or at the end of time? What I wanted from Hawking was some touch of the miraculous.

I followed him down to London the next day, where he had organized an Einstein centennial celebration at the Royal Society. Hawking was less interested in spending more time talking to me than in seeing his colleagues and friends from around the world. We spent two days playing hide-and-seek. Finally, when it was over I cornered him and Page on the Royal Society steps. I asked him again the question I asked in the elevator. "After we get a real quantum theory of gravity, will there still be singularities?"

"Yes, I think so," he allowed, and then repeated what he had said before. There would always be singularities in the complex mathematics he was using to derive the theory. "One can find ways of going around these singularities, but the fact that they are there is important and they give rise to important physical effects."

I was confused already. You could get rid of them mathematically. "But physically is the singularity there?"

"This comes back to the question of whether there's a unique space-time picture," he answered blandly. "I think that if you take quantum mechanics seriously, you cannot have a unique space-time picture."

"Does that mean that in some cases the singularity might not be there?"

"What does it mean to say that the singularity is there?" responded Hawking. "All you can ask is what are the probabilities of making certain observations. I don't think you can actually observe a singularity."

Penrose had said the same thing when I asked him, because the singularity was always in the future, like next Tuesday. "It's always in front of you or something," I offered up lamely.

"Yes, but I'm not sure even then whether or not it hits you. If it hits you, then you can't even really ask the question about whether it's there."

"When it hits you it's too late."

"If you can ask the question," Hawking continued, "then that apparently means the answer is no, that you haven't hit the singularity."

"This isn't very helpful," I confessed. This was a funny conversation to have with a man who had rejected biology because it was too fuzzy and indeterminate. Maybe I was asking the wrong question. I decided to take another tack. What about our destiny? What about the singularity at the end of time? "Is everything going to shrink down to zero in the big crunch?"

"The question of whether the universe will collapse is an unresolved question."

Wait a minute, wait a minute. I began to have a sinking feeling about what he was going to say next. But I had to ask anyway. An electron was one thing, but the universe was something else entirely. Surely it had to go one way or another. You were either dead or not dead. The universe either existed or it did not. He didn't mean to say, did he, that the universe could end for some observers and not for others?

Hawking smiled. Yes, he told me. Yes.

7

THE BIG BANG

Thick snow swirled through the Sawatch Range in central Colorado. It glided like a soft glacier down the Roaring Fork valley from the Independence Pass, filling the nooks and crannies between 14,000-foot peaks with fluff, snagging in the lodgepole pines like torn cotton. It obliterated Aspen Mountain, known to skiers as Ajax, which rises out of the center of the town of Aspen like a giant on a bended knee. Drifting through the town's Victorian streets, it muffled the sounds of early morning traffic and the noisy cheer of breakfasting skiers. It had been so far a very dry winter in the Rockies.

On the far side of town from the mountain, where cottages gave way to meadow and rushing canyon streams, the snow drifted on the flat roofs of a collection of small cedar buildings housing the Aspen Center for Physics. Inside the largest of the buildings two dozen physicists and astronomers wearing ski clothes sat at classroom desks in a small auditorium listening bleary-eyed to lectures and arguing about the universe.

One man, tall and slender, all legs, with straight dark hair, small features, and a perpetually quizzical look on his face, slouched lazily in the front row with his chair turned sideways so that his back was against the wall. His legs extended out like a pair of skis, and the speakers had to be careful not to trip over them. He took no notes; instead he engaged in a running banter with the speaker, butting in with a comment, a criticism, or a joke whenever he felt like it. "Ahh, that's truth and beauty," he responded to one particular idea. As if the whole seminar were a two-person conversation taking place in his own living room or office.

His name was Philip James Edward Peebles—"Jim" for short—and he had a right to feel at home. His name didn't get in the newspapers much, for he was a cosmologist's cosmologist. In a sense, anywhere in the universe that cosmologists gathered was his office; in a very real sense they were all his pupils and disciples. Most of the questions they had come to ask—why are there galaxies? what is the universe made of? why is it homogenous, or is it?—were questions he had taught them to ask, while he sat to the side gently scorning their answers.

Like some fractious church, the subject of cosmology enfolded different streams of endeavor. One was the classic observational cosmology of Sandage, what he sometimes called practical cosmology or cosmography, measuring the universe with the calipers of the 200-inch looking for two numbers that described the expansion. Another was the theoretical quest of Hawking, sitting and thinking about grand principles that attended its creation and destruction. Peebles was the author of a third style of cosmology, less grand but grittier and more ambitious in its aims. He and his followers called themselves physical cosmologists; they were less concerned with the fate of the universe than with its origin. They wanted to move from describing how the universe *is* to explaining *why* the universe is the way it is—why there are galaxies and matter. In short, the universe was a physics problem. By the eighties most of the people in the world who called themselves cosmologists were physical cosmologists.

Jim Peebles remembered the day he had become a cosmologist. It was a hot day in the summer of 1964, hot as hell. The leafy colonial lanes of Princeton—where Einstein had once walked and where Wheeler now held sway with his mesmerizing vision of gravitational apocalypse—seemed wilted and hazy. It was the summer of the Mississippi civil rights drive and the Beatles' ride to the top of the music charts. The world trembled on the brink of several explosions; one of them was going to happen in New Jersey.

He had no inkling of being picked out by fate. He spent the evening as he did every Friday, tucked into a desk drinking beer and eating pizza while the mysteries of physics were unfolded on a blackboard. Like many young physicists who fancied themselves pretty bright, he had been mucking around all spring trying to quantize gravity—and getting nowhere. In another year his fellowship would be up and he would have to figure out where to go next, but for now life was pretty good.

At the time Peebles was twenty-nine years old; he and his wife, Alison, his college sweetheart, had two small daughters. His life seemed almost too normal for a driven field like cosmology. An outdoorsman, he had a personality like water running downhill and easy manners. He never sat, he sprawled. His curiosity also sprawled—Peebles's idea of a good time was trying to figure something out. It was that nose for a good time, combined with a laconic disregard for scholarship, that was to make one of physics' greatest triumphs also one of its most frustrating comedies.

Peebles had been born and raised near Winnipeg on the plains of Manitoba, the youngest and only boy of three children. He discovered a mechanical bent as a youth and enrolled in the University of Manitoba intending to be an engineer, but he says the engineers there were a relatively dull lot. The physicists seemed to be having a lot more fun and threw better parties. So he switched to physics—where he excelled. He would later recall the quality of teaching as being equal to that of the Ivy League, although its range was narrower. He graduated, for example, without ever having a course in quantum physics. A professor who had attended Princeton steered him south to graduate school. It was a bruising experience.

"I went from being a big fish at Manitoba to a small fish at Princeton," he said. "I wasn't particularly happy as a graduate student much of the time. I was low man on the totem pole. I didn't know what was going on. I was the best student of the year at Manitoba. I came to Princeton and I was the least well prepared. I was totally lost. I sat in on courses in which I didn't understand the introductory sentence." He grinned tightly and without mirth.

"But that is what graduate school is like."

It was in those foundering days that he had his first dry unsatisfactory brush with cosmology. It didn't appeal to him; in fact, it repelled him. He had first encountered it in graduate school, studying for his general Ph.D. exams. Princeton had no graduate course requirements, let alone a course on cosmology; students were supposed to pick up the subject on their own. "One of the topics on the general exam is general relativity," Peebles recalled. "And commonly, questions in general relativity are on cosmology. What turned me off was the thought that the universe could be as simple as the standard big bang models suggested."

It was at that same time, however, that Peebles fell in with Bob Dicke, another of the grand old men of physics at Princeton. Dicke was a sort of Renaissance man. Trained as an atomic physicist, he had shifted into radar during the war and invented techniques that later served as the basis for atomic clocks and lasers. Among his inventions was the Dicke radiometer, which was a sensitive receiver for very short wavelength (microwave) radio waves. In the fifties Dicke had turned his attention to gravity. As an alternative to general relativity, he invented a theory in which the gravitational constant changed over time; in that theory, gravity would get weaker relative to the other forces as the universe got older.

On Friday evenings Dicke ran a small seminar that featured wide-ranging discussions of physics, usually accompanied by beer and pizza. A friend of Peebles's belonged to Dicke's group. "I came along to some of the meetings and I got trapped by all the interesting things he was doing. And so after a fairly short foray into particle physics—I only wrote one paper on particle physics—I fell into Bob Dicke's orbit." He and Alison became Friday night regulars and counted themselves, he says with a grin, among the more vigorous partiers.

Peebles enjoyed the wide range of physics that was required to test Dicke's ideas. For his thesis he figured out how to tell whether a number known as the fine-structure constant, which characterizes the strength of electromagnetic forces, had been changing over cosmic time, by studying the radioactive decay in various meteorites formed billions of years ago. The answer was that the constant did not change. "The remarkable thing is—physics way back then was awful close to physics now," he said.

This sort of work led Peebles gradually into astrophysics and also ended his career as an experimental physicist. Once he showed Dicke his plan for a complicated experiment. The plan was in the form of a flow chart. Dicke couldn't make heads or tails out of it. Finally he asked his young graduate student, "Have you considered theory?"

Peebles got his degree and stayed on in Dicke's group as a post-doctoral fellow. A postdoc gets an office and a stipend, typically for a two-year term, and has no duties except to hang out and do research. It's the most crucial time in a scientist's career, a chance to build a track record before the youthful fevers of dedication and inspiration get diluted by adult responsibilities and distractions.

Dicke often lectured about the beauty and mystery of the expanding universe and on that fateful summer Friday he brought it up again. Like many scientists he was unhappy, however, with the singularity, the idea of the universe starting from nothing. According to Peebles he kept complaining about it. "Consider the universe as it is today," Dicke would say. "You can trace back to the universe as it was yesterday, and then the day before. But in the conventional big bang, you run into the day zero where things stop."

Dicke's preferred solution was an oscillating universe. Rather than coming out of nothing on day zero, he suggested, the universe could be rebounding—"bouncing"—from being squeezed during a previous collapse and big crunch. That solved the creation problem—at least for this immediate cycle—but it raised another problem. According to the astronomers the matter in this universe had started out in a pure and simple form as hydrogen, with the heavier elements being cooked later on in stars. What had happened to the heavy elements that had been made in stars during the previous cycle of expansion and collapse?

Dicke concluded that during the compression phase between big crunch and new big bang the universe must have gotten hot enough to decompose atomic nuclei and erase all traces of the previous era of cosmic history—a billion degrees would do. Matter that hot, according to the laws of thermodynamics, would radiate high-energy gamma rays. The very early universe would have been an intense fireball.

Dicke wondered what had happened to the heat radiated from that fireball. He likened the expanding universe to a box with mirrored walls. The gamma rays inside could not escape as the box got bigger. Thus there were the same number of photons inside now as when the universe was young, but their wavelengths would have increased along with the size of the universe. In effect, the expanding universe would cool as it got bigger, like gas exploding in the exploding combustion chamber of an engine. By the present era, Dicke figured, the temperature of the universe should only be a few degrees above absolute zero.[1] The leftover thermal radiation from the primordial fireball would have been stretched in wavelength all the way to the other end of the electromagnetic spectrum, and would have been transformed into radio waves.

But not just any radio waves. They would be coming from everywhere, because the universe is everywhere, and their spectrum—the distribution of intensity with wavelength—would show the characteristic humped shape of thermal, or black-body, radiation: The energy would be smeared across a broad range of wavelengths, the intensity peaking at a particular wavelength and then falling off rapidly at greater and lesser wavelengths. It would be easy to recognize in the radio part of the spectrum if it could be detected at all.

"Wouldn't it be fun?" he asked his group, "if someone looked for this radiation?" He gave Dave Wilkinson and Peter Roll, two young radio astronomers, the job of thinking about how to do it. Then he turned to Peebles. "Why don't you go and think about the theoretical implications?"

"Which was an inspiring remark," Peebles recalled wryly, "because there were so many to think about." What would have been the effects

---

1. Absolute zero is the temperature at which all random motion stops and a so-called perfect gas would lose all pressure and volume. According to the third law of thermodynamics ("You can't get out of the game") nothing can be colder. It corresponds to $-459°F$ and $-273°C$. Physicists prefer to use the more rational Kelvin scale, which defines absolute zero as zero degrees and goes up from there. It uses the same temperature gradations as Celsius; thus water freezes at $273°K$ and boils at $373°K$.

of this radiation on the big bang? How would it interact with matter? How would you describe the radiation?

At that point in the summer of 1964 Peebles and Dicke and the rest of the crew were poised for the Nobel Prize. They were on the track of the spore of the creation—not some subtle manifestation of it, such as the slope or straightness of a line on some graph of galactic velocities—but the thing itself, the big bang, still rattling and echoing around in this cage of a universe.

Peebles had a vice, however, and it was about to cost him dearly. Like many bright people he was a little sloppy about his homework and had never developed the habit of hitting the books. "I was never strong on the literature. Still am not strong on the literature; I have to have something pointed out to me, often, before I'll recognize that someone else could have done something. It's so much more fun to think things through on your own than it is to read someone else's paper."

Invited to mull over cosmic radiation, Peebles didn't go to the library. He went home and thought.

The first thing that occurred to him was that if the early universe was very hot it would have been like a pressure cooker. In fact, it was like a star. "And I vaguely knew that when stars exploded, one predicted that the abundances of the elements would evolve in a way that is computable."

Stars burn hydrogen into helium. So Peebles set out to calculate how much helium the big bang would have produced. He came up with a figure that established that about 25 percent of the mass of the primordial universe would be turned into helium. Without knowing it, Peebles had solved a major cosmological mystery. Helium was one element that the stellar nucleosynthesis models of Hoyle, Fowler, and Burbidge and Burbidge might not be able to account for.

"I didn't know any astronomy," he explained sheepishly. "There was a helium problem, and it was simply that helium is very abundant. In our sun it is something like twenty-five percent of the mass. It seems to be equally abundant in very old stars, and you had the puzzle: Where did this helium come from?

"I'm a terrible astronomer. There are wild gaps in my knowledge. There were even papers I could have read explaining in detail why it was a puzzle that helium is so abundant and so uniform.[2] But that wasn't any motivation for me. I had no idea what helium was doing. I just did the calculations and got a large helium abundance." He laughed. "But I didn't know whether that answer was good or bad. I compared it to the helium

---

2. Including one published by Hoyle and Roger Tayler in *Nature* in 1964, in which they concluded that a big bang had to result in a universe at least 14 percent helium.

abundance in Jupiter because the first paper I had written in astronomy was on the structure of Jupiter, so I knew about helium there. And there was not a discrepancy. In my naive way I thought that was a triumph."

Peebles also estimated that the temperature of the leftover radiation suffusing the universe today would be about 10°K; it should produce a signal in the microwave part of the radio spectrum and exhibit the characteristic black-body curve of energy against wavelength. He wrote all this up and sent it to the *Physical Review* early in 1965. The editor sent it back. "It was rejected on the very good grounds that I hadn't looked at the history of the subject," Peebles groaned, "and I should have been aware that George Gamow had plowed all this ground along with Alpher and Herman years before. I was promptly told to go look up the old stuff, and see what was new."

If Peebles had done his homework, he would have learned that he and Dicke had reinvented George Gamow's old idea of a cosmic fireball. Recall that during the forties, when the expanding universe was still controversial, Gamow and his collaborators at Johns Hopkins University had attempted to explain the origin of the chemical elements by thermonuclear synthesis in a hot primordial fireball. Gamow's scheme had worked great for helium—the next simplest element after hydrogen—but had failed to produce heavier elements. Astronomers accepted instead Hoyle and Fowler's assertion that the elements were cooked in stars, and simply neglected all of Gamow's theories. In so doing they may have thrown the baby out with the bathwater.

Gamow's group had also suggested that the cooled remnant of the big bang radiation should still be around. In a paper published in *Nature* in 1949, two of his disciples, Ralph Alpher and Robert Herman, calculated that the present-day temperature of the universe should be about 5°K. Inexplicably, nobody followed up on this prediction; by the sixties it had been forgotten.

Dicke was one of those who had completely forgotten about Gamow's work in this regard. Not only that, but he had also forgotten that when he was at MIT during the war he had already done an experiment to measure sky background radiation, if there was any. He detected none, which led him to conclude that if there was any, it was less than 20°K.

Radio and microwave technology had improved greatly since Dicke's MIT days, and he realized that fireball radiation of only a few degrees might be detectable today, allowing physicists to measure the actual temperature of the universe and thus make serious computations about the original fireball. At Dicke's urging Wilkinson and Roll were assembling their apparatus on a Princeton rooftop to do just that.

Meanwhile, however, unbeknownst to Peebles and Dicke, the

putative big bang radiation had already been detected. A few miles away over in Holmdel, New Jersey, a pair of radio astronomers from Bell Labs were at that very same moment scratching their heads about a faint om- nidirectional hiss coming out of a radio antenna they were calibrating and planning to use for radio astronomy. In fact for years there had been buried in the data published by the phone company hints of some sort of anomalous low-level signal intruding on their own—too faint to be of any practical significance. "For many years people at Bell Labs sat on this anomaly under the reasonable assumption that it was something in their instruments, something they just couldn't explain," said Peebles. It was only while Peebles was boning up on his history, that word came along the grapevine that a pair of radio astronomers over at Bell Labs had some funny noise in their system. It turned out to be the big bang calling.

Bell Labs had a glorious history in accidental astronomical dis- coveries. It was in Holmdel, in the thirties, that an engineer named Karl Jansky built an antenna to investigate the sources of static on long-distance lines and discovered there was static coming from the center of the Milky Way, thus inventing radio astronomy.

The discovery of the cosmic background radiation was partly another such accident, but it was also a tribute to the pride and fussiness of the two young radio astronomers involved, Arno Penzias and Robert Wilson. They were a classic pair: Penzias aggressive, talkative, and im- aginative; Wilson thoughtful and meticulous. They had been recruited in the early sixties to modify a special horn-shaped antenna built to bounce signals off the *Echo* satellite so that it could send and receive microwave transmissions from the new *Telstar* communications satellite. Their re- ward, once this was accomplished, would be the opportunity to use the antenna for radio astronomy.

The Holmdel apparatus resembled a giant alpenhorn on its side; its 20-foot-square opening caught the microwaves and then funneled them to a new state-of-the-art receiver and amplifier, whose heart was a ruby crystal. Problems developed, however, when Penzias and Wilson tried to calibrate the antenna's gain by measuring its response to an airborne radio transmitter. There turned out to be a constant unexplained signal—a hiss, a hum—coming out of the receiver no matter how the antenna was positioned, even if it was pointed at empty space, no matter when the tests were done.

The temperature of this anomalous signal was about 3 degrees above absolute zero.[3] It was minuscule, but it was persistent and annoying.

---

3. Radio engineers typically characterize the strength of a signal in terms of temper- ature, by comparing the intensity at some wavelength to the intensity of thermal radiation that would be emitted by an idealized black body in the same wavelength band.

Penzias and Wilson were sure it was noise in their system. No known cosmic source broadcasted at the wavelength they were tuned to, 7.35 centimeters (4080 megahertz), and none could be so uniform and constant. They tore their hair for about a year trying to get rid of it. They took apart the electronics, shooed pigeons and shoveled pigeon droppings out of the horn, and taped over rivets, with no luck.

Finally, in the spring of 1965, Penzias and Wilson gave up. If the signal was real, which they doubted, it was important—whatever it might be. But they didn't want to make fools of themselves in case they had overlooked something in their troubleshooting, so they decided to bury their announcement of it in a twenty-page technical report on their calibration procedures. While they were writing it, Penzias was tipped off to Peebles's paper, which was circulating as a preprint, and called up Dicke for a copy.

After he read it, he called Dicke again. You'd better come and take a look at this, he said.

Dicke, Roll, and Wilkinson drove down to Holmdel. It didn't take them long to conclude that Penzias and Wilson were first-rate astronomers and that they had indeed probably tuned into the fading remnant of the primordial big bang itself. Peebles recalled that when Dicke came back he was pleased that old cronies from his days in the radar business still remembered him.

Penzias and Wilson were thrilled: At last, an explanation for the troublesome hiss. They weren't lousy engineers after all. The cosmological implications of their discovery went right past them at first. Neither of them took cosmology too seriously; Wilson, in fact, was a fan of the steady state theory.

Peebles, on the other hand, was dumbfounded. "I was very excited," he recalled. "I felt a little disbelief—you mean this is actually working out? It was a long shot to begin with, a highly speculative idea that you can detect radiation left over from the big bang, and I think that all of us—well, I can't speak for the others, but my feeling was one of relief, that there was something in this after all.

"It's an awesome topic. It happens so seldom that you venture an inspired guess and it works out, on a scale so far removed from what we already knew. But it worked!"

The Princeton group hurried to write a quick note explaining the theoretical and cosmological consequences of the microwave radiation discovery while Penzias and Wilson wrote one on the actual observations. They were printed back-to-back in the *Astrophysical Journal Letters*. Cautious to the end, the Bell Labs pair restricted themselves to a dry description of the radio data, eschewing all interpretation. "A possible explanation for the observed excess noise temperature is the one given by Dicke, Peebles, Roll and Wilkinson (1965) in a companion letter in this issue." Before sending it in, they did yet another test, driving around the Holmdel site with a portable radio generator.

Before the papers appeared, science reporter Walter Sullivan got wind of the discovery, and it was announced on the front page of the *New York Times* that astronomers had discovered the explosion that had given birth to the universe. When Penzias and Wilson read it, the ramifications of their discovery finally sunk in.

In Colorado, Gamow, by then retired, read the same newspaper accounts with an equal appreciation of their ramifications and with mounting indignation. Neither the Princeton nor the Bell Labs papers mentioned the Gamow group's prediction of big bang radiation. He wrote an indignant letter to Dicke (who admitted he should have known better), pointing out various places the prediction had appeared in print. His indignation and sense of injustice never faded. Years later at a conference on the microwave radiation, Gamow said, "If I lose a nickel and someone finds a nickel, I can't prove it's my nickel. Still, I lost a nickel just where they found one." Disgusted, Alpher and Herman left physics.

The brouhaha about the authorship of the cosmic background radiation was a black eye for the scholarly gentlemanly traditions of cosmology, and perhaps explains why Penzias and Wilson got the Nobel Prize in 1978 for its discovery, but nobody got a prize for predicting it.

Peebles was rather defensive about the idea that he and Dicke had just recapitulated Gamow's work. "I don't think anyone previously had realized that one could do an experiment to measure this stuff," he argued. "In fact I asked Gamow that very question: Had people recognized the experimental possibilities? His answer was clearly no. It was clear they didn't realize that this stuff could be detected now."

Steven Weinberg, a Harvard particle physicist who wrote a bestselling book, *The First Three Minutes,* about cosmology, took a more philosophical view of the failure of astronomers to follow through on Gamow's original prediction. It was, after all, part of a discredited theory—the attempt to ascribe all element production to the big bang. "This is often the way it is in physics," he concluded. "Our mistake is not that we take our theories too seriously, but that we do not take them seriously enough. It is always hard to realize that these numbers and equations we play with at our desks have something to do with the real world. . . . The most important thing accomplished by the ultimate discovery of the three-degree radiation background in 1965 was to force us all to take seriously the idea that there *was* an early universe."

The steady state theory practically vanished overnight. But there was no huge rush to embrace the idea of the early universe—more of a stately parade. Although Penzias and Wilson's discovery was hugely suggestive, it was only one measurement at one wavelength of the obscure and difficult microwave band of the electromagnetic spectrum. The cosmic background radiation was presumably smeared out across the spectrum, and only by measuring it at many different wavelengths could astronomers be sure that its intensity distribution conformed to the classic black-body hump. Physicists were a long way from showing or knowing that the cosmic radiation had the black-body spectrum of a primordial fireball, or that it really did fill the sky.

In the summer of 1965 Wilkinson and Roll got their Dicke radiometer working and were able to detect the background at a second wavelength, and it gave roughly the same temperature. Measuring the microwave background became a cottage industry, a subbranch of radio astronomy. Observations mounted and points were filled in on the long wavelength side of the spectral distribution, which showed the intensity rising as wavelength decreased, as a black body would. Peebles was encouraged. The actual peak of the hump and the shorter wavelength radiation on the other side was harder to detect and measure because it fell in a difficult part of the far-infrared or submillimeter-wavelength region of the spectrum. Many of the instrumental techniques were classified, and observations had to be performed above the atmosphere, by sounding rockets and balloons. Nevertheless the characteristic curve of a black body eventually emerged and showed the universe with a temperature today of 2.7 degrees above absolute zero.

While some astronomers concentrated on confirming the spectral characteristics of the cosmic background, others investigated its spatial characteristics, searching for variations from one part of the sky to the next. If the microwave radiation really was the remnant of the primordial fireball, then its distribution in the sky ought to reveal how matter and energy were arranged during the earliest moments of the universe. Astronomers searched in vain for any departure from smoothness, any hint of structure in that early fireball, but the microwave background proved to be astonishingly uniform and bland. From pole to pole, North, South, East, West, the temperature of the sky background deviated hardly a whit from 2.7. There were no hot spots or cold spots, no ghostly trace of the merry clumps and chains of galaxies that would decorate the sky 15 billion years later.

Finally, in 1977, a team of Berkeley astronomers, flying a radiometer in a U-2, found a minute variation in the microwave background, but the variation itself was so smooth and regular that it only reinforced the impression of miraculous blandness. The sky, they found, was three-thousandths of a degree warmer in the direction of the southern end of the constellation Leo and correspondingly cooler in the opposite direction. The pattern, they pointed out, was exactly what would be produced by the Doppler effect—the same effect that caused redshifts—if the earth and Milky Way, and presumably the whole local group of galaxies, were drifting through an absolutely uniform bath of radiation. The microwaves would look slightly hotter in the direction we were going and slightly cooler in our wake.

All this went some ways to quell Peebles's suspicion of cosmology, but it took a long time. "It wasn't until ten years after the Wilson-Penzias discovery that I really felt comfortable with this idea," Peebles confessed. "You know that there was a long period in which there were substantial anomalies from the rocket measurements. They were pretty good measurements, they were awfully good people, and one would have been very justified in saying that what they were showing was that this idea was not right. I guess I couldn't put a finger on a day when I felt rather strongly that this really is thermal radiation left over from the big bang. It was a rather gradual flow of opinions that many of us went through, from one of hope that this might be right to cautious optimism, to a pretty strong belief that this is the way it has to be. Which is where I am now and where I think most people are."

Peebles became a kind of paleontologist, using the microwave radiation to explore the putative early universe. The exciting thing about the microwave discovery was that by knowing the temperature of the universe now, he could calculate the temperature at any time in the past, even, say, when the universe was only a minute or a second old, when protons were being fused together with neutrons to form helium. The radiation (and the universe) would behave just like a gas in a cylinder, which grew hotter as it was squeezed.

Another paleontological tool at his disposal, he realized, was the abundances of the light elements. The percentage of helium in an ancient globular cluster star was a relic of processes that took place during the first few seconds of time. Now that he could calculate the temperature back then, he could go back and calculate the other properties of the early universe, jiggling the numbers until today's helium abundance came out right. Temperature, density, pressure—these were all quantities that physicists dealt with. Before the discovery of the microwave background, cosmology had been astronomy, but afterward, in the fall of 1965, it became physics. Peebles rewrote and expanded his rejected paper on cosmic radiation and helium production.

"So I guess by then I was already aware of the helium problem," he explained, his voice arching. "But I must say that the main driver in doing these computations wasn't so much to solve an astrophysical problem with helium, but rather to use the production of helium as a probe for cosmology. Because one recognized if you adjusted the cosmological theory, you would adjust the predictions, perhaps very dramatically, of what was produced during the big bang. That seemed to me to be more exciting."

Potentially one of the most sensitive "probes," Peebles found, was deuterium, a heavy isotope of hydrogen important on earth for nuclear weapons work. Normal hydrogen atoms consist of an electron circling a proton; the deuterium nucleus contains a proton and neutron. Deuterium was a way station in the building of a helium nucleus, which normally has two protons and two neutrons. The denser the early universe was, the more deuterium got used up in helium synthesis and the less pure deuterium survived to emerge from the big bang thermonuclear furnace. Small changes in the density of matter, Peebles saw, would lead to enormous changes in the final cosmic abundance of deuterium. The trouble was, no one knew what the cosmic deuterium abundance was, or how much was subsequently burned or produced in stars.

With this work, Peebles had begun what would be the style of a new science: rolling back the universe to earlier and earlier times and hotter and hotter temperatures and applying whatever physics—if any—were applicable at those energies and densities to calculate the consequences for today's observable universe.

Having pointed the way into the primordial fireball, Peebles did not go in. "I wrote two papers and then I moved on to other things, and I think that was the right decision. Since I'm not an astronomer I don't care to go into all of the details of how you estimate helium abundance in stars, how you correct for helium produced in supernovae, how you make a detailed comparison between theory and observation. I think that was well left to others."

The job of making detailed calculations of nucleosynthesis in the big bang and comparing the results to reality passed to California, where it was eventually taken up by Fowler, Hoyle, Robert Wagoner, and a never-ending line of postdocs that Fowler anchored like a giant galaxy at Caltech. Within a few years, aided and abetted by new spacecraft techniques for making delicate astronomical measurements, they found that when they adjusted big bang parameters of temperature, pressure and density to make the correct amounts of helium, the same calculations also predicted the correct abundances of deuterium and lithium—even though helium was 25 percent of the universe while lithium was less than a billionth. The big bang theory apparently worked.

Peebles, watching from afar, was proud. "It certainly looks as though there has been a great triumph here. You can compute what happened when the universe was one second old—it's spectacular! Again, if someone had told me that when I was a graduate student, I think I would have laughed. The universe is too complicated. But sometimes, the universe is amazingly simple, like the hydrogen atom—it's hard to get it wrong." He laughed.

In his own work Peebles went the other way and became involved with complication. He sought the origin of galaxies, how the universe had gone from the smooth blandness of the primordial fireball to the bright knots of galaxies and stars. He was quietly, methodically promoted through the ranks, becoming a full professor in 1972 with a big office next to Dicke's; he never had to leave Princeton. He and Alison had another daughter. He began teaching the first course on what he called "physical cosmology" to distinguish it from the Sandagean pursuit of Hubble constants and curved space. He was constantly invited to symposia and conferences, but he had a healthy suspicion of the usefulness of such gatherings. "If we met half as often, we wouldn't be that far behind," he said. One symposium he remembered chiefly because the institute sponsoring it had hired a new chef.

His favorite such occasions were small gatherings where he could badger the speaker, and none more so than the workshops put on at the Aspen Center for Physics, a small, private research center which started out as a summer retreat for theoretical physicists. One of its founders was a former nuclear physicist who owns much of Woody Creek—just down the valley from Aspen proper. Run, in the words of one regular, on "a shoestring" provided by the National Science Foundation, the physics center was another of the small jewels in Aspen's glitzy cultural crown. In the summer physicists came for extended stays to sit around picnic tables in the sun and talk and write papers together. In the winter they came to ski. Before and after the slopes were open they gathered in their ski clothes to talk physics. The dialogue would continue all day on chair lifts and in warming huts.

Skiing was the sport of the cosmologists. Peebles skied better than anybody with such long legs should be expected to ski. He carved his way down the mogul-packed upper slopes with deceptive speed, his body curved into a long maroon question mark. He'd start out slow and then suddenly he'd be halfway to the bottom. I hadn't skied in twelve years and I had to hustle to catch up. When he hit the bottom where the slope flattened slightly and ran out down an undulating gorge, he shouted, "It's Suzy Chapstick time!" and bent into a tuck to schuss the rest of the way home.

8

THE GALAXY MAKERS

Why should there be galaxies in the universe? It was an ancient problem, as old as Hubble, who had hoped that his tuning-fork diagram of galaxy types might turn out to provide some clue to their evolution and origin. Somehow the elegant simplicity of the big bang, the smooth, undifferentiated fireball, had congealed into the jeweled clumps of galaxies and stars that spangle the modern universe. But when and how? What physics made giant elliptical galaxies so uniform, for example, that a Sandage could dare to use them to plumb cosmic distances? Was there any sense or pattern behind the organization of matter in the universe, or was it just random chaos?

Two trails of thought occurred to Peebles on the day back in 1964 when Dicke told him to go off and think about cosmic radiation. The first was that the early universe would be like an exploding star, transmogrifying elements; that led to the thriving science of big bang nucleosynthesis. While a younger generation of astrophysicists sought to plumb the big bang by studying the intricacies of nucleosynthesis and the abundances of trace elements scattered among interstellar dust grains and in the vapors of stars, Peebles concentrated his energies on the second question. In effect turning from the microcosm to the macrocosm, he sought to understand the origin of galaxies.

He was one of two men, one in the West and one in the East, who sought answers to those questions and who became the poles of opposing world views. They would find themselves, in a few years, trying to examine the big bang by studying the largest and most obvious aspects of the universe—the wheeling galaxies themselves. That was the question that raged on the chair lifts of Aspen.

What would be the leading preoccupation of physical cosmologists began with a simple thought. It occurred to Peebles, as in fact it had occurred to Gamow, that it was too hot in the early stages of the universe for galaxies to start to form. The fireball radiation would exert enormous pressure that would blow matter around like dust in a furnace. Any clumps that tried to coalesce would be blown apart.

"I realized that this [was] an awfully significant result because it [set] a bound on when you can form galaxies," Peebles recalled lazily. His long Levi-clad legs arced from his leaned-back chair up over his desk and down the front of it, ending in comfortable-looking hiking boots, a foot from my face. What he had determined was that no primordial lumps would be able to grow until the universe had expanded and cooled to the point where the force of gravity pulling clumps of matter together outweighed the pressure of radiation keeping them apart. That time could be calculated, giving the galaxy theorists something at last to sink their teeth into. If no one could now say when or how galaxies did form, at least Peebles could now say when they *did not* form. It was progress and it whetted his appetite for the subject.

Peebles set out to explore how galaxies could grow out of this inferno. Soviet theorists E. M. Lifshitz and I. M. Khalatnikov had shown a couple of years earlier that if there were any lumps in the distribution of matter in the early universe, they would grow. Gravity amplified enthusiastically every primordial ripple of density in the expanding spacetime. As the universe expanded and the radiation cooled, incipient clouds of gas or particles would grow and then eventually begin to condense under their own weight into stars, or galaxies of them. If this were true, then the great clouds of galaxies on the sky today would be descended from minute seeds in the original fireball, and studying the organization of the galaxies might reveal clues to the origin of the universe.

Conversely, the organization and large-scale structure that emerged from gravity's amplifier to make the modern universe depended on what kind of lumps or ripples existed at the beginning. Was the primordial universe inclined to small lumps over large lumps? Was it lumpy on all scales, with lumps within lumps? Peebles suggested that the universe favored small lumps. If the lumps were like waves on a pond, the waves with the shortest wavelengths would have higher crests, and they would grow the fastest. In his scheme of things, the smallest objects would be the first to collapse and bear light. A hierarchy of collapsing and clustering of ever larger conglomerations of matter would occur. In 1966 he used the microwave background data to calculate that the combination of cooling and expanding would select clouds with a mass of about a million suns as the first to collapse and form recognizable astronomical objects.

As Peebles tells it, Dicke walked in one day and said, "Hey, wait a minute. You ignorant physicist, don't you know that the mass of those objects you're computing and their densities are remarkably similar to those of globular clusters?" Peebles didn't, but that news was exciting. Globular clusters were very old and distinctive objects; their origin and role in the galaxy were enigmatic. Now Peebles had discovered that they might be the building blocks of galaxies.

Dicke and he proceeded to dash off a paper. But, as Peebles remarked, "It's characteristic of my style of operation that I wrote a rash early version of a paper that was roundly rejected because I had missed some good physical points."

After being sent again to the library, Peebles rewrote the paper, but his troubles still weren't over. Chandrasekhar, the editor of the *Astrophysical Journal,* couldn't find a referee who both understood the paper and believed it. Finally he took it under his own wing, only to discover that they had slipped in some humor, which he made them take out.

"We had remarks about the life of one of these [primordial gas] clouds being nasty, brutish, and short," he explained, referring to the English philosopher Thomas Hobbes's famous description of life before the institution of government, "and he didn't like that. It's not much of a joke. And it was an important point. We had to explain why, if the first objects to form were these gas clouds, most of the mass of the universe wasn't tied up in globular star clusters. We had to argue that the chances were very high that the gas clouds would be disrupted by collisions before they could form themselves into star clusters. That was the remark about the life of a cloud being nasty and brutish and short." He concluded wryly, "The paper was received with not much excitement by the astronomical community."

Peebles went on to formulate and advocate what he called the hierarchical or "bottom-up" theory of galaxy formation. Galaxies would assemble themselves out of small clouds of stars and then swarm together into small clusters. The clusters would in turn gather into even larger groups. The longer time went on, the larger the large-scale structure of the universe would get.

It sounded good, but was it true? Did nature work like that, and was there a way to find out?

*   *   *

At this point Peebles's career was influenced by an accident. On his way back from a year's stay at Caltech in 1969, he arranged to stop off for a month at Los Alamos National Laboratories in New Mexico, home of the atomic bomb, partly to visit astrophysicists there but mainly to have a break in the cross-country drive. "Los Alamos has, of course, enormous computers," he explained. "It always has the state of the art. Since I was there for only a month, and I had all these computers around, the one project that seemed feasible and profitable would be to use these big computers to do simulations.

"I think the computer was a CDC 6600," Peebles said airily, frowning at his own lack of memory. "I didn't pay much attention to what it was, except that it was more powerful than any I'd seen, and it was many magnitudes larger than any I could have gotten at a university computer center. In those days there were big computers at universities at the computer centers, but you paid in research money for computation time, so you didn't use large gobs of it. At Los Alamos I could let it run all night.

"So I did it. I remember with great pleasure that month. Alison was a little bored because Los Alamos is really a very isolated slow community. But when I wasn't working we were traveling, and New Mexico is a wonderful state."

The kind of exercises he had in mind were called N-body simulations, and they were to play a prominent role down the road in cosmology. The goal was to mimic how gravity would affect the arrangement of the contents of an expanding universe. Peebles began the calculation by sprinkling 2000 points, each representing, say, the mass of a galaxy, around in an imaginary box that represented part of the universe. Knowing the location and velocity of each point, the computer determines gravitational force exerted on each point by all the others, lets the particles drift off under the influence of these forces for the computer equivalent of a few million years, finds out where the particles wind up, and displays the result. Then it would redo the computation based on the new distribution, and keep going. At the end Peebles would have a sort of movie of galaxy mass points moving around, slowly clustering. If he were lucky the distribution of points would look something like the maps of galaxies on the sky.

And in fact they did. The resulting pictures showed dots slowly gathering themselves into a smooth, round cloud. "That led to a pretty nice model for the way something like the Coma cluster of galaxies evolved," said Peebles. "It's still a pretty nice model—close to reality, probably."

Back in Princeton Peebles continued doing simulations of the expanding universe with graduate student Ed Groth, who greatly refined Peebles's crude computer algorithms. But simply looking at pictures and comparing them to maps grated on Peebles's physicist mentality. He thirsted for numbers. He began to wonder if there was some quantitative way to compare the galaxy distributions in his simulations to those in the sky.

Another travel adventure spurred him on. During a visit to the University of Toronto to lecture about the microwave background radiation, Peebles fell into a discussion with Sidney van den Bergh, the Canadian Sandage, about whether the galaxies were really distributed homogeneously through the universe—as Hubble had claimed and Sandage believed. A Caltech graduate student named George Abell (subsequently a UCLA professor and author of a famous textbook) had spent years scouring the big glass plates of the Palomar Sky Survey for clusters of galaxies. Van den Bergh had a map of these so-called Abell clusters in his office, and the clusters appeared gathered in clumps and long chains.

"Look, even at this depth, how lumpy things look," said van den Bergh, pointing to the map.

Peebles replied, "How do you know that that distribution isn't simply a random distribution of points? Are these clumps real or are they accidents?"

"I can't tell," said van den Bergh. "They look real to me. Why don't you check it out?"

"I think I will," Peebles responded, "sounds like a fun project."

On the airplane flying back home he thought of a way to apply statistics to the galaxy and cluster distributions and began to work it out. "I remember," said Peebles, laughing, "that sitting next to me on the plane was a lady who didn't say anything to me the whole flight, and I was occupied, scribbling away, working this out; but at the end of the flight she leaned over and whispered to me, 'Young man, you're getting all your homework done, aren't you?' "

The technique he began to invent that day is known as the correlation function, and it became one of the most powerful tools of cosmology. The correlation function was a measure of the sociability of galaxies. The idea was simple; the math, horrendous: Find a galaxy in the sky. Draw a little circle around it and ask what were the odds of finding another galaxy in the circle. Now draw a larger circle. What were the odds of finding another galaxy in the annulus between the two circles? And so the correlation function just gave the probability of finding another galaxy within any given distance of the first galaxy. A computer, Peebles realized, could easily do all the circle drawing and counting on the results of a simulation as well as on a map of the sky, yielding numbers that could be used to compare reality to hypothesis.

Moreover, he thought gleefully, the correlation function was a start toward understanding the physics behind the original ripples in the fireball that were the template on which the clusters of galaxies had been laid down.

"Well, it seemed to me to be a wonderful thing to be able to say what the initial character of density fluctuations was," Peebles said. "Also, of course, I came to see an amusing challenge—there were all these galaxy catalogs floating around with which people had done relatively little because until you had high-powered computers there wasn't much you could do. Computers at that time were not new, but new enough that no one else had recognized, well, there's a gold mine here to be mined, until I came along."

He began casting around for suitable galaxy samples to run through his computerized mill. Suitable samples were hard to come by. To make his statistics sound, Peebles needed to know the locations in the sky of tens of thousands of galaxies. Gathering such large lists was to be a perpetual problem in cosmology. At the time, Sandage had measured the redshifts of only a few hundred. Somehow—he doesn't remember how—Peebles heard about an extraordinary archive of data at Lick Observatory in northern California. Between 1947 and 1954 the director of the observatory, Donald Shane, and a student, Carl Wirtanen, had taken a series of 1256 photographs of the entire northern sky through the observatory's 36-inch Crossley reflector onto 17-inch-square glass plates. Then, dividing the sky into cells 10 minutes of arc (a third the diameter of the full moon) across, they had counted the numbers of galaxies visible down to the 19th magnitude in each cell. They tallied more than a million galaxies. From the galaxies' magnitudes, Shane and Wirtanen estimated distances of a billion light-years for the farthest ones.

Peebles found out through Dicke, who knew Shane, that Shane still had the original data sheets; Peebles called him up. "He's a very friendly person. He invited me to come visit him in Santa Cruz and stay at his place in the redwoods outside town in his little summer cottage. Which was really a treat." Peebles came back with a microfilm of the data sheets.

"I'm so impressed with the fact that Shane had worked and had taken these data at a time when one didn't have sophisticated computers. They had no reason to think that anyone would ever look at the data with any care. By the time that they were completed I don't know if there was a computer that could have held that catalog in its memory. I doubt it. And yet they took such scrupulous pains in setting the thing up with statistical controls, that when computers did become available we were able to do an awful lot with that catalog."

Peebles hired keypunchers to enter the data onto cards and got back a pile of punch cards bigger than he could carry. A series of graduate students and postdocs did the grunt work of checking the data for accuracy and programming, and reducing them. "Whenever I can avoid it, my students or friends, colleagues do the computations," he explained. "I guess that's another aspect of my impatience, and I don't really enjoy sitting staring at something like a terminal."

The payoff didn't come until about 1975, when Peebles and his crew were finally able to compute the correlation function. In a nutshell, the correlation function was just a measure of the length, or angular scale, on which galaxies like to congregate, and what the Lick data showed was that—up to a limit—galaxies like to congregate on *every* scale. The best place to find one galaxy was right next to another, the next best place was a little farther away, and so on; the likelihood fell off smoothly with distance, out to a distance of about 50 million light-years, where the function suddenly plummeted.

What did this mean? On small scales—smaller than clusters of galaxies—nature was organized relentlessly, like a set of Chinese boxes. Any clump of galaxies that you looked at would be composed of smaller clumps, and within those smaller clumps would be smaller clumps still. A photograph or map of the universe would look the same on any scale—up to a point.

That point was when the scale of your photograph exceeded 50 million light-years, the distance at which the correlation function abruptly seemed to "break." Galaxies separated by more than that magic distance were apparently physically unrelated. "That, of course, is of interest," murmured Peebles, "because if you have a characteristic length, you have something to get your teeth into. There must be some physics there."

Peebles thought the break represented the skirmish line in gravity's ongoing attempt to arrange the primordial contents in the universe. His idea of how the universe was structured reminded me of the old radical's prescription for political and social change: the individual organizes the family, the family organizes the block, the block organizes the city, the city organizes the state, and so on. This process would take generations; you could measure time by the level to which the organizing process had reached. In Peebles's bottom-up theory of clustering, the original distribution of matter was determined by minute random inhomogeneities in the big bang. As time went on and the universe expanded, clouds of gas condensed around the denser parts of the universe; galaxies formed and gathered themselves in groups. The groups gathered into larger groups, and the hierarchy swelled outward with time. The break in the correlation function indicated to what scale the clustering tendency or organizing process of the universe had advanced. Everything smaller than 50 million light-years had already been rearranged; everything larger than 50 million light-years still followed the original primordial fault lines.

So Peebles had perhaps fulfilled his dream of making contact with primeval physics through the large-scale structure of the universe. Understanding that physics and proving his rather intuitive interpretation of the correlation function would turn out to be a heartbreaking and treacherous task, but in the meantime the break served as a convenient signature to use in evaluating computer simulations. Statistical cosmology became a buzzword and a growing industry in university departments and in the literature. The correlation function was objective and easy to use.

"They're convenient—they're reliable," explained Peebles. "That is the main reason why these correlation functions have been so easily used. But they're crude statistics; they don't tell you a lot about the way galaxies are distributed. For example, they don't tell you if galaxies are distributed in sheets or bubbles or chains. One can see lines in the Lick map, but one can see lines in noise, the eye is such a good pattern maker.

"Anyway, despite the limitations these correlation functions have some virtue—I think overwhelming virtue: You can measure the suckers."

Correlation functions made it into the technical lexicon, but it was another aspect of the Lick project that made it into the popular culture. Peebles and his colleagues transformed the Lick galaxy counts back into a visual map of the sky with each 10-minute cell depicted in white or various shades of gray, down to black, depending on how many galaxies were in that cell. The map was published as a poster, called "One Million Galaxies," by Steward Brand of *Whole Earth Catalog* fame, and thus found its way into libraries, dorm rooms, and various dens of countercultural iniquity.

To gaze at this map is to be reminded of all the unclaimed richness of the universe, for which dry statistics either cannot or will not provide testimony. It goes without saying that the whole sky map is not a uniform gray. There are great clots of white and black, and whorls and chains that snake through the firmament like turbulent smoke. The practiced eye can pick out the Coma cluster; the Virgo cluster, which lies in the same general direction in the sky, is so spread out that it makes no impression here. What about the rest—the windy strings and sheets? Are they real, or just a trick of the eye, the same eye that saw canals on Mars? Is this the texture of creation, the primordial pattern?

A hint of the primordial anything was usually enough to set an astronomer's or physicist's blood on fire, but Peebles's style, laconic and cool, was to distance himself from the white heat that motivated other cosmologists. His papers were often flip; his lectures occasionally took the form of grading his opponents' theories as if he were a sort of master schoolteacher. "God didn't intend me to be a cosmologist," he told me once, riding a ski lift in Aspen. "I just think of myself as a physicist, no different than somebody doing acoustics."

I asked him if he had any prejudices or preferences regarding the fate of the universe, the kingpin question that made most cosmologists sweat. Was it open and destined to expand forever, or was it closed and doomed to collapse and crunch itself?

"I think that it's a dull question," he said. "If someone gave me on a tablet of clay, the answer and the numbers, I would be disappointed. I would throw it away, because the great discoveries are not going to be a final number but the method you come to apply to learn that number. What good would it do us to know? The q-noughts or whatever it is? Those numbers in isolation don't mean much to me. They're just numbers, not the ball of wax you untangle to get the numbers. Actually, when I got into this subject, as you noticed, I entered as a skeptic thinking that surely the universe is not this simple. I certainly did not have any dreams of discovering the value of q-nought. That seemed the furthest thing from my mind. The big question was, Is this cosmology close to the truth?

"And I guess that's still my question. We've come a long way in this subject and I now believe that the big bang cosmology is close to the truth. It's a miracle, that the human mind could come up with something that describes the evolution of the universe over these enormous spans of time. Eventually we'll know whether the universe is open or closed. And I suspect at the same time we'll have a pretty good idea why." He laughed and crossed his legs. "But I certainly don't have any preferences. Open or closed—either one is too awesome to prefer."

Peebles could be called "the father of physical cosmology," but on the other side of the world, operating out of a shabby wooden barrackslike building in Moscow that housed the Institute for Physical Problems, there was a rival for that title.

Yakov Boris Zeldovich was everything Peebles was not. Peebles was rangy and sardonic; Zeldovich was short and fiery. He had never even gone to college. Peebles was analytical; Zeldovich intuitive. Peebles was criticized for being aloof with his students and not helping them get jobs; Zeldovich dominated and guided their careers like a godfather. Peebles concluded that the universe was built from the bottom up; Zeldovich concluded that it had been built from the top down.

My first sight of Zeldovich was in an ancient, paneled lecture room one March morning in 1986 at the Sternberg State University in Moscow, where he ran a prestigious seminar two Mondays a month. On this particular day, Carl Sagan, who was in town on Halley's Comet business, had been invited to speak about nuclear winter. About 120 scientists and academicians were jammed into the room. They were all wearing suits, except for Zeldovich, a muscular, bald man about the stature and shape of a fire hydrant, who was in a gray sweater with a red sawtooth pattern. At the time Zeldovich was seventy-two and had just married for the third time. He was bouncing around, robust and cheerful, like a man thirty years younger. He greeted Sagan effusively, complaining that Sagan had not brought along his wife, Ann, and then he looked through me with impassive gray-blue eyes when I tried to introduce myself.

Sagan talked for three hours—a Soviet seminar being serious business—and for much of the time Zeldovich didn't even seem to be paying attention. He sat in the front row like a ringmaster, whispering to colleagues, passing notes, thumbing the stack of magazine articles I had given him, sometimes speaking out loud in Russian. I thought he was being rude, but it turned out he was arguing with Sagan's translator. Finally, halfway through the morning he dismissed the translator with a wave of his hand, leapt up on the stage, and took over the translating himself.

When it was over he gave me his business card—one side printed in Russian, the other in English—which identified him as the head of the theoretical department of the Institute for Physical Problems of the USSR Academy of Sciences and member of the U.S. Academy of Sciences and Britain's Royal Society as well as a Hero of Socialist Labor. He scrawled his phone number on it and told me to call him at six the next morning. I was to learn later that it was a famous Zeldovich habit to ask people to call him early in the morning with the answers to puzzles, but when I telephoned he was confused. He answered the phone in Russian. He didn't remember my name. "Who is it? Who is it?" he kept asking. I explained slowly, fighting against the panic of losing my brief connection with the man whom Sandage had described as "the Einstein of our profession," that we had met at Sagan's lecture. A pause.

"Is this Sagan?"

Finally his voice brightened. "Ahh, the *correspondent*."

I liked the sound of that. Four hours later I had followed his instructions and the address on the English side of his card to the shadow of the Yury Gagarin monument—a statue of the world's first cosmonaut standing atop a soaring pillar with the earth lying at his feet like a soccer ball. I didn't know where to go from there, except across the river of traffic on busy Leninsky Prospekt, at which point I was grabbed by a cop for jaywalking. The cop jabbered away in Russian, oblivious to my protests that I didn't understand. I gave him my passport; he took it and kept talking sternly.

Finally I handed him Zeldovich's bilingual business card. The cop's eyes widened, and he smiled. He gave me my passport back and pointed down the street to a cluster of low white buildings.

Zeldovich was waiting for me in a small, gloomy first-floor office tucked behind a staircase. He was wearing the same sawtooth sweater as the day before. His office had no windows, one lamp, and a worn couch. On the wall was a framed photograph of a laughing Zeldovich in a T-shirt imprinted with the legend "2.7 2.7 2.7 2.7 2.7 . . . ," the temperature of the cosmic microwave radiation. He was in a twinkly mood, and no sooner had I settled on the couch than he raised an index finger in the air and started giving me a physics lecture.

"I think that the need to understand the surrounding world is something very deep in the human mind," he pronounced. "It's just perhaps the most distinguishing thing between animals and humans.

"At first glance there is an immense difficulty. You see very different animals, the heavens, the stars. You see very different things. And then two questions are naturally posed. What is the present situation and from what did it come?

"Now it is microphysics which gives us the understanding of all the, um, different shapes. We understand that the different shapes of what we see are different structures of elements. From a few, uh, different stones you can build very different structures. So first you come to the idea that all is built from a hundred chemical elements, and you go deeper and you see that the elements themselves are atoms. And then you go deeper and you see the atoms are made from nuclei and electrons. And the nuclei from protons and neutrons. Then you go even further and you see that even protons and neutrons are not elementary but are built from quarks."

Whatever Zeldovich himself was made of, it was sturdy stuff. His whole life was a testament to smarts and moxie. He had been born in Minsk in 1914. "I was always the shortest one in my school class," he said. His formal education was hampered by the civil strife in Russia following the Great War, which closed the schools, and by the fact that he was Jewish. He remembered staying home for several years and studying with a tutor. By the time he was ten or twelve, he recalled, he had "come into science." He talked to his father about what to do. They agreed that math was only for geniuses, and that there was nothing much left to discover in physics (which has periodically gone through such times, when its practitioners thought it was "finished").

Yakov Borisovich therefore decided to pursue chemistry. Unable to attend high school or college, he got a job as a lab assistant at the quaintly named Institute of Processing of Useful Ores. When he was seventeen he was sent on an errand to the Physical Technical Institute in Leningrad and beguiled the chemists there in a learned discussion about nitroglycerin. They invited him back. A little finagling and adjustment of budgets ensued. The upshot was that he was traded to the Leningrad institute in exchange for a vacuum pump.

The diminutive assistant quickly established himself as the brightest person in the lab. He absorbed knowledge like a sponge. Within five years he had written and defended a dissertation to become a candidate of science, the equivalent of a Ph.D.

He became an expert on the behavior of gases and particularly on combustion, a line of work that led him to the bomb. Zeldovich wrote a paper on chain reactions and nuclear fission in uranium that was published in the open scientific literature—the last unclassified paper on the subject—just before World War II. During the war, he worked on jet propulsion. The late forties and fifties are a big hole in the official Zeldovich résumé, years he only refers to obliquely as work on uranium fission. Zeldovich was a charter member of the Soviet atomic bomb project. After that he was part of the effort to build the hydrogen bomb, where he was joined by Andrei Sakharov, later the famous dissident and Nobel Peace Prize winner. Zeldovich and Sakharov were the key designers of the Russian hydrogen bomb. Each of them received the Order of Lenin and was three times named a Hero of Socialist Labor.

In a secret bomb-research camp called The Installation, Sakharov and Zeldovich grew close. They lived next door to each other, they talked several times a day, and competed to solve problems both profound and trivial. "Sometimes we played games, as it were," Sakharov later wrote in *Nature,* "competing in the speed and elegance of the solutions (the one who solved it first ran to tell the other at any hour of the day or night)."

Sakharov said, "In science, Zeldovich was a humble man (although the manner in which he sometimes took part in discussions, defending what he considered to be the scientifically irrefutable, could give a somewhat different impression). He was almost childishly delighted when he had managed to achieve some important piece of work, or had overcome a methodological difficulty by an elegant method, and felt failures and errors keenly."

Sakharov was not so sanguine about Zeldovich's randy personal life, which included numerous what he called "below the waist" affairs. Zeldovich woke him up in the middle of the night once to borrow money for his mistress, a prisoner who worked at the Installation. Pregnant, she was about to be released and exiled to the Far East where she bore him a daughter in an ice-floored cottage.

When Sakharov began to speak out politically about disarmament and human rights, Zeldovich, determinedly apolitical, hung back. He joined Sakharov in fighting a Khrushchev proposal that everybody, including scientists, put in some time on collective farms—an idea that would ruin China for a generation. And when a delegation from the National Academy of Sciences asked him once to sign a letter criticizing Sakharov he threw them out of his office.

Sakharov was eventually exiled to Gorky for his outspoken heresies. He felt abandoned by Zeldovich. "A coldness," Sakharov complained, had crept into their relationship. He felt Zeldovich was a bit of a coward. Zeldovich's friends say that he was comfortable supporting Sakharov from a distance when he was protesting matters of principle. But when Sakharov started going on hunger strikes to obtain exit visas for his family, he had gone too far. Zeldovich had little use for Yelena Bonner, Sakharov's wife, whom he blamed for distracting his old friend. He muttered pointedly about the fact that Hawking had never let anything distract him from physics.

The nuclear weapons work gradually made Zeldovich a physicist. The field, he decided, wasn't used up after all. In the fifties he was writing papers about particle physics from the secret city in which the bomb makers were sequestered. The high-powered life of a bomb physicist— chauffeurs and bodyguards—was beginning to chafe. "When uranium fission became more engineering then physics," said Zeldovich, "I was allowed to work on astronomy."

Research in the Soviet system is conducted by the USSR Academy of Sciences through a series of institutes, most of which are in Moscow. Zeldovich moved to Moscow, taking a post at the Institute for Applied Mathematics and also the Space Research Institute.

Zeldovich took the big bang more seriously, right from the start, than his Western colleagues. He had come to conceive of the universe as a gigantic particle physics experiment. In particle physics, elementary particles were accelerated to enormous energies and then slammed together, annihilating them in a microscopic fireball so that physicists could search the debris flying outward from that fireball for new or more fundamental particles. Zeldovich thought that by examining the relics flying outward from the big bang—galaxies, stars, rocks, and people—science could find clues to the fundamental laws of physics and the creation of the universe. "Sometimes," he said, "I say the universe is a poor man's paleontology. We have to work on footprints. One cannot build a foundation in cosmology. The fundamental theory in physics is not settled."

The search for clues led him to Gamow's papers from the forties on nucleosynthesis in the big bang and the prediction that the heat from that explosion should still be around. "Gamow made very useful errors," Zeldovich said. He realized that this primordial radiation should be detectable as radio waves and even suggested in the early sixties a radio telescope that could do the trick—a horn antenna owned by Bell Labs in Holmdel. Dicke and Peebles never saw his paper. Neither did Penzias and Wilson. Shortly thereafter Zeldovich misread a table of cosmic element abundances and concluded that helium was only 10 percent of the universe instead of 26 percent. That was so at odds with Gamow's model that Zeldovich concluded that the so-called hot big bang was wrong.

"For one year, I believed in a cold big bang." Meanwhile the microwave radiation was discovered where Zeldovich had suggested looking for it—in Holmdel. "Now the hot big bang, like the existence of atoms, is established forever," he said smiling.

By the middle of the sixties he had assembled a group of astrophysicists and general relativists around him that rivaled Wheeler's at Princeton and Sciama's at Cambridge. Kip Thorne, Wheeler's student, frequently commuted to Moscow, and so did Hawking. "Half the seminal ideas in relativistic astrophysics in the sixties and seventies came from that group," said Thorne. Because Moscow was the center of Russian science, the circle of disciples around Zeldovich tended to grow year after year until it became unwieldy. According to Thorne, Zeldovich would then engineer a feud with someone so that he could break up the group and start over.

Zeldovich's disciples seemed drawn to him by his magnetic personality, his intuitive feel for physics, and his gift for relentless but gentle pedagogy. Typical was the testimony of Sergei Shandarin, a muscular-looking young theorist who met Zeldovich as a graduate student during seminars at the math institute and was soon drawn into the circle. There were about fifteen people in the seminar, recalled Shandarin, who went the first time with a friend. "I felt something quite different about science with him. It was not just boring scholarship. I had never seen a person who understood all subjects.

"He gave away ideas, he didn't care what happened to them next. The group was a dictatorship. You learn that is the best way; you could do worse picking your own problems. Zeldovich expected you to solve problems immediately. He would call at six in the morning for the answer. Many people were embarrassed. Once I said, 'I didn't do it yet.' Zeldovich put the receiver down. Then he phoned again a few minutes later."

Physics was Zeldovich's life. His second wife was a physicist, and his children and their spouses were all physicists. He had a big blackboard installed in his living room. Every morning he arose at five o'clock to work alone and receive calls from colleagues and students. The living room was also full of medicine balls and barbells. Physicists who stayed in his apartment got dragged into weight-lifting sessions. Zeldovich played tennis until he was seventy. Shandarin remembered Zeldovich challenging him and another physicist two on one. After the game was over and they were going up to Zeldovich's apartment, a three-story walk-up, Zeldovich suggested a contest to see who could leap the farthest up the staircase.

His disciples paint the picture of Zeldovich as a kind of Zorba the Cosmologist, a drinker and dancer for whom there were no barriers between life and science. Zeldovich used to put on his medals when he was going out for a bout of drinking so that the Moscow police, notoriously hard on drunks, would not hassle him.

Joe Silk, a Berkeley theorist, described how he had met Zeldovich at a conference in Prague. "At that time the Russian cosmologists were not nearly so well known as they are now. He dragged me off with a whole group of his students, and they took us to a swimming pool in Prague, and we spent an hour or two swimming around and discussing things. At that time I guess he was about in the early sixties but with the physique and abilities of a much younger man. He really is a life and soul type. At another symposium in Poland there was a banquet afterward, there was a band playing, and so he would go around the tables and he was dancing with everybody, women, men.

"Then they put on a cabaret and it turned out to be a striptease artist, for some crazy reason, who did the show. And, of course, as soon as the band started playing Zeldovich started joining in with her, trying to dance with her in the middle of her act. Everyone started bursting out laughing uproariously. He is just a tremendous extrovert."

Zeldovich's status as almost a cult figure in cosmology was enhanced by the fact that he could not travel out of the Eastern bloc, because of his previous national security work. It was a sore point with him. He kept a map of the world on the wall of his office and every time he got an invitation to a conference he couldn't attend, he'd stick a pin in the map.

It was Silk who put Zeldovich on the course of the "top-down" theory of galaxy formation, which was to set him at odds with Peebles and divided cosmology between East and West. In his Ph.D. thesis Silk had examined the effect of the fireball radiation on density fluctuations in the primeval fireball, extending Peebles's work but coming up with a new result. Silk concluded that the intense radiation pressure would tend to iron out irregularities in the primordial gas, smoothing out any small-scale features. The result was that all lumps smaller than about 10^{13} solar masses would be erased. That was the mass of a hundred Milky Ways—a medium-size cluster of galaxies. The building blocks of the universe were not small, they were large.

In 1969 Zeldovich asked himself what would happen to such a gas cloud as the universe expanded and cooled. How did it turn into stars? It was sort of a large-scale combustion problem. He began by realizing that the cloud would probably not be perfectly spherical. It was more likely to be cigar shaped, and it was probably spinning. When the cloud cooled it would collapse lopsidedly; the gas would compress fastest along its shortest dimension. The cloud would form an elongated pancake. Material slapping together from either side of the pancake would make a shock wave that would spread out, heating the thin sheet of gas and breaking it into fragments. These fragments, said Zeldovich, were what subsequently condensed into galaxies.

In other words, according to Zeldovich, the largest structures in the universe formed first and then broke up to make smaller objects.

Zeldovich realized there would be observable consequences of his top-down, or "pancake" process. The galaxies and clusters of galaxies would still be assembled in the pancake formations in which they were born. On the sky all these pancake clouds of galaxies would form an intersecting pattern of ribbons and filaments, which was not unlike the impression one would later get from looking at the Lick million-galaxies map. "Like a net," he explained, "or the shadows on the bottom of a swimming pool."

One day at a conference he happened to be staring at the bottom of a swimming pool and noticed that the ribbonlike patterns of light and shadow cast by surface waves and swells was identical to the kind of pattern galaxies should make in the sky. The mathematics of the two situations, he realized, were analogous. Suddenly, says Shandarin, Zeldovich was seeing the same pattern everywhere. He saw it again at a lecture when the speaker's slides were left in the project for too long and began to blister. Zeldovich was never happy until he had a simple, visual way to explain an idea.

He decided his crew should make a model by which they could produce the ribbonlike patterns and analyze them. Like Peebles, Zeldovich was always looking for quantitative ways to analyze qualitative information. The key to success was producing a surface with smooth random waves on it. When light shined through, the random bumps would refract it into cosmiclike ribbons. Inspired by the melted slides, he suggested first that Shandarin get some plastic and burn it.

Zeldovich called up the next day and asked if Shandarin had done it yet. He reported that he had tried, but burning didn't work. Okay, said Zeldovich, he had another idea: Pour a big mound of epoxy and stir waves in it. The epoxy would set with a smooth random lumpy surface.

Shandarin reminded Zeldovich that it was Sunday, and he couldn't buy any epoxy. I have some glue, come to my apartment, Zeldovich said. So Shandarin took a train across town, picked up the glue, and went back home. "Again we didn't succeed."

Two days later Zeldovich hired an artisan to etch random dots on a piece of glass with a laser, oil, and acid. This time it worked—the plate cast perfect filamentary shadows. Zeldovich was happy. "A swimming pool keeps changing," explained Shandarin. Now he had something he could show off and study as well.

By the mid-seventies the lines were drawn between the bottom-up theory of Peebles and the top-down theory of Zeldovich. Peebles said that galaxies formed first and then slowly assembled themselves into clusters. Zeldovich said the opposite: that the clusters formed and then broke up into galaxies. It was a little like arguing what came first, the chicken or the egg?

Behind the screen of his consistent good humor, Peebles was intransigently opposed to the Zeldovichian universe. "I was certainly willing to accept pancakes as one possibility, but by no means the only one," he said mildly. "The debate was over the question of sequence of formation. Did galaxies come before clusters or the other way around? Of course, the subject goes a long way back. There were debates about this in the thirties. It was something that Hubble wrote about, and also Lemaître. My prejudices, I think, were molded on the picture of small forming before large. It seemed to fit the galaxy statistics that we had, at least in our minds.

"But then one also had some direct bits of information that seemed to be awfully compelling. For example: We're in the outskirts of a cluster, Virgo, that looks to me to be still growing, still forming, but we're in a galaxy that certainly is very old and we can see the galaxies are, by and large, old, full of old stars. So on the face of it, it's manifest— our galaxies formed before this particular cluster. And if this cluster is a late former, why wouldn't clusters in general be formed after galaxies?"

And yet the vague clues that had been gleaned by then about the large-scale structure of the universe—the mysterious Lick map, the strings of Abell clusters dotting the sky—were suggestively redolent of Zeldovich's ribbonlike swimming pool shadows.

What did nature really say? If Peebles was right, the sky was full of little lumps of light getting larger; if Zeldovich was right, it was a mesh streaked with sheets of galaxies and empty volumes in between. Just as the debate between the steady state and the big bang had energized classic observational cosmology, the opposition between the bottom-up and top-down theories of galaxy formation gave form to the infant discipline of physical cosmology. Now that there was controversy, it was a science. Sides could be chosen up, theories and observations made to clash.

What *did* nature say? It would be years before they found out that Peebles and Zeldovich were both right.

II

FERMILAND

*Only when men shall roll up space like a piece of leather
will there be an end of sorrow apart from knowing God.*
                        —Svetasvatara Upanishad

9

THE LONG MARCH

While physicists began to explore the big bang, the astronomical side of cosmology, the quest that Allan Sandage called the search for two numbers went on. Sandage kept going to Palomar, month after month, year after year, climbing into his heated observing suit, riding the clanky elevator to his lonely cage in the telescope, amassing the data needed to fulfill his cosmological mandate.

The path that Hubble had laid out had two parallel but independent tracks. One was to measure the deceleration parameter, q_0. That number would tell him the shape of the universe, the curvature of space-time, and whether the universe was closed and would end some day in a crushing collapse, or whether it was open and the galaxies would just keep drifting off forever. The second was to remeasure the Hubble constant, H_0, the number that told him how fast the universe was expanding, and how big and old it was. The distance scale was slower, drier work, and it was not until the late sixties that this second campaign started to bear fruit.

In any expanding universe H_0 was just the ratio of a galaxy's redshift velocity to its distance. In principle, then, the Hubble constant was easy to measure: Just ascertain the distance to a bunch of galaxies and measure their redshifts. In practice it was the hardest thing in astronomy, making the Hubble diagram look like doodling on the blackboard. It was knowing distances, the third dimension, that had made astronomy a quantitative science, had allowed science to break through the dome of the sky, and had made modern cosmology possible. Without it the stars and galaxies were just smears of light. How could a man— Sandage or Hubble—gauge the distances of little smears of light that he could never touch, never view from any other angle?

Very poorly, if history was any guide. Hubble's original value for the Hubble constant had been 530 kilometers per second per megaparsec, which meant in English that a galaxy would be flying away 300 miles per second faster for every megaparsec (roughly every 3 million light-years) that it was farther away. By 1956 Sandage and Baade had lowered the Hubble constant by a factor of 3 to 180, which meant that a redshift velocity increase of 300 miles per second corresponded to a distance increase not of 3 million light-years, but of 9 million light-years. It meant the distances between the galaxies were three times as large as had been thought and that the universe was correspondingly three times older. By the early sixties the accepted value was 100; Sandage, following up on the promise made in his prize lecture in 1956, wrote a paper suggesting that it could be as low as 75, based on a study of the apparent brightness of globular clusters in the Virgo cluster. According to the latter value, the universe could be 13 billion years old.

"First of all, astronomy is almost an impossible field, and anyone who doesn't—" Sandage's voice broke off, exasperated at the mere thought of his burden. "You know how Herschel got the distance to the first stars?" he asked. "He said, suppose all the stars are just exactly like the sun. We know the apparent magnitude of the sun, and the apparent magnitude of the stars. Therefore the mean distance to the stars is such-and-such. Well, that's a great assumption, and it's almost right."

In his epic observing campaign in the twenties and thirties, as the reader will recall, Hubble had used Cepheid variable stars as his "standard candles" to find the distances to the other galaxies in the local group. Cepheids—named after Delta Cephei, the first example known— pulsate, and as they pulsate their brightness changes. Over time the record of a Cepheid's output—its light curve—goes up and down like a saw-tooth. What made Cepheids perfect distance indicators was that their absolute magnitudes were proportional to their pulsation periods. That meant that the absolute magnitude and thus the distance of a Cepheid— and of the group of stars or galaxy it was part of—could be read off the sawtooth light curve.

When Sandage started working for Hubble, one of his first as-signments had been to snap photographs of nearby galaxies to search for Cepheid variable stars. He had kept doing it over the years. Nearly every-one who had ever worked at Santa Barbara Street had chipped in with photographs at one time or another. Sandage figured that it took at least thirty or forty plates of a galaxy over a ten-year span to have enough data to find the Cepheids and determine a distance.

By 1962 there were hundreds of plates stacked in Sandage's office awaiting the truly tedious task of sifting their billions of stars for the telltale sawtooth light curves of Cepheids. Bowen, Mount Wilson's director, told him he could hire an assistant. Sandage went looking and found a man named Gustav Tammann.

When Allan Sandage came into his life, Tammann was sitting in a café in northern Italy sipping coffee. A Swiss astronomer descended from what he described as "the lowest level of nobility" in czarist Russia, Tammann is tall and debonair, with a cleft chin and a shy-little-boy smile, and speaks in a tiny, precise voice with only the vaguest trace of a German accent. He and Sandage were at a three-week summer school, one of those occasions where illustrious scientists get a free stay somewhere nice in exchange for a series of lectures to a small number of colleagues and carefully selected graduate students. During the course of their mutual stay, Sandage had gotten a good look at Tammann in action during the scientific sessions and evidently liked what he saw. When they met for coffee, Tammann asked his advice about a proffered job; Sandage responded by offering Tammann a job with him, reducing the distance-scale data.

"It takes a very special skill," recalled Sandage years later. "It takes incredible patience. It takes a certain long-range point of view also; it takes a certain attitude. These people are really very rare. Henrietta Swope, Baade, Hubble, they all had it. This man clearly was unique and good."

They were an incongruous and perfect pair. If Sandage approached cosmology with a religious intensity, ascending to charismatic exuberance and plunging into brooding curses of self-doubt, Tammann maintained the even keel and good humor of an aristocrat at sport. Tammann was charming and waggish, waving around a long gold cigarette holder, even on mountaintops, where he preferred not to go, but if obliged went dressed in a suit and topcoat. With his dandyish ways, he had an air of high manners and low mischief. It was hard to imagine him ever raising his voice; it was easy to imagine that his smile was often a smirk. He admitted he was a clumsy observer, but he could do what he called the "dirty craftsmanship" of photometry.

Tammann's grandfather had fled Russia for Germany in 1906. Tammann had been born in Göttingen, but after his father died his Swiss-born mother took the family to Basel. Gustav had announced his astronomical interests to the family at the age of six when he drew zodiac signs for everybody on a blue tablecloth. It had, he recalled, a dramatic effect on the adults. He went to the University of Basel and hung out at the law school, his sister being a lawyer, before committing himself to astronomy.

When they met, Sandage was already a great man, and Tammann deferred to his preeminence completely. He said he saw his job as holding Sandage's coat and otherwise easing the way for Sandage to accomplish what destiny had decreed for him. When the critics needed attacking, Tammann would do it; when Sandage was depressed, Tammann cheered him up. "I accept him as the leader," said Tammann in his little Swiss-watch voice. "For some reason he accepts me as a collaborator."

Tammann arrived in Pasadena in February 1963, about the time Schmidt was deciphering the quasars, expecting a couple of years' worth of work.

The dirty craftsmanship for which he had signed on was by far the longest and least-romantic part of the cosmological process. "We spend our lives looking at black marks on a photographic plate," he sighed. The instrument of this terror by boredom was a machine called a blink comparator. The blink machine held two plates of the same section of sky—exposed at different times but perfectly aligned, star for star, on the identical scale—and projected them alternatively on a screen. *Blink:* There was the galaxy M81 on February 12, 1950. *Blink:* M81 on March 18. In a month a lot of things could happen in a bowl of a hundred billion stars—some swing around and eclipse each other, some explode or flare, a relative handful of others—the Cepheids—jittery with age, beat rhythmically in and out of prominence against the creamy swirl of galaxy. In a month a hundred billion things remain the same, burning in a steady pace toward the thermonuclear grave.

In this two-frame motion picture the discrepancies between two frozen moments leapt into animation. The screen was filled with blinking stars. Each was a suspect Cepheid, but was more likely an asteroid, an emulsion defect, or some other false alarm, to be arrowed on a working print. Plate after plate, the process went on: New candidates were added while one-night stands and mistakes were stricken from the list. At the end what was left were the repeaters—real Cepheids, recurrent novae, binary stars, irregular variables.

Tammann's work began next door to the Milky Way in a group of galaxies dominated by the bright spiral M81. There were seventeen galaxies in that group, thought to be the nearest neighbor to our own local group.

Hubble himself had launched the quest for the new Hubble constant on his first run on the 200-inch telescope, on November 9, 1949, when he exposed a plate of M81, which lies just off the lip of the Big Dipper, in Ursa Major. In photographs M81 looked like a dropped egg

spinning in soup—all nucleus bordered by a smooth, filmy ring of nebulosity. Nearby is another spiral known as NGC 2403, which looks like
a smeared Z in the sky with a diffuse center and splotchy slashes for arms.

It had been common wisdom since Hubble's time that M81 and
NGC 2403 were at the same distance; the distance to either one would
suffice for both and for the whole group of galaxies. In 1954 Sandage
had made an abortive effort to find variable stars in M81 and come up
with forty-one candidates, but he had made a mistake with his comparison
stars. Therefore, he told Tammann to try M81 first, but it was, Sandage
says, "hard as holy Ned." Tammann stared and stared but the Cepheids
just did not present themselves. So they put away M81 and took up NGC
2403 instead.

Tammann began to get results, but he hated the blink machine.
The differences, physical differences, between the plates drove him crazy;
if one was grainier than the other, for example, the sky would seem to
explode. Variations in the seeing would change the contrast between the
stars and the sky and make them appear to be brightening when they
weren't.

He abandoned the machine and just held the plates up to a
window, searching for changes in tiny memorized star fields sector by
sector with a small magnifier. "I don't think Allan liked it that I was
doing it that way," he said slyly.

There were forty plates. After a year and a half of Tammann's
comparing in this fashion, a few dozen stars remained arrowed on the
working print.

The next step in the game was to chart the magnitude changes
of these suspect stars. Around each of the numbered stars, a cloud of
presumably stable stars had been marked. Spanning the range from gray
dots barely distinguishable from the photographic grain on the plate
emulsion to swollen black blotches, they were the scale on which the
inconstancy of the Cepheid star would be measured, timed, and testified
to. Tammann didn't have to know anything about those comparison stars
other than that they were, in fact, constant to do the next part, which
was to go through the entire series of plates, twenty years in the life of
a galaxy and find the comparison star that was closest in magnitude to
the Cepheid at each moment. The result was a history of the relative
changes in brightness of the Cepheid.

"One of the most marvelous experiences is to have a series of fifty
plates and you see a Cepheid," Sandage said brightly, his fingers dancing
along a tabletop. "You estimate its brightness. You don't have to have
numbers; here are three comparison stars: $a, b, c,$ the Cepheid brightness
is a third of the way from a to b, so you call it a.3. So you make these
estimates and then take these plates from over ten years, you phase them

all and get the period. And when you're successful there's not a single deviant observation in the whole batch. You've gotta get the period exactly right, but you've touched reality.

"And you've got its period and you just know you got it. There's no feeling like it in the world."

Eventually, finally, *a*.3, *b*.2, and the rest had to become numbers. How bright were the comparison stars? To Sandage that was the crux, the real battleground of it all, and he treated it like guerrilla war. Magnitudes could be measured photographically, as Tammann was doing for the Cepheids, from the sizes of dots on a plate, by comparing them to the dots made by stars that were known quantities. But that was getting to be old-fashioned.

The way of choice, as of 1963, was called photoelectric photometry. The photometer was just a fancy light meter that measured the amount of light coming through a small round aperture in which the star is centered. The photometer was equipped with standardized filters that isolated different parts of the spectrum called "UBVRIJK," for ultraviolet, blue, yellow (or visual), red, and three bands of infrared. With a set of these measurements a star's spectrum could quickly be characterized, quantized, and compared to other stars.

So for every one of the arrowed variables and novae on the M81 and NGC 2403 plates, Sandage had to go to the telescope and take the photoelectric measure of the calibration stars around them. It was easy to lose the game here, he thought. A half-magnitude error in the brightness of some obscure star could swell or shrink the universe by 30 percent. Hubble himself had bungled his calibration stars.

Waving his cigarette holder, Tammann tagged along to Palomar to help Sandage with the excruciating job of measuring stars. Sandage's perspicacity and tenacity in this tedious business impressed him. "Allan Sandage has the best sequences in astronomy," Tammann gushed. "He goes incredibly faint." He recalled watching Sandage offset the telescope from visible stars to measure the light from stars he couldn't see in the eyepiece, stars so faint that sometimes he couldn't even tell at first if the photometer was registering a signal. None of the Cepheids were visible in the eyepiece; they spent their whole time bouncing around in invisibility. Sandage had to go into the shadows after them. The two astronomers would spend a third of the night, Tammann remembered with a groan, just measuring and remeasuring standard stars to monitor the clarity of the atmosphere and make sure no invisible cloud was casting a shadow on the universe.

Tammann usually stayed down in the control room while Sandage submitted himself to the spartan marathon of the prime focus cage. Sometimes, when the night had gotten long, Tammann would stand in front of the microphone and pour water back and forth between a pair of glasses loudly enough so that it could be heard up in the cage. "Hey, Allan, do you have to go to the bathroom?"

"YOU SON OF A BITCH."

Recalling those labors, Sandage complained, "If Columbus had known how far the New World was, none of his sailors would have sailed. If we had known how hard it would be to get the light curves in NGC 2403 we would not have tried. We simply would not have tried. And if Gustav and I had known the disaster we would subject ourselves to by the criticism of all those people—and the criticism still continues—why, we would have gone into a Trappist monastery fifteen years ago."

The light curves the two men did finally obtain looked chopped, like a row of icebergs. Their bottoms were under the foggy background, in the land of the plate limit—Hubble's shadowed realm—where star images were no denser than, and thus indistinguishable from, the noise of random grain concentrations in the plate emulsion. Sandage and Tammann struggled mightily to make sense of the amputated and noisy data.

Finally, when one more cosmic landmark was nearly free from the shadows of ghostly error, a single correction remained—a correction that could not be made. Air and dust, on Palomar and in the lanes of the Milky Way, redden and dim starlight, like Hollywood smog reddening a sunset, making the star and its home galaxy appear farther than they actually are. Sandage and Tammann had tables of data to help them correct for Milky Way dust and their measurements of standard stars to help them correct for the local atmosphere, but there was no way for them to estimate by how much the star's light had been meddled with in its home galaxy. Usually they would analyze the "color" of the star—the ratio of its magnitudes in different spectral bands—to see if it was abnormally red. If so, the star's light had probably been dimmed and reddened by dust in its own galaxy and, therefore, was giving a false distance reading. This time, though, the data were so poor that Sandage didn't trust the colors to tell him anything about the stars or dust in NGC 2403. He couldn't tell if that had been reddened or not. He assumed not.

So at last they were ready to publish a distance for NGC 2403 and its group. By now it was 1967, almost twenty years since the first data had been taken, by a man who was now dead. Their paper bore marks of the struggle and echoes of Hubble's magisterial tone. One remarkable passage sailed off the pages of the *ApJ* like a long cosmic groan—a report from the shadows:

> We wish to comment on the extreme difficulty of work on variable stars at the exceedingly faint levels encountered here. . . . It is often impossible to prove if an actual change

of intensity between two plates of different quality has taken place for a given star. Often, inspection of a large fraction of the plate material is necessary before the variability of the faintest stars can be proved. Because of these factors, Hubble's early detection of Cepheids in galaxies outside the Local Group must be considered a remarkable achievement.

In what would be a trend running through all of Sandage and Tammann's work, the distance they obtained for NGC 2403 and its group, 10.6 million light-years, was significantly beyond Hubble's original estimate. If this held up, the universe was clearly going to be even larger and older, and the Hubble constant lower than anyone had suspected. But NGC 2403 was only the first, nearest galaxy, barely outside of the local group. The answer was one datum point, the first definitive distance, in a chain that would lead to the Hubble constant and the expansion of the universe.

It was then that Barry Madore came to town, a tall, lean Canadian, who was a graduate student of van den Bergh's, the observational cosmologist at the University of Toronto. Looking at Sandage and Tammann's data, he discerned a discrepancy between the red plates and the blue ones. It seemed to him as if the Cepheids had been greatly reddened by the dust in NGC 2403 after all. Not all the diminution of their light was due to distance, he concluded, which could mean that NGC 2403 was closer than Sandage and Tammann's figure of 10.6 million light-years. Madore put it at more like 7.1 million light-years.

Sandage invited Madore to Pasadena. They talked for two days. Sandage protested that he and Tammann had stated in their paper that the colors of NGC 2403 Cepheids were no good. If the data weren't good enough to deduce the reddening, Madore insisted, they weren't good enough for anything, including a distance. Suddenly, on the third day, as Madore tells it, Sandage laid into him.

"If you publish we will respond," Sandage said frostily and then dismissed Madore.

"I was devastated," said Madore years later. He published anyway, in a British journal, *Monthly Notices of the Royal Astronomical Society,* not the *Astrophysical Journal,* which is the main arena of cosmology. Sandage and Tammann did not answer. Publicly they ignored Madore's criticism, but it gnawed at Sandage's gut. He was Hubble's successor—he had spent fifteen years riding in that cold cage and staring at the black dots, and now a callow graduate student had come with no data of his own. Had come to say, after all that, that Sandage was wrong.

He took the criticism personally. "I was not prepared for that at the time," Sandage said. "The training didn't include all of that, that's right. If you have a result, why shouldn't everybody agree with it? Everyone believe it?

"So then the human psyche comes in," he said scowling sternly, referring to his own stung feelings, "and we were criticized very, very severely after spending seventeen years on the problem, by a young man who was just not even quite out of graduate school. He came and spent time with me and I told him he was wrong and he went and published a devastating paper against Tammann and me."

Bloody but unbowed, Sandage and Tammann marched outward in their quest to measure the distances of farther and farther galaxies. The farther they went the harder it got to find Cepheids. They are fairly bright stars, but Sandage and Tammann could not use them to get very far out in the realm of galaxies. Beyond 10 million light-years Cepheids began to fade into the background mists of the billions of stars in which they lived. Sandage needed brighter candles to survey the depths to the giant clusters and beyond, where galaxies themselves became soft dots on the plate, where what he wanted to know was the distance to eternity.

The farther out he could measure the distances of galaxies, the more accurately he could measure the Hubble constant, because the local bumpings around of individual galaxies would have smaller and smaller influence on the data. The farther out he could measure, the more he would be certain that the Hubble expansion was a universal effect, the same in every direction, the more he would know it was the supreme law. Sandage conceived of the Hubble constant project as rungs on a ladder, or steps outward through concentric zones of the universe. Each rung was a different distance indicator that depended for its justification on the data acquired on the rung below, in other words a kind of astronomical leapfrog. "There are all kinds of problems," said Sandage. "You calibrate things you can and then find their twins in association with brighter objects."

The higher and farther outward he climbed, the brighter and more unreliable the standard candles he was forced to use became. Cepheids were the bottom rung. For the next rung he used the flame-colored patches of hot ionized hydrogen gas that dot the arms of spiral galaxies. Hubble had mistaken these hydrogen clouds for bright red stars in his original distance estimates, when in fact they were actually the birthplaces of whole clusters of stars. These clusters burn their way out of the clouds that had condensed to form them like chicks pecking out of a shell. In the nearby galaxies whose distance had been calibrated by Cepheids, Sandage and Tammann found a correlation between the sizes of the hydrogen clouds and the brightness of their parent galaxies. They found a similar relationship between the size of a galaxy and the brightest blue stars in it. These two correlations formed parallel rungs that allowed them to extend their distance measurements to galaxies beyond Cepheid range. And so on, outward. Calibrate something; find its twin and use it to calibrate something brighter or bigger. On the third rung, galaxies themselves were standard candles.

Meanwhile there was a similar march inward through the galaxy. The Cepheids themselves still had to be calibrated, which meant finding an independent measure of their distance. The trouble was, most of them were in the large Magellanic cloud or in very distant star clusters in the galaxy. Astronomers could only estimate the distances of these clusters by comparing the magnitudes and colors of their stars to those in the so-called main sequence. Take the yellow stars, for example, and assign them the luminosity of yellow stars like our own on the main sequence. Of course that meant that the main sequence itself had to be calibrated.

This backward chain of inferences came to rest in the Hyades, a V-shaped grouping of stars west of Orion that forms a backdrop for Aldebaran, the red eye of the bull. It was the nearest star cluster whose distance could be determined independently. The Hyades stars were all moving across the sky; in the course of historical astronomy, their charted coordinates, on average, had migrated about one three-hundredth the width of the full moon. If the cluster were at infinity, the stars would have moved on parallel tracks, but in fact, the tracks converged, the way a flock of geese seems to get smaller as it recedes toward the southern horizon. That geometric distortion allowed astronomers to calculate its distance as around 135 light-years. That was the true foundation of the distance scale, yet it was an unstable one. The Hyades were abnormally high in heavy metals, and many astronomers wondered, therefore, if they were a fit comparison to other stars. Some astronomers questioned whether the Hyades were a bound cluster at all, or were drifting apart or still contracting, throwing a monkey wrench into all that immaculate geometry. Every time the Hyades twitched, the galaxies shuddered, and the universe grew larger or smaller. And the Hyades twitched often.

To get his one simple number, the Hubble constant, Sandage had in effect to keep up with all of astrophysics. If the process seemed somewhat haphazard and less than elegant, Sandage was not apologetic. "You have to have the answer," he explained. "You use everything that's available to you. It's like the clutching of a drowning man. He clutches at the nearest straw to him. Distances of galaxies are almost impossible to obtain. You can say, 'Aha, all these problems,' but what else *can you do?* You use the best method that's available to you at the time, or you just throw up your hands. Either that turns you on or it doesn't. And if it turns you on and you're willing to do it, you get some kicks, but it's questionable. Churchill said if you're not down in the arena fighting, if you're not down there—" his voice broke off. "I feel terribly strongly about this that it's very easy to criticize if you're not down there fighting the lions."

For Sandage and Tammann one of the major cruxes of their entire ladder was to get the distance to a giant spiral galaxy known as M101, which lurks just above the handle of the Big Dipper. Sometimes called the Pinwheel, M101 is one of the most beautiful galaxies. It looks like a hippie starfish in the sky, its loose splayed arms bedecked and gaudily laden with pink clouds of hydrogen, ragged pockets of dark, rich dust, and blazing clouds of young blue stars. Its small, yellowed nucleus and bright, open arms marked M101 as an Sc-type spiral of the giant class. Next to the giant ellipticals at the centers of rich clusters, giant Sc galaxies were the most luminous "normal" galaxies in the universe. Van den Bergh had organized galaxies into luminosity classes, and had suggested that the giant Sc's seemed to be uniform enough to serve as standard candles.

M101 was the nearest giant Sc. If Sandage could only measure the distance to that galaxy and determine its true luminosity, he would have a yardstick that he could apply as far as the 200-inch could see. The Hubble constant would fall into their laps.

Sandage and Tammann began by scrutinizing M101's splattery star lanes for Cepheid variables. None showed up, which meant the galaxy was at least far enough for distance to dim the Cepheids into the plate fog. Next they looked for supergiant stars in M101, but couldn't find anything brighter than 21st-magnitude, again indicating that the galaxy was far away. The spiral arms were full of bright hydrogen clouds, but Sandage and Tammann's calibration of the hydrogen clouds did not extend to galaxies as large and bright as M101.

So they abandoned the main galaxy and turned to the smaller galaxies surrounding it: round dwarfs and ragged irregulars. Sandage and Tammann measured the diameters of the hydrogen clouds in those smaller galaxies. From them the astronomers determined that the distance to the whole group, and presumably to M101, the kingpin Pinwheel, was 22 million light-years, farther than Hubble had ever dreamed.

That meant that M101 was as luminous as 100 billion stars. But was this truly a standard candle?

Continuing the cosmic leapfrog game of twinning and calibrating, Sandage and Tammann measured the sizes of the hydrogen clouds in M101 and used them to extend the relation between the sizes of the clouds and the brightness of their parent galaxies to the giant ScI class. The relative sizes of the clouds in more distant giant Sc galaxies marched in lockstep with the redshifts, which suggested that the clouds were indeed a good standard yardstick. When Sandage and Tammann computed the distances of those galaxies, the giant spirals turned out indeed to be roughly similar, about as luminous as 100 billion suns. They would make adequate standard candles. The goal was in sight.

Already, with the pivotal galaxy M101, some things were becoming clear. The galactic distances that Sandage measured were larger than all the previous estimates, which meant that the universe was larger and, therefore, older than anyone had suspected. The Hubble constant was not 100, or even 75; it was more like 50 kilometers per second per megaparsec. The universe could be as old as 20 billion years. That was comforting; new estimates of the ages of the oldest stars and of the galaxy from the ages of radioactive elements had been giving answers in the teens of billions of years. How closely all three of these "clocks" agreed when Sandage and Tammann were finally done would tell them how much—if at all—the universe had slowed down and whether it would keep expanding forever.

All that remained were the tedious days and nights sitting and analyzing the redshifts and magnitudes of these little starfish smears in the sky. Giant spirals could be identified out to redshifts of 21,000 kilometers per second—ten times the putative distance of the Virgo cluster.

The Hubble constant chased Sandage into middle age. In 1970 his mother died of lung cancer, and Sandage quit smoking. His father remarried a year later. "Until about 1968 or 1969 Sandage was more mellow," recalled Tuton, the former night assistant on the 200-inch. "Then he became more and more driven."

"It's like building a house," Sandage went on, his voice flat. "It's an engineering thing, putting nails in, it's not the excitement that you read about. You look at plates, measure amplitudes, stars are flickering up and down below the plate limit. You get their periods and lo and behold you get their distances. What's so great about that? It's marvelous to get a distance, because it's almost impossible to believe that you can do it. But there's nothing mysterious about it."

In 1972 Sandage and Tammann used their growing mound of data to get a preview of the answer to the question of the fate of the universe— was it going to expand forever or not? And it was not the answer Sandage had been expecting all those years. It came in response to a group of astronomers from the University of Texas who had written a paper in support of a heretical motion about the organization of the universe, namely, that every structure in it was part of something larger, which in turn was part of something else still larger, endlessly. This meant that there was no such thing as a representative volume of the universe and that the Hubble constant would always increase with distance. Their idea flew in the face of the common wisdom espoused by Hubble and Sandage that galaxies were spread more or less evenly through space with occasional large clumps, and that the Hubble constant was measurable.

Sandage sprang into action. "For some reason it was one of the very few times I've felt compelled to answer the critics," he said. He sent Tammann and Eduardo Hardy down to the Mount Wilson library where they spent eighteen hours counting all the galaxies in the Zwicky catalog, some 30,000 individual entries. As Sandage suspected, the statistics—the numbers of galaxies at different magnitudes—supported the classic Hubble view of the distribution of galaxies. "The paper took two days to write; it really was another of those marvelous, exciting two or three days," recalled Sandage.

During that week it dawned on Sandage for the first time that if there were large concentrations in mass in the universe, such as the huge Virgo cluster, then the extra gravity associated with those clusters would make the universe expand lopsidedly. The galaxies in the vicinity of large clusters like Virgo would tend to get pulled toward the cluster. It presented him with a conundrum. Sandage thought he knew that the universe was not expanding lopsidedly. All his data seemed to indicate that the Hubble constant was the same in the direction of Virgo as away from it, as if the cluster's gravitational pull did not exist. The Hubble flow was "noiseless." How could that be? The only answer he could see was that gravity was unimportant in the universe. The fact that the galaxies could serenely sail right past big clots like Virgo with nary a jitter or a course deviation implied that there wasn't enough gravity in the universe to slow the galaxies and eventually call them all home. Space was so flat, and q_0 was so low, that the clusters made no difference. The galaxies didn't know the others were there. They were already free. And the universe would expand forever.

And yet, other studies, especially those involving his beloved Hubble diagram, still pointed to a closed universe, suggesting that space was highly curved and would swallow itself like a giant black hole at the end of time. The data pointed both ways; nature seemed to be talking out of both sides of her mouth. There was still no clear indication of the fate of creation, Sandage concluded, but the time for a final decision was drawing close.

10

THE ENDLESS GOOD-BYE

The fate of the universe was the $64,000 question in cosmology. "The fate of the universe"—by what possible audacity could a person even say those words? What did they mean? Which was more absurd: The notion that the universe went on forever, or the notion that it didn't? The idea that one might be able to predict the future condition of everything that existed—yea, or existence itself—might seem preposterous and a little foolhardy. As preposterous, perhaps, as the idea that the universe could have had a beginning, and yet the skies were ringing with testimony of its fiery birth. The cosmological question, as Sandage called it, had riveted the attention of astronomers and their public ever since Hubble had begun cosmology by discovering that the universe was expanding. Would it continue to expand forever, the galaxies departing each other for good? Or would it slow down gradually and eventually recontract, the galaxies swarming back together in a final cataclysm that was a backward replay of the explosion that brought it into existence? Would existence wink out into a black hole at the end of time? If it did, would the universe rebound Phoenix-like from its own crushed ashes? For twenty years, every astronomical discovery, every new paper or newspaper Sunday supplement article came with those questions tagging along—what did this mean about the fate of the universe?

Sandage had hoped that the answer would come from the Hubble diagram, on which the redshift velocities of nearly identical galaxies at vastly different distances—so-called standard candles—were compared to see if the cosmic expansion was slowing down and by how much. The Hubble diagram seemed to indicate that the universe was closed, that it would one day, some 60 billion years from now, collapse. In his heart Sandage wanted to believe that result, that he was getting The Answer, but his faith was buffeted as the hideous complications of the Hubble diagram were pointed out.

On the other hand, preliminary data from his and Tammann's project to measure the Hubble constant showed the universe expanding evenly and uniformly in all directions, both toward and away from imposing concentrations of galaxies, which implied that there was not enough gravity in the universe to affect the expansion. The galaxies would sail on, lonely and free. Which was it? The answer seemed buried in shadows and ghostly errors.

It was long dry work, this systematic combing of the cosmos. Sandage had lost his enthusiasm long before the end was in sight. In the early seventies the end was drawing near. But while he fiddled with his Hubble diagrams and continued his long march with Tammann through the cosmos, a new generation of cosmologists had grown up, less inclined to wait decades for someone else to get the answer.

In 1967 Sandage took a trip to the University of Texas to give a talk about cosmology. Before he could speak, a young woman, a graduate student, stood up and told the audience that everything they were about to hear was wrong. Sandage was stunned and outraged—an outrage he never was to forget. "It was typical of him to recall with exaggeration," said Tammann. "She was a graduate student. Allan was already Allan."

The woman who had announced herself into Sandage's life so unforgettably was Beatrice Tinsley. She was short with dark, curly hair and resembled the Peanuts character Lucy, in temperament as well as looks. Tinsley was only the loudest of many astronomers who argued that Sandage's standard candles—giant elliptical galaxies—were not standard enough to reveal the fate of the universe. On one level the argument between Sandage and Tinsley was a technical dispute about which kind of stars generated the most light in elliptical galaxies—red giants or normal hydrogen-burning stars. On another level it was about the fate of the universe. She and Sandage were to spend the next ten years dueling.

Tinsley had taken a circuitous but determined route to cosmology, and once there she was not to be denied. She had been born Beatrice Hill, the daughter of an Anglican clergyman, in Cambridge, England, during the Blitz, and had been raised in New Zealand, where her father gradually became involved in politics. Writing about her later, her father described her as a genius in Newton's sense, of being able to take infinite pains.

Her two loves were music and mathematics. At the University of Canterbury in Christchurch, she played in a chamber music group four nights a week and fell under the spell of physics, which, she felt, combined the seductive elegance of pure math with a certain amount of practicality. She learned, she said, "to question everything." She emerged in 1962 with a master's degree and married Brian Tinsley, an atmospheric and auroral physicist. When he got a job at the Southwest Center for Advanced Studies in Dallas, she followed him to the United States.

She brought with her a scholarship from New Zealand. Having decided to go into astrophysics, she used it to wheedle a position with the mathematics group of the Dallas center. She arrived just in time for the first Texas Symposium and the parade of Wheeler, Schmidt, Greenstein, Gold, and Hoyle wrestling with the issues of quasars and singularities.

In a short while she found life in Dallas scientifically and culturally stagnant. Although she loved to travel around the country and took delight in its rambling possibilities, she was appalled at the civil rights situation in the South. A feminist slightly ahead of her time, Tinsley also did not fit into the life of a scientist's wife in Dallas and caused a minor scandal by refusing to host a tea when it was her turn. She preferred to hang out with the other scientists, the men, but even that soon palled, relativity was still a mathematical game to the center theorists.

When she and Brian failed to conceive a baby, she enrolled in graduate school at the University of Texas, 200 miles away in Austin. The astronomy department there was being built up by the energetic Harlan Smith, who had hired a French cosmologist Gérard de Vaucouleurs, among others. Tinsley commuted twice a week from Dallas and commenced a thesis. As her specialty she picked the evolution of galaxies. It was here that she crossed swords with Sandage.

The bone of contention between them was the Hubble diagram. Sandage had concluded that the Hubble diagram indicated that the universe was closed. That conclusion, however, depended on how much the diagram had to be corrected for supposed tendency of elliptical galaxies to redden and dim as they got older and the brightest and bluest of their stars burned out. How much did these galaxies change? Not enough, Sandage decided, based on observations and his knowledge of the H-R diagram, to change the overall conclusion.

Tinsley was not impressed by Sandage's result; she was not alone in feeling that it had been based on lazy thinking. In her thesis she tackled the problem theoretically, asking how various collections of stars would evolve and how their collective properties would change as they aged. The answer was quite a bit, enough to change the verdict of the Hubble diagram from a closed universe to an open one.

Tinsley's thesis was published in 1968, after a long struggle with an "anonymous referee," with the cosmology section greatly curtailed. Sandage rejected her conclusion, say Tinsley's friends. He simply treated it as if it didn't exist. Tinsley was wounded by Sandage's reaction but she was a combative person. It made her work all the harder, and she launched into the problem of galaxy evolution.

She and Brian had adopted children, and she spent the years after her thesis at home in Dallas taking care of them and doing cosmology on the side. She got involved as well in antiwar activities and joined Zero Population Growth. She kept her hand in enough to plague Sandage, who once told a colleague that he opened each new Tinsley paper with trembling.

Then, in 1972, to add insult to injury, Tinsley came to Caltech and Palomar-Mount Wilson.

The person who brought her there was a Caltech professor named Jim Gunn, who would become the second member of what would soon become a young cosmological gang of four. "Sandage didn't like Beatrice from the very beginning," he recalled one afternoon in the large corner office he now inhabits at Princeton. "She was right and she knew it. She was assertive."

Gunn, a wiry figure, now baldish with long brown curls cascading down the back of his neck, a bushy beard, and squarish wire-frame spectacles, had come to Caltech as a graduate student in 1961, after studying physics at Rice. He met Sandage when he was working on his Ph.D. thesis about the background light in the universe and whether its grainy pattern could be used to discriminate between the cosmological world models. Gunn found Sandage fascinating—"He's a religious man, a peculiarity among our breed"—but their relationship was distant. "Sandage was well known not to be interested in students," he said in an Oxbridge accent that belied his rural Texas roots. "I never thought about working for him. Mount Wilson never took students, anyway."

After two years in the army, which he spent assigned to Caltech's Jet Propulsion Laboratory (JPL) building spacecraft instruments, Gunn went to Princeton. By 1970 he was back at Caltech teaching Robertson's old course. While at JPL he had built a small electronic spectroscope for studying planets that turned out to be perfect for looking at dim galaxies as well. With that he was pitched into the front line of observational cosmology, a discipline in which progress was driven largely by improvements in instrumentation. The giant 200-inch had sparked one revolution;

electronic light detectors that were more sensitive and versatile than photographic plates would spark the next. Gunn, who was enamored of high technology, could design and build state-of-the-art instruments, observe with them, and do theoretical calculations when it rained. He became one of the few astronomers, in an age of ferocious specialization, who could do everything.

Once, when asked who was the best astronomer in the world, Sandage had replied, "Well, young Jim Gunn is doing pretty well. If he keeps it up, he could be number two."

In 1970 Gunn had served notice that nothing was sacrosanct any more, neither in cosmology nor in Pasadena. He and Beverly Oke, another Caltech astronomer, decided to take up the elusive Hubble diagram. They applied for time on the 200-inch telescope to perform modern observations of the bright elliptical cluster galaxies. "Sandage wasn't doing anything in that regard," said Gunn. "We didn't consider it a direct assault."

Sandage, of course, did, just as he had when Greenstein rushed into quasars. When he objected to Gunn's getting 200-inch time to pursue what was in essence Mount Wilson and Sandage's project, he was overruled. After all, it was in the best traditions of science that everybody's work is checked by someone.

But, of course, it was not in the Gentleman's Code to do it at the same observatory.

The rival project went ahead. Gunn had been following Tinsley's work and, knowing they needed a theorist to unravel the effects of galactic evolution on their measurements, arranged for her to spend several months in Pasadena in 1972, giving her a new lease on life after her domestic confinement. They became close friends as well as professional allies.

There were now two groups doing cosmology in Pasadena, trooping to Palomar, that did not speak to each other except in the angry language of referees' reports and peer reviews. Sandage eventually published an eight-part series of papers on the Hubble diagram. He and Tinsley made contradictory statements about the evolutionary correction in the Hale Observatories annual report.

The next member of the rebel cosmological team was recruited in Cambridge, England, in the summer of 1972 when Gunn, who was visiting Hoyle's institute for the summer, spent some time hiking and talking with David Schramm, another summer visitor. Schramm, a hulking outgoing redhead and the Caltech wrestling coach, was a graduate student of Fowler's. He was an expert on nucleosynthesis and interested in the big bang.

Peebles had pointed out, recall, that the amount of deuterium (a heavy isotope of hydrogen) produced in the big bang was extremely sensitive to the density of nuclear matter in the early universe. The definitive calculations had been done by Hoyle, Fowler, and Robert Wagoner in the late sixties. Only recently, however, had space techniques made it possible to measure the abundance of interstellar deuterium, which in turn made it possible to estimate that density: the mass density of the universe when it was a few minutes old. The results strongly suggested that the universe only had about a tenth of the mass needed to generate enough gravity to ever recollapse itself. The implication was that the universe would expand forever. Schramm was enthusiastic about using the nucleosynthesis data to resolve the cosmological question.

Gunn was impressed. Work he had been doing with a Caltech postdoc named J. Richard Gott on the dynamics of galaxies and clusters pointed to a similar conclusion: that the galaxies *in toto* were neither massive nor numerous enough to arrest the expansion of the universe.

In the fall both Schramm and Tinsley were at the University of Texas, talking about doing a paper, when Gott came to town for a colloquium. The four of them decided to combine forces and write one paper, reviewing the growing preponderance of evidence that the universe was open, that Wheeler's big crunch, the collapse and snuffing of the laws of physics, was not in the cosmic cards after all. Most of these data were not new, and much of them were not even the young scientists' own work, but it was the first time anyone had assembled all the evidence together and dared to stand behind it. "We were sort of young Turks wanting to upset the establishment," said Schramm, chuckling. "One of the motivations was to show that the best way to solve cosmology was not the Hubble diagram."

It was a formidable quartet. Gott, a moon-faced Kentuckian who speaks in a slow drawl, was the youngest and also perhaps the deepest. A pure theorist and expert on relativity and the life of Einstein, he specialized in concocting strange universes. In one celebrated example, he theorized that at the moment of the big bang the universe had split into three parts: a universe of ordinary matter going forward in time; a universe of antimatter going backward in time; and a universe of tachyons—hypothetical particles that go faster than the speed of light. None of these universes could ever connect with one another.

Schramm was the physicist, the representative of a new breed who regarded the universe as a physics problem, and the most aggressive. Gunn, the closest of the four to being a classic "Sandagean" astronomer, was the most versatile. As for Tinsley, "Beatrice," sighed Gunn fondly with a faraway look in his eyes, "Beatrice was the glue."

She did most of the writing. The paper was called "An Unbound Universe?" If Sandage's prose tended to sound like cosmic groans reeking of dry authority, Tinsley's tone had the snap of youthful impertinence. It began with a quote from the Greek philosopher Lucretius:

> Desist from thrusting out reasoning from your mind because of its disconcerting novelty. Weigh it, rather, with a discerning judgment then, if it seems to you true, give in. If it is false, gird yourself to oppose it. For the mind wants to discover by reasoning what exists in the infinity of space that lies out there, beyond the ramparts of this world. . . . Here, then, is my first point. In all dimensions alike, on this side or that, upward or downward through the universe, there is no end.

Tinsley and her comrades made the case for a new way of doing cosmology. Instead of asking, How is space-time curved?—a task involving dauntingly long-distance astronomy with all its mystical uncertainties and Hubble diagram perplexities—they asked, Is the universe heavy or dense enough to drag itself back down to oblivion?

According to the Friedmann equations of the expanding universe there was a critical density of mass and energy—equivalent to about one hydrogen atom per cubic meter. If the universe were denser than that, it would eventually stop expanding and collapse; if it were less dense the galaxies would fly out forever. In equations, the ratio of the actual density of the universe to this critical density was denoted by a capital omega, the last letter of the Greek alphabet, symbol of kingdom come. Written it looks like a horseshoe: Ω. Mathematically, omega is just twice q_0. An omega of exactly 1.0, the critical density, corresponded to the universe with q_0 equal to 0.5: flat space, a universe balanced on the edge that would stop expanding after an infinite amount of time, the galaxies flying apart like rocks thrown just at escape velocity. If omega were less than 1.0, the universe would expand forever. If it were greater than 1.0, gravity would slow the expansion to a halt one day far in the future. Then the universe would recollapse and eventually swallow itself in the big crunch. In effect their paper enshrined omega as the new Holy Grail of cosmology, replacing the quaint and tiresome q_0.

To measure omega an astronomer need only stake out a suitably representative volume of space—the larger the better, but there was no need to survey all the way to the quasars—and somehow add up all the mass inside. Gott, Gunn, Schramm, and Tinsley (listing themselves alphabetically) showed that however you added up the mass in the universe, omega came out to be pitifully small—much, much less than 1.0—and, therefore, way too small to close the universe.

One way to measure omega was to add up all the light from the galaxies; the mass of all the stars needed to produce that much light only amounted to 0.01 of the critical mass. A more powerful way to estimate the masses of galaxies was by analyzing their motions and the way they tugged on each other in pairs and clusters. That method included the gravitational contributions of black holes, dim stars, and any other dark stuff in galaxies; it increased the imputed masses of galaxies tenfold (apparently, most of the material in galaxies didn't shine), but still left omega at only 0.1, the four reported, far short of the density needed to close the universe.

The final way to estimate omega was by measuring deuterium. Schramm's line of thought was the centerpiece of their argument. The beauty of deuterium was that it made no difference to the calculations what form the matter was in *today*—whether it was in a black hole, dust, or a blazing star—as long as it had passed through the thermonuclear cauldron of the big bang. Just by determining the relative abundance of deuterium in some representative sample of the cosmos, astronomers could calculate directly what the density of matter—ordinary matter, that is, neutrons and protons—had been when the universe was a few minutes old. The deuterium abundance gave the same answer as the galaxies: that omega was only 0.1.

"The verdict is not yet in [on the fate of the universe]," they concluded, "but perhaps the mood of the jury is becoming perceptible."

It took them longer to get the paper published than it did to write it. They sent it first in 1973 to *Nature,* where it was rejected as "inappropriate." It was finally published in the *Astrophysical Journal* in December 1974.

If Tinsley and her friends were right and the universe was open, then Sandage had been beaten to the greatest prize in cosmology, the end of Hubble's road. But he hadn't been beaten by much. The race was in fact virtually a dead heat. Sandage's route to the answer was pure astronomy in the classic Hubble vein. In 1961, he had invented an alternative to the slippery uncertainties of the Hubble diagram; he called it the time-scale test. It was a way of comparing two independent "clocks" to see how much the expansion of the universe had slowed down. Globular clusters were the first clock. Globular clusters were made of the oldest stars in the galaxy, maybe the universe, and figuring out how to date them had been his first big accomplishment. Suppose, he said, that you knew the ages of the oldest stars. Because the universe couldn't be younger than its inhabitants, he would then know that the universe had to be at least that many billion years old.

The second clock was the expansion of the universe. If he knew how fast the universe was expanding he could calculate how long it would have taken the galaxies to spread out to their current distances from each other at the present expansion rate. That would give him a second estimate of the age of the universe. The universe, like a person, had surely grown faster when it was young than it is growing now. Therefore the second clock, based on today's expansion rate, should always give a greater age than the globular cluster clock. By comparing the readings of these two "clocks," Sandage could determine how much gravity had slowed the expansion. For example, suppose the globular clusters proved to be 5 billion years old, while the so-called Hubble time was 10 billion years. Then Sandage would know that expansion must have been slowed down by at least half since the birth of the universe; at that rate it would soon coast to a complete stop and then reverse. Knowing those two "ages," he could solve the equations for q_0.

In effect, having launched the Hubble constant project in quest of one of the sacred numbers of cosmology, at the end Sandage would harvest them both.

Sandage and Tammann, based on their study of the spiral galaxies, finally concluded that the Hubble constant was 57 kilometers per second per megaparsec, plus or minus 15 percent.[1] Shortly thereafter an adjustment to stellar theory nudged the Hyades star cluster farther out, which made everything else in the universe farther and bigger and brighter. The Hubble constant became an even 50.

The latter value made the so-called Hubble time 20 billion years, which was how old the universe would be if it had expanded at the same speed, without slowing down, ever since creation. If that was the oldest the universe could be, what was the youngest it could be? Surely not younger than the stars. According to Sandage's most recent calculations, the oldest globular cluster stars were 14 billion years. He added a billion years or so for the galaxy to form and figured out that those two ages were pretty close. The universe had hardly slowed down at all. And it wasn't going to slow down. Ever.

On March 1975, these words, written by Sandage, appeared in the *Astrophysical Journal*.

> From this analysis we conclude that if H_0 has the value
> given . . . and if the age of globular clusters is about 14 bil-
> lion years, then H_0 has the value given . . . and if the age of
> globular clusters is about 14 billion years, than q_0 cannot be
> as large as 1. It may be as small as 0.03, in which case (a)

1. The same amount of uncertainty that Hubble had allowed for a constant ten times larger.

most of the mass is in galaxies, (b) the Universe has happened only once, and (c) the expansion will never stop.

In the Tinsley and Gunn camp, despite all their bravado, there was quiet relief that the old man had come up with the same answer as they had. "It's a terrible surprise," Sandage told *Time* magazine.

What about the contradictory evidence from the Hubble diagram that the universe was closed? In the eyes of most astronomers, Tinsley had won that argument. The Hubble diagram based on giant elliptical galaxies had fallen. The crux of the argument came down to a technical disagreement about what kind of stars produced most of the light in elliptical galaxies. The answer would determine whether the galaxies got radically dimmer with age, as Tinsley claimed, necessitating a drastic correction to the Hubble diagram and its message of a closed universe. Sandage said they were normal hydrogen-burning stars; Tinsley said they were red giants—one of the stages of evolution that stars pass through after they exhaust their hydrogen fuel and leave the so-called main sequence.

In 1974 the astronomer Jay Frogel, now at Ohio State, made infrared observations of elliptical galaxies and found in their light the unmistakable signature of red giant stars. Tinsley had been right.

Tammann got a preprint of Frogel's paper in the mail. "I was about to give a talk," explained Tammann. "I called up Allan and asked, 'Can I say the universe is open?'

" 'Of course.' He wasn't angry or even sad. It just shows how flexible he is when he is presented with the data. Up to then, q-nought was always 0.5 or more. It changed overnight."

Sandage, in fact, had fulfilled his quest nobly. His disappointment was tempered by the realization that the final result was a great vindication of the cosmological framework in which first Hubble and then he had toiled for decades. The Friedmann equations worked. The time scales matched—that was a miracle. The answer didn't mean as much to him as the fact that there *was* an answer. So what if it wasn't the answer that the theorists thought was the prettiest? We lived in a Friedmann universe. The simplest model worked. But it didn't feel like God.

"So the universe will continue to expand forever," Sandage said, "and galaxies will get farther and farther apart, and things will just die. That's the way it is. It doesn't really matter whether I feel lonely about it or not."

The open universe gave a new meaning to loneliness. It opened vast, unbelievable realms of time and space to informed speculation—gulfs that made all of cosmic history to date an indistinguishable splinter.

If the closed, oscillating universe punctuated physical existence with a big crunch like a pair of clashing cymbals, the open universe was like the piano chord struck at the end of the Beatles song "A Day in the Life"— a single brief jangly burst of light and sound, quavering and ringing as it slowly fades out into the blackness.

In this scenario, apocalypse (such as it is) will happen in stages. The first thing that will happen is that the stars will burn out and the galaxies will fade. The sun will run out of hydrogen in a mere 5 billion years and swell up into a red giant, turning the earth into a black cinder. In 100 billion years the Milky Way will be a cemetery full of stellar corpses: black holes, neutron stars, and white dwarfs. Within a billion billion (10^{18}) years, these will all be conglomerated into a single giant black hole at the center of the former galaxy. In 10^{27} years, all the galaxies in a cluster will have merged into one supergalactic black hole; the universe will consist of black holes rushing away from each other through dead cold space. In 10^{100} years these black holes, billions of times as massive as the sun, will evaporate. Nothing will remain but feeble, dilute pools of particles and radiation separated by trillions of light-years.

What else might happen depended on how seriously you took the more exotic theories of physics. For example, some speculative theories of particle physics predicted that protons, the fundamental building blocks of atoms and thus of all ordinary matter, were unstable and would fizz away radioactively after about 10^{30} (a million trillion trillion) years or so. There would be nothing but lightweight junk particles left in the universe. Some scientists envisioned the formation in the far far future of pseudo-atoms composed of these junk particles—electrons and positrons circling each other at a distance equal to the radius of today's universe. No matter how huge and empty the universe got, they pointed out, it could never be completely empty. By virtue of the uncertainty principle there would always be vacuum fluctuations; space would at least be microscopically alive even if there was no one to look at it.

What were the prospects for life in such a universe? The imaginative Freeman Dyson suggested that civilizations could sustain themselves almost indefinitely by tapping the energy that could be extracted from a rotating black hole. Of course, it would prove futile if the protons in their bones meanwhile decayed. Still, there was nothing in principle, he concluded, to keep life from going on forever if it could adapt itself to the infinitely slow rhythms of the far future.

Wheeler, the eternal prophet, found this prospect of eternal decay, a one-way universe that was born and never died, distasteful aesthetically and emotionally, as well as intellectually. It was unsymmetrical. It was like being trapped in a football stadium with the home team hopelessly behind in a fourth quarter that would never end. A universe that could never die could never be reborn. There was no second chance for creation. Where was the frame that gave meaning to existence? Wheeler insisted that the extra mass that would raise omega and close the universe would be found somewhere some time.

He was not alone. The preponderance of evidence in favor of an open universe was impressive, and yet. . . . One night during the 1974 Texas Symposium, *New York Times* reporter Walter Sullivan took a group of astrophysicists to dinner and asked them to vote on whether they thought the universe was open or closed. The vote, he reported, was unanimous for a closed universe. After 1974, looking for the other 90 percent of the universe was to become one of the major themes of cosmology.

Tinsley was one of the few astronomers who *liked* having an open universe. "It may be 'bad science' to like the universe being open because it feels better, but there is in me a strong delight in that possibility," she wrote her father. "I think I am tied to the idea of expanding forever— like life in a sense—more than spatial infinity."

On another occasion she wrote, "I don't think it is weakness to be motivated by emotions. What else is the driving force, or the inspiration to think of useful theories? Only if emotional attachment to one's own theory makes one blind to alternatives is it bad."

Another who did not join in the quest for the missing mass was Sandage. "For some reason I would like to have it closed," he admitted haltingly. "Yes, but to think the universe happened only once—that makes it even more mysterious, in a sense. It's outside the realm of science, what happened before the first microsecond. Why it got itself into that, how it got itself into that state?

"But it's no more mysterious than noting the tremendous complexity of the chemical balance of the human body. You cut yourself— and why is it the white corpuscles know exactly where to go to close the wound? That's a miracle. And I don't believe that's due to progressive selection of the fittest. It's just too fine a mechanism. I don't know what I'm saying now, I don't know what the next sentence is.

"I don't mean that points to the existence of God, whatever that means. Newton's laws are God, in a sense. But I find it all so rational and so amazingly beautiful and so mysterious."

That same line of thought often ran through Sandage's conversations, that the world was too magical to be an accident, although in his milder moments he admitted that he didn't know enough about evolution to be shooting his mouth off. In 1975 he was eating lunch with Graham Berry, the director of the Caltech news service, when he got on the subject of religion in his usual enthusiastic manner. The couple at the next table started following the conversation. Finally the man got up, introduced himself as a minister, and asked if he could join them. He thought Sandage was a minister, too. Sandage was thrilled.

"I don't know what I would call myself," he said in 1977, describing the strange nexus between science and religion into which his pursuit of the stars had stuck him.

"If you believe anything of the hard science of cosmology, there was an event that happened that can be age dated back in the past. And just the very fact that science can say that statement, that cosmology can understand the universe at a much earlier state and it did emerge from a state that was fundamentally different. Now that's an act of creation. Within the realm of science one cannot say any more detail about that creation than the First Book of Genesis.

"Well, I think that the whole rationality of the universe is a mystery. The fact that Newton's and Einstein's equations work is one of the world's great mysteries. And in that sense I'm very religious."

Reaching the end of a quarter-century trail gave Sandage no relief. To Tammann, Sandage was at his worst when he finished something. The frenzy of completion would fend off the larger pressure for a while, but then, as soon as it was gone, what Tammann called the "debt of nature" would come crashing back in on Sandage. He would start pestering Tammann for work that was due on the next paper.

Sandage plunged into new projects. He had to finish measuring redshifts of the 1300 brightest galaxies in the sky, which had been compiled in a catalog by the astronomers Harlow Shapley and Adelaide Ames at Harvard in 1932. In collaboration with a pair of younger astronomers, Jim Westphal and Jerome Kristian, he set out to redo the Hubble diagram, remeasuring the magnitudes of elliptical galaxies with a sensitive new television system—one of his few ventures into high technology.

The pressure and pace invariably left his collaborators by the wayside. Tuton remembered Sandage badgering Kristian during their attempts to extend the cursed Hubble diagram. There was never enough time for Sandage at the telescope; every second counted. Kristian was too slow. He would be up in the cage trying to make some measurement and Sandage would be prowling restlessly down in the control room calling up over the intercom. "What's taking so long, Jerry? Are you done yet? What's going on? C'mon, talk to us, Jerry."

Only Tammann seemed to be able to endure Sandage's blasts year in and year out. Once, even he rebelled against the accelerating workload, and curiously it was a comfort to Sandage. He had always worried about drowning Tammann, but now he knew that Tammann would fight back.

Deadlines were a particular burden to the two of them. Writing,

Sandage moaned, was "turning blood into ink." Every calculation was hell. Sandage often copped out on appearing at conferences, but to Tammann's chagrin, never on delivering a paper once the deadline was agreed to. "It makes life hard and, therefore, has to be done," said Tammann. The debt of nature could never be repaid.

"Allan is so talented, so powerful," Tammann said marveling, "and so unhappy because he can never accomplish what he sees as his duty, what people expect of him." He recalled Sandage complaining once, why hadn't it been given to him to discover a great law of nature, as it had been to Hubble? "Allan Sandage wanted to be Hubble. Hubble wanted to be Einstein. And Einstein wanted to be a peacemaker," he sighed. "The great ones pay."

The competition from astronomers outside Pasadena was increasing. The Mount Wilson monopoly had ended. In 1974 a 158-inch telescope began operation at the Kitt Peak National Observatory outside Tucson, run by a consortium called Associated Universities for Research in Astronomy (AURA) and funded by the National Science Foundation. A couple of years later AURA and NSF built a twin telescope in Chile to open up the southern skies. The Europeans followed suit. Putting a large telescope in Chile had actually been Mount Wilson's idea; Hale had dreamed of a 200-inch in the Southern Hemisphere. Carnegie, Mount Wilson's owner, had lost a tug of war over Ford Foundation funding for the telescope when the head of the Ford Foundation moved over to become the director of the National Science Foundation, AURA's patron. Carnegie finally built a 100-inch telescope at Las Campanas in Chile. The new telescopes were being equipped with electronic detectors like the charge-coupled device (CCD), a silicon wafer on which light was collected as electric charge and then read out by a computer. A good photographic emulsion might capture 5 percent of the light falling on it to blacken grains; a CCD gobbled 50 to 80 percent of the available light. One of these transformed an ordinary telescope into a giant. Now anybody could reach the edge of the universe, the depths of redshift and time where even the quasars fell off, the era before which there might not have been galaxies.

Even with all this new equipment, observatories could not keep up with the rising demand for telescope time. In the wake of the space program, quasars, black holes, and the big bang, astrophysics had become a glamour field. New astronomy departments were starting every year, staffed by freshly minted Ph.D.s and by physicists flooding across interdisciplinary lines to join the Great Work. In the face of this population pressure Sandage was somehow able actually to increase his observing time. In 1977 he spent 105 nights on various mountains. He particularly liked the Las Campanas telescope, which could photograph more than two square degrees of sky in one shot, 41 times more than the 200-inch could image.

Sandage didn't make it easy for the younger astronomers and the physicists to get to know him. His door at Santa Barbara Street was usually closed, and he virtually stopped going to meetings and conferences. In his office he sat with the heritage of Hubble and Baade and Humason stacked around him. Asked a question he was liable to wave at the locked file cabinets and plate boxes and answer aloofly, "The answer's in there."

There were two Sandages, really. According to legend a young acolyte might first meet Uncle Allan, the charismatic teller of tales with a dry wit and religious intensity about the creation event, a puller of legs and nudger of elbows in the Lyndon Johnson style, kidder of the young ladies, drinker of Manhattans, speller of bromides and Bondi theorems. Sooner or later there would be the clash of scientific opinion, and the disillusion and disappointment would set in; the acolyte had failed him. The jollity would slide off like an old snakeskin. Out came That Son-of-a-Bitch Sandage with a dry-ice voice, slumped shoulders, cold half-lidded blue-gray eyes, and all the warmth and empathy of some patch of sky next to Virgo, lowering the boom, writing devastating referee's reports on one's paper or proposal, telling one not to publish, delivering edicts. We at Mount Wilson . . .

It was rumored in astronomical circles that it was Sandage who had driven Gunn out of Pasadena to Princeton, where he became a full professor and MacArthur "genius grant" fellow. Gunn denied it. The principal reason he left, he said, was to get out of building instruments all the time (he was drafted to help build one of the cameras for the space telescope) and to do more theoretical work.

"Allan does not like competition. He established the discipline of observational cosmology and he thinks it should be his. I think he feels, and I can sympathize," Gunn said frowning, "that many sloppy and shoddy papers are rushed to print. He is seldom nasty in person."

Tinsley's star also soared in the wake of the unbound universe, but according to Schramm her personality changed shortly thereafter. An uncle had been on the DC-10 that crashed taking off in Paris in 1976, killing all aboard. Before, said Schramm, she seemed happy. Afterward she was more driven, more determined to do what she did best.

Eventually Dallas became too small. Yale and Chicago were pursuing her with professorships, while the head of the University of Texas, Dallas—which is what the old Southwest Center had become—failed to answer her letters asking for a promotion. Feeling her children were old enough to do without her, she divorced Brian and went to Yale, where she moved in with Richard Larson, a mild-mannered astronomer she had known in Pasadena.

At Yale she became a role model and champion for the tiny but growing band of women in astronomy. Tinsley never lost her combative edge and had difficulty accepting the fact that people respected her and took her seriously. "She never lost the feeling of fighting the world," said Larson.

Tinsley continued to wage war against the Hubble diagram, railing against its usefulness as a cosmological tool in several papers. The last straw for the credibility of the Hubble diagram came when the Princeton theorists suggested that everybody had been on the wrong track: Giant elliptical galaxies, they said, did not get dimmer with age—as everyone had thought—but might actually get brighter by swallowing smaller galaxies that wandered too close. Subsequently, photographs of giant elliptical galaxies with two or more bright central spots in them—presumably the half-digested nuclei of victim galaxies—gave credence to the notion of galactic cannibalism.

In 1977 Tinsley organized a major symposium at Yale on the evolution of stars and galaxies. Sandage was invited but didn't show up. That same year a lump showed up on her thigh that turned out to be a melanoma. Tinsley's years at Yale were shadowed by a fight that she finally couldn't win. By 1980, despite surgery to remove the original tumor, the disease had spread through her body, and she spent the last year of her life checking in and out of the Yale infirmary undergoing radiation and chemotherapy while holding seminars in her room. Toward the end, in 1981, she wrote a poem:

> Let me be like Bach, creating fugues,
> Till suddenly the pen will move no more
> Let all my themes within—of ancient light,
> Of origins, and change and human worth—
> Let all their melodies still intertwine,
> Evolve and merge with growing unity,
> > Ever without fading,
> > Ever without a final chord . . .
> Till suddenly my mind can bear no more.

11

FERMILAND

The headquarters of the Fermi National Accelerator Laboratory rise from the Illinois plains 30 miles west of Chicago like a pair of concrete archer's bows standing back-to-back with glass stretched between them. Outward from its lonely spire a collection of berms and sheds and roads scroll a series of interconnected loops and spirals across the prairie. The largest of them, known as the Main Ring, marks the location of a 4-mile-long underground vacuum pipe in which protons are whipped by a small city's supply of radio energy to the speed of light and then smashed together in the effort to create new forms of matter.

It was to this environs that David Schramm came after the publication of "An Unbound Universe?" and played a leading role in a process that had begun with Peebles and Zeldovich: the transmogrification of cosmology into particle physics and vice versa.

Schramm, the nuclear physicist of the gang of four, had most strongly pushed the idea of the abundance of primordial deuterium as a barometer of the density and fate of the universe. Schramm knew that this line of reasoning was the wave of the future. The answer—that the universe was open and would go on forever—was less important than the technique: studying the subatomic physics of the early fireball to understand the large-scale properties of the universe 15 billion years later. In fact, in less than a decade the answer, for Schramm and other brave theorists, would change. "A universe with omega of 1.0 is the only one that makes any sense," Schramm, of all people, told me. The evidence would come not from the sky, but from elementary particle physics. The towering Fermilab high-rise and its underground ring were to come to symbolize the union of those who studied the very small, the subatomic world, and those who studied the very large, the universe. Schramm, a natural hustler and organizer, was to build his empire on that blurry scientific boundary.

At six foot three and 230 pounds, there was maybe 20 pounds too much of Schramm. He wore eastern-style cowboy boots and a tie with a picture of Saturn on it that looked a little lost on his chest. His forehead was sunburned and furrowed; a peak of wavy red hair swept over his brow. His pale eyebrows made him look old around his blue eyes. He moved and talked with the exaggerated delicacy of a giant who is aware and a little in awe of his own destructive power. He had a habit of softening his more assertive or outrageous statements with a soft chuckle that a fawning Chicago newspaper columnist once called a "cosmic giggle." I have never seen him raise his voice or have to. Nobody, it seems, would want to see him mad.

Schramm had been raised in a middle-class neighborhood in St. Louis, the oldest of three sons of a librarian and a pilot whose career was cut short by an eye injury before he could attain glory in World War II. Inheriting his father's love of airplanes, young Dave devoured encyclopedia articles about aviation and decided early on that he was interested in doing something in science, but there were no role models for him to follow. He never did any homework. Science and math came easily, but he was afraid his ability in them marked him as different or weird.

"Mostly I was interested in girls, athletics, and hanging out," he said. "I tried to worm my way into social groups by being a star athlete." Schramm succeeded. He was an all-state tackle on the football team and the Missouri State wrestling champion his senior year, as well as a member of the debate team. The summer after his senior year he married his high-school girlfriend, Melinda. "My parents were not really pleased," Schramm recalled, "but they still supported us."

In the meantime a pair of uncles who had contact with the "outside world" convinced him to raise his sights beyond a midwestern jock university and apply to the Massachusetts Institute of Technology (MIT). Schramm, with perfect SAT scores, got in. "I didn't realize I could go to places like Harvard or Stanford," he said. "I didn't even apply to those other places."

MIT, a maze of gray stone industrial neoclassic buildings with numbers instead of names, sprawls along the Charles River in Cambridge, a few miles downstream from Harvard. The names of great scientists and engineers are chiseled along the cornices. From across the river in Back Bay, Boston, it looks like a small, grimy city punctuated by tall columns and weighty domes reminiscent of Rome's Pantheon. The main thoroughfare on campus is a quarter-mile-long central corridor that runs, with one jog, from the pillared entrance on Massachusetts Avenue to the cubical little humanities building dangling like one toe off the end of the octopus complex.

One of MIT's more curious and revealing flourishes resides in an old cafeteria known as Walker Memorial. It stands beyond the humanities building, its columned entrance facing the river. Inside, the walls are covered with frescoes of classical and biblical scenes. One panel next to the entrance is particularly striking. It shows a tall, handsome, white-haired man in a lab coat standing before two smoking urns. A dog slinks around the base of one urn. In the foreground, figures representing industry, government, and the military look up from a table in awe. Angels hover in the background. Above this display of might and promise are the words ET GRITIS SICVT DII [sic] SCIENTES BONVM ET MALVM, "And you will be like gods, knowing good and evil."

They are, of course, the Devil's words.

When Schramm arrived in 1963, the Institute's gray walls would buzz once an hour with throngs of purposeful humanity—mostly white and male. MIT called itself a universe polarized—or, as the joke would have it, "paralyzed"—around science. "You get a peculiar attitude," Schramm said. "Most of the people majoring in humanities at MIT are in humanities because they were flunking out of science. You get the attitude that people in the humanities are dumb. It takes a long time to get rid of that."

Married, and soon with a son, Schramm found himself in a place where for the first time he had to work hard, but the reward was a straight-A record. Along the way he played rugby and managed to become the New England heavyweight wrestling champion. "I didn't know what it meant to be a scientist," he said. "I had no idea it meant a Ph.D. I thought it was sitting around the lab soldering with the graduate students. I knew I wanted to be in physics." A year of working in a nuclear physics laboratory and an astrophysics course taught by the gifted Philip Morrison put Schramm in the groove. The big bang radiation had just been discovered. At MIT, astrophysics was a quantitative discipline that eschewed descriptions of constellations and galaxy types in favor of the hard-core calculations of the stars and the early universe as thermonuclear pressure cookers. In a word: Peebles-type cosmology, the universe as a physics problem.

Schramm went on to Caltech graduate school with the ambition of combining nuclear physics and astrophysics. That meant he was a student both of Willy Fowler's and of Gerry Wasserburg, who among other things ran a laboratory known as the Lunatic Asylum, which was built to analyze the moon rocks brought back by astronauts. He became an expert on exploding stars and the multitude of nuclear chain reactions and decays within them that strewed exotic radioactive elements across the galaxy. One of Schramm's first jobs under Wasserburg was inventing a way, mathematically, to untangle the different generations of long-lived radioactive nuclei like plutonium and thorium that contributed to the present-day abundances and their decay products. The oldest thorium had to come from the first stars in the Milky Way galaxy to explode, so Schramm's method was a way to date the galaxy. According to that and other nuclear chronologies, the galaxy and thus the universe were at least 10 billion years old, a figure close, in fact, to the ages that Sandage and others were deriving from globular clusters. It made a name for him at Caltech, which was a great relief, because he was used to being a big man on campus. "Maarten Schmidt and Richard Feynman knew my name," he recounted happily. "They would ask me questions about the age of the universe."

The atmosphere around Fowler's lab was enjoyable for Schramm. There were parties on Friday night after the seminars. "It was fun to watch the faculty members get drunk and chase women," he said grinning smugly.

Although Schramm was a theorist, he was required to get experimental experience, and so he worked in the Lunatic Asylum. Wasserburg's group specialized in measurements so sensitive that people worried that silver atoms drifting out of the fillings in their teeth would contaminate the data. Schramm found the life of an experimenter tedious. "You work hard on trivial things, but it's good for theorists to do an experiment. You learn how hard it is to do science, and about the uncertainties in the numbers."

Meanwhile wrestling continued to claim his allegiance. Schramm placed second in the Olympic trials for Greco-Roman wrestling in 1968. Unfortunately, only the winner got to go to Mexico City. On the way to the trials, he says, he beat five former national champions and realized they weren't so tough. "It was good for the ego." For the next few years Schramm was a national force. "There were four of us around the country who were pretty even." He thought of shucking physics and becoming a wrestling coach. Then, in 1972, with the Munich Olympics approaching, Fowler offered to take Schramm along with him on his yearly summer visit to Hoyle in Cambridge.

It wasn't hard to choose. "I had a one-in-four chance of going to Munich and a one-in-one chance of going to Cambridge," Schramm explained. The Schramms arrived in Cambridge to find that their lodgings were not ready and so moved in for a week with the Gunns, old friends who were also visiting that summer. During the week Schramm and Gunn began the conversation about the fate of the universe that led to the formation of the gang of four.

Schramm had gotten his introduction to cosmology in a course on general relativity taught by Kip Thorne, Wheeler's old student. During the term he wrote a paper on big bang nucleosynthesis, which was a hot topic around Caltech in those days. The other application of nuclear physics to astrophysics was, of course, to determine what had actually occurred during the big bang, when the universe was hot enough to fuse hydrogen into helium and perhaps a few other light elements. Peebles, reinventing some of Gamow's ideas, had made calculations along these lines in 1965. When Peebles dropped out of nucleosynthesis to study the formation of galaxies, Fowler, inspired by Hoyle, began to tinker around with the early universe, using a computer code written by his postdoc Robert Wagoner.

Fowler was interested in how closely his computations could match the observed amount of helium in the universe, but he noticed that minuscule amounts of deuterium (heavy hydrogen), lithium, boron, beryllium, and helium-3[1] also dribbled out of the big bang calculations. Fowler kept track of them, but mostly for bookkeeping purposes, because the amounts produced of these elements appeared to him to be cosmically irrelevant. For example, in big bang models of a universe with critical density, less than one hydrogen atom in a million had an extra neutron in its nucleus, making it deuterium, but in seawater about 1 in every 10,000 hydrogens was deuterium. In 1962 Greenstein, Fowler himself, and Hoyle had concluded that this deuterium, and lithium as well, was produced in young protostars; during the sixties most astronomers believed them.

Then in 1969 while they were on the moon, the *Apollo 11* astronauts unfurled an aluminum window shade to capture particles blowing off the sun. The results suggested that the ratio of hydrogen to deuterium in the sun and presumably the rest of the cosmos was more like 100,000 to 1, in the range of the big bang calculation. The seawater ratio, in fact, was an anomaly caused by chemical effects. About this time Jean Audouze, a student at the French Institute for Astrophysics, and his thesis adviser, Hubert Reeves, showed in a paper that protostars could not produce deuterium and lithium after all. Audouze heard about the Apollo results,

1. The identity of an atom—what kind of element it is—is determined by the number of protons in its nucleus, or its *atomic number*. Hydrogen, with a nucleus of a single proton, has atomic number 1. Helium, with two protons, has atomic number 2. The different species or isotopes of a given element are designated by their *atomic weights*—the sum of the numbers of protons and neutrons in their nuclei. Regular helium has a nucleus of two protons and two neutrons and thus an atomic weight of 4. Helium-3 has two protons but only one neutron in its nucleus.

which were analyzed by a Swiss named Johannes Geiss. On a trip to Caltech Audouze told Schramm about it, and Schramm pounced like a hungry bear.

"We got all excited," he said. Audouze and Schramm realized that the *Apollo 11* results meant that the big bang calculations *were* giving the right deuterium counts, as well as those for helium. At about the same time, calculations by Hubert Reeves showed that cosmic-ray collisions could account for the production of some trace light elements, like lithium-6 and beryllium-9, that neither the big bang nor the stellar evolution could match. Everything was falling into place. That gave Schramm and Audouze confidence that they could use the big bang calculations and modern element abundances to find out about the early universe. Deuterium, too, was a primordial element, and splattered on the aluminum foil was a relic of the universe when it was only a few minutes old.

In the meantime David Black, an astronomer at NASA's Ames Research Center, had reached a similar conclusion about the deuterium abundance, based on analyzing the gases trapped in meteorites. Black knew what Peebles had found out: that the big bang deuterium abundance was wildly sensitive to the density of mass and energy in the primordial fireball—to *omega,* in other words, the key to the fate of the universe. In a short paper in *Nature* in 1971, he pointed out that the deuterium abundance as implied by the meteorites fit a model of the fireball in which omega was very low, only a tenth of the value needed to close the universe; it meant that the universe was open. That was the first published shot in a revolution.

A year later Schramm, Audouze, Fowler, and Reeves wrote a long paper for the *ApJ* detailing the creation of the light elements. They noted that the deuterium had most likely been made in the big bang and that, if so, its abundance could measure the fate of the universe.

Cosmology had taken a big step. Whereas the helium abundance was a test of whether the big bang had happened at all, the trace elements like deuterium and lithium were sensitive barometers of primordial conditions and their abundances could be used to probe details of the big bang. In the production of helium during the big bang, deuterium was an intermediary step. The denser the universe, the sooner in time the fusion process began and the more helium was produced and the less deuterium was left over. So the more deuterium there was in the modern universe, the more open the universe was. According to the aluminum foil, there was ten times too much deuterium for the universe to be closed. Later in 1973 the measurement was backed up by ultraviolet observations with the *Copernicus* satellite, which could see the notch that interstellar deuterium took out of the light of distant stars, and concluded that 2 out of every 100,000 hydrogen atoms were deuterium—evidence of an open universe.

Not all cosmologists were so enthusiastic. "We won't know if you're right until we measure q-nought," grumbled one Texas senior astronomer to Schramm.

Schramm could not understand such conservatism. The universe was a pretty simple reactor; the computations represented less a mathematical leap than a philosophical one. In that regard he was fearless. He was in love with recklessness, and now, as far as he was concerned, he had the cosmological mystery in a hammerlock. He didn't have to stumble around the dark trying to decipher the light from some galaxy at the edge of time. The signature of the big bang was all around him—it was in him, in his blood and his beard. Fowler's basement nuclear laboratory was as good a telescope as the 200-inch—and it was cleaner. Big bang nucleosynthesis gave the answer.

Schramm talked the results up shamelessly that summer to Gunn and to everybody else. In the fall he moved to the University of Texas and met Tinsley. The article "An Unbound Universe?" ensued. When Sandage's own measurements supported the gang of four's conclusion that q_0 was negligible and that the universe seemed likely to expand forever, Schramm's confidence was vindicated.

Robert Wagoner, now a Stanford theorist who sort of resembles the character in the MIT fresco—tall and prematurely white-haired—had written an elaborate computer program for the big bang calculations with Hoyle and Fowler. His code became a kind of cookbook of the elements: hydrogen, helium, deuterium, lithium, boron, and beryllium were products of the recipe. Matching the recipe's output to the real universe became a demanding observational game. But as more and more data were painfully wrenched from the sky, the numbers looked better and better. The simplest big bang model worked. The relative abundances of the different elements all fit together in one consistent model. Each part could be challenged, Schramm conceded, but together all the observations added up in a viselike, cross-braced whole. "We're just saying the simplest explanation works," said Schramm proudly. There was a cosmic history. All the way back to a hundredth of a second, when the observable universe was about the size of the moon.

In principle it was simple, based on today's three frigid degrees, to extrapolate the temperature and density of the universe back to any moment in time, back to a millionth or a billionth of a second after time began. Both fell in a simple mathematical way as the universe expanded. Theoretically, at the exact moment of creation—the singularity—the universe was a pinprick of energy of infinite temperature and density. According to the laws of relativity and quantum mechanics, elementary particles—like electrons and quarks—could form in pairs with their antiparticle mates out of this intense energy field. At the same time these particle-antiparticle pairs were meeting and mutually annihilating each other, dissolving back into the radiation bath. The hotter it was, the more massive were the particles that could be created. So the universe at any time in the first seconds of the big bang was a stew of radiation and elementary particles being created and destroyed.

As of 1974, the standard model began when the universe was one one-hundredth of a second old and had a temperature of about 100 billion degrees K. (Above that temperature, particle physicists did not know enough about the forces of nature to predict how the particles that would be created would interact.) At that ripe old age, the universe would have been 4 billion times as dense as seawater. It would be a sea of electrons, positrons, photons, and funny massless particles called neutrinos bumping, annihilating, and reforming, along with a much smaller supply of protons and neutrons left over from an earlier, hotter moment of creation and annihilation. It was still too hot for any of these particles to stick together into nuclei.

When the universe was about fourteen seconds old and the temperature had dropped to only 3 billion degrees K, it became too cool even to manufacture electrons and positrons—the lightest known particles—out of the ambient energy. As a result their numbers began to decline precipitously as annihilation proceeded unbalanced by creation. In fact annihilation was ruthlessly efficient in ridding the universe of particles; out of the original population of electrons and protons only one particle in a billion would survive the first few seconds. All the matter that now exists in the universe, all the crystalline shapes and blazing stars and chains of galaxies, is based on less than a trace of what once existed, is based on barely more than nothing. Once the annihilation-creation circus had ceased, the universe, in terms of mass-energy, was essentially pure radiation.

As soon as there were protons there was hydrogen—whose nucleus consists of a single proton—in the universe. At the age of 100 seconds the expanding universe had cooled to a billion degrees K, and it was suddenly cool enough for protons and neutrons to stick together in deuterium nuclei. An orgy of nucleosynthesis ensued. During the next few minutes the deuterium gobbled free neutrons and protons in a complex chain of reactions to build lithium and helium. By the end of that time, when it was too cool for fusion to proceed any further, about a quarter of the mass of the universe had been transformed to helium; the rest was hydrogen and a tiny, tiny amount of lithium. The universe was then about one-tenth as dense as seawater. It consisted of bare atomic nuclei and free electrons flying around independently of one another.

That state of affairs stayed the same as the universe kept expanding for the next few hundred thousand years, until the temperature had fallen all the way to around 4000°K, the surface temperature of a cool red star. At this point it was cool enough for atoms to form: electrons fell into orbit around the nuclei. The capture of electrons into the atoms made the universe suddenly transparent to the background radiation, which was then at the wavelengths of visible light; the whole sky would have glowed like the inside of a furnace. Subsequently, matter no longer felt any pressure from the radiation. It was only then, as Peebles had realized back in his original paper on the big bang and galaxy formation, that whatever lumps existed in the primordial stew could begin to grow by the force of gravity, attracting other matter in the slow, inexorable process of becoming galaxies.

As the universe kept expanding and cooling, the cosmic radiation lengthened in wavelength and fell into the infrared and invisibility. The universe went dark for a billion years. Under cover of blackness the galaxies grew—or the superclusters pancaked and fragmented—and clouds of gas formed and swirled and condensed and condensed some more. Waves of stars formed, exploded, collapsed into tight whirling clumps, and collapsed again. Giant black holes rent space along fault lines laid down perhaps in the first microsecond of time, surrounded by debris. The long epoch of galaxy birth was ended by the glint of quasars, as the black holes, finding themselves anchoring glittery wheels of stars, began to eat up the detritus of birth. The stage was set for slower chemical and biological evolution.

The standard model was a triumph. It explained the elements. The problem with this model would be the problem with all the models that superseded it: What happened during the stretch of time it described depended on what had happened in the uncharted moments before. No matter how far into the past cosmologists blazed their way, it seemed that the most crucial interesting events had always occurred earlier.

At about the time that "An Unbound Universe?" was published, Schramm, then twenty-eight, moved to the University of Chicago as a tenured associate professor. The university, where Amos Alonzo Stagg coached football, the first Heisman trophy was won, and Enrico Fermi built the first nuclear reactor underneath the football stadium, is an island of Ivy League intellect amid the mean sprawl of the South Side of Chicago. The community hews closely together. As an urban midwesterner, Schramm was right at home. He bought a large townhouse on the edge of campus. Within a few years he and Melinda got divorced. "She had to keep switching around taking courses here and there," he said of their traveling undergraduate and graduate days. He sighed. "We missed out on a lot." She wound up getting a law degree in Chicago.

Meanwhile Schramm had begun his love affair with particle physics, and a strange elementary particle known as the neutrino that was to play a rich role in the lives of cosmologists.

Neutrinos were among the most paradoxical members of the zoo of elementary particles that were discovered after the war. Produced during radioactive decay, they supposedly had neither charge nor mass and they traveled, consequently, at the speed of light. Their only interaction with the world (besides gravity) was by means of something called the "weak" force, which causes some kinds of radioactive decay. It was so weak that, according to calculations, a typical neutrino could pass through a billion miles of water unhindered—stars and planets were transparent to them. Neutrinos were produced by the trillions every second in stars like the sun, and copious quantities of them were left over from the big bang, but this constant cosmic rain was invisible. They were the ghost riders of the universe. If particles were like people, then neutrinos were the CIA spooks of the subatomic world—aloof and slippery, gone before you knew they had been there, slipping through borders like moonlight through a window. No prison could hold them.

This bizarreness reached the popular culture. The novelist John Updike wrote a poem about them in the *New Yorker* titled "Cosmic Gall":

> Neutrinos, they are very small.
>   They have no charge and have no mass
> And they do not interact at all.
> The earth is just a silly ball
>   To them, through which they pass,
> Like dustmaids down a drafty hall
>   Or photons through a sheet of glass.
>   They snub the most exquisite gas,
> Ignore the most substantial wall,
>   Cold-shoulder steel and sounding brass,
> Insult the stallion in his stall,
>   And, scorning barriers of class,
> Infiltrate you and me! Like tall
> And painless guillotines, they fall
>   Down through our heads into the grass.
> At night, they enter at Nepal
>   And pierce the lover and his lass
> From underneath the bed—you call
>   It wonderful: I call it crass.

Schramm's involvement with neutrinos came from his work on supernova explosions. According to some theories, it was neutrinos—of all the unlikely actors—that did the work of blowing apart a star. The way the process worked was like this: Supernova explosions began when a massive star ran out of nuclear fuel and its core collapsed, shrinking in about a second from a glob of iron as big as the earth to a nugget of densely packed neutrons about the size of Manhattan. Almost all the energy of that free-fall inward came out in the form of neutrinos, created when protons and neutrons in the shrinking mass were squeezed together, 10^{58} of them. There was power in numbers; despite their apparent disdain for other matter, the neutrinos, as they flooded out of the star's dense center, exerted enough pressure to blow the outer layers of the star into space.

The success of this scenario depended on a feature of the weak interaction called neutral currents, which were predicted by new theories of the weak force but which had not been observed by the early seventies. Without these currents the neutrinos didn't supply enough oomph to kick apart the poor star. Without them, in fact, theorists had no good explanation for how stars explode.

Schramm decided he'd better learn about neutral currents and weak force physics. "Because of neutral currents I got in touch with particle physicists," he explained. Soon he found himself tracking information back and forth between the two communities of scientists. "I became known among astronomers as an expert on particle physics. Heh, heh, heh."

It was his fascination with neutrinos and the big bang that led Schramm to one of the deepest problems in physics, one that he dared to suggest could be solved by cosmology: Just how many kinds of elementary particles are there?

Particle physics is the branch of science charged with the mission of discovering the fundamental constituents of matter and the laws that govern them. As exemplified by the 4-mile-long vacuum tunnel then beginning operation at Fermilab out in Batavia, west of Chicago, this was an expensive and violent pursuit in the late twentieth century; physicists and countries leapfrogged one another building bigger and bigger machines to smash atoms and particles together with more and more force to see what came out. The task was frustrating; the harder they banged particles together and looked at the results, the more so-called elementary particles physicists found.

Neutrinos were a key to the proliferating families of matter. The original neutrinos were paired with electrons. The two particles were produced together, as for example, when a neutron (which is radioactively unstable outside of the nucleus) decayed into an electron and a proton.

Along with the electron a neutrino would come flying away. Some rare instances of radioactive decay, however, spit out not an electron but a muon, which is like an electron; it has a negative charge, but is 200 times more massive—an electron with a weight problem, went the joke. Muons had first been discovered in cosmic rays, somewhat to the consternation of physicists. "Who ordered that?" Columbia University Nobelist Isador Rabi is supposed to have remarked. They seemed to have no obvious role to play in creation; they were just junk particles.

The muon was also accompanied by a neutrino, but it was a different neutrino from the one that came with the electron. It seemed somehow to remember that it had been born with a muon; in subsequent interactions downstream it eschewed ordinary matter and interacted only with muons. It was called the muon neutrino; together with the muon it seemed to constitute a second family of matter. When an even heavier electronlike particle called the tau was discovered, physicists figured that there must be a third neutrino, a tau neutrino to go with it.

Three families of particles. One family comprised matter as we know it. What were the others for? Most of the theories of physics suggested that there was a relation between the number of neutrino types and the number of quarks, which were imputed to be the building blocks of protons and neutrons. How many more neutrino families were there waiting to be discovered as the accelerator physicists pushed to higher and higher energies? And why?

In the spring of 1976, Gunn visited Schramm in Chicago. Gunn had also done some theoretical work on neutrinos, so they were on his mind, as well. The two of them were fooling around with the big bang equations on the blackboard when they noticed something they hadn't seen before: there was a factor in the equations for the number of neutrino types. The more kinds of neutrinos there were in the universe, the more helium would be produced. Schramm and Gunn realized they could turn the equations around. Knowing how much helium had been made in the big bang, they could calculate how many kinds of neutrinos and thus how many families of elementary particles there could be. For a universe of 25 percent helium, the answer was seven or less. Which meant that no matter how large and energetic physicists made their accelerators, someday there would be a limit to their quest. The roll call of building blocks of matter would not go on forever; in fact it might be almost over.

Their reasoning was subtle. The big bang was full of neutrinos, as well as other particles. Being naturally aloof, neutrinos would not directly participate in nucleosynthesis, anymore than spectators do in a football game. Their numbers would have an indirect effect, however. Theoretically, they didn't have any mass, but because they were moving at the speed of light, they did have energy and thus they added to the mass-energy density of the universe. The more neutrinos there were, in other words, or the more kinds of neutrinos, the more mass-energy there was in the universe, and thus the denser the universe was. And the denser the universe was, the more helium was produced. To put it the other way around, the less helium there was in the universe, the fewer families of neutrinos there could be.

Schramm and Gunn wrote a draft of a paper, revealing how cosmology could be used as a constraint on the queen of sciences, particle physics. As it happened Schramm had been invited to a summer workshop on big bang nucleosynthesis in Aspen. When he arrived, he went to see the organizer of the workshop, his old friend Gary Steigman, and showed him the neutrino paper. Whereupon Steigman whipped out a paper *he* had written reaching the same conclusion.

Neither Steigman nor Schramm and Gunn, it turned out, were the first to see the connection between neutrinos and helium. Hoyle and Tayler had pointed it out in 1964, and Zeldovich's collaborator V. F. Shvartsman had also hinted at it in 1969. In science, however, having an idea is only half the battle. The reward for a new idea is not applause but argument from people who take you seriously enough to try to destroy you. The glory and honor justly go to those who are willing to stand up for an idea and commit themselves and their tenure prospects to the ego-grinding process of convincing their colleagues, pushing it on the colloquium circuit, and generally making noise about it. Schramm, Steigman, and Gunn weren't the first to think of neutrinos in the big bang, but they were prepared to make a lot of noise about it.

Steigman and Schramm were kindred souls. Steigman, tall, curly haired, and athletic, with a Hollywood smile, was also a tough city kid. He had grown up in the Bronx, in the shadow of Yankee Stadium, passing up the entrance exam for the famous Bronx High School of Science once because he wanted to be "normal," and going on anyway to study physics at the City College of New York, Cornell, and New York University.

He had gotten into astrophysics when his NYU thesis adviser suggested he look into the problem of why the universe seemed to be made exclusively of matter and not antimatter. All the known laws of physics prescribed that matter and antimatter had to be created and destroyed together, in equal amounts. That meant either that there should be vast amounts of antimatter somewhere in the universe—antimatter galaxies, maybe—or that all the matter and antimatter should have destroyed each other completely during the big bang, and the universe today should be made of nothing. Clearly the latter was not the case, so what had happened to the antimatter? Was the universe unbalanced? Steigman never quite solved the matter-antimatter problem. In a paper he showed symmetric cosmologies didn't work and that the universe must have become unbalanced in favor of matter very early on. In the meantime he became an expert on the big bang and an early convert to the idea that particle physics and cosmology were related.

After getting his degree in 1968 he spent some time at Hoyle's institute in Cambridge and one night met Fowler there. They drank too much and got into a heated argument about physics. The next day Fowler decided Steigman had been right, sought him out, and gave him a standing invitation to come to Caltech. He accepted and shared an office with a cheerful hulk named David Schramm, a disciple of Fowler's who was learning how to grunt and crunch through the chain of nucleosynthesis reactions that were the lifework of stars and certain moments of the universe. They became friends but did not work together.

When he left California, Steigman found himself knocking around. He spent five years at Yale and, after losing out to Beatrice Tinsley for a tenured position, landed at the Bartol Research Foundation at the University of Delaware. He spent his summers at the Aspen Center for Physics, which had long been a favorite summer colony of the Caltech particle theorists.

Steigman enjoyed the life of a physicist. As he explained once, happily munching canapés at the Aspen home of one of the center's directors, "This is a way you can enjoy a high lifestyle without a high salary." He kept urging his former officemate Schramm to come to Aspen.

Finally Schramm showed up for Steigman's nucleosynthesis workshop. In preparation for the meeting Steigman had gone over the big bang calculations and made the amazing discovery that if you knew the helium abundance of the universe, you could practically read off the number of neutrino species. The number of families of elementary particles in the universe, he thought, was not a bad mystery for a "half-astrophysicist"—as the renowned theorist Murray Gell-Mann used to call him —in a flaky field like cosmology to solve.

Steigman and Schramm combined their papers, including Gunn, and sent it off to the *Physical Review Letters,* whose editors turned it down, stating that it was probably correct but not very interesting. The paper was published in *Physics Letters* in 1977. From the start, Schramm and Steigman considered it a possible ticket to Stockholm as well as a beachhead for cosmologists in particle physics, and Schramm in particular campaigned aggressively to convince the world of its importance. Saying there could be no more than seven families of elementary particles was a definite statement, and although it might sound like a very safe one, because only three families were currently known, it really wasn't.

"The trend was for more numbers of neutrinos as accelerators went to higher energy," explained Schramm in his soft voice. "We said this trend wasn't going to continue. There was no statement from particle physics on the number of generations. It could be a thousand. For the first time cosmology was giving something back to physics."

Schramm got invited to give talks to particle physicists, but met a lot of skepticism. "People still wanted to hear about it," he said, "but they thought it was amusing. Their tone was sort of 'Why don't you come and entertain us?' Heh, heh, heh."

To be a cosmologist, Schramm argued noisily but patiently in his gentle giant mode over and over, you had to know particle physics. The future of both cosmology and of particle physics, he thought, was in the slim hundredth of a second that separated eternity from the start of the "standard model"—the big bang nuclear physics that he and Steigman would defend at the drop of a hat as if it were holy scripture. Every time physicists succeeded in turning the energies of their particle accelerators up a notch, they were blasting a little further back in time, re-creating for a fraction of a millisecond the conditions when the universe was a little hotter and a little younger. At that pinnacle of time and energy, every new particle discovered, every shade of force or dimension that was proposed could cast long shadows down the corridors of cosmic history.

And physicists, he contended, should pay more attention to astronomy. The universe itself, after all, with all its features large and small—the galaxies, stars, atoms, the very protons and electrons in our fingernails—was a relic of historical events, most of which had happened at energies forever beyond the wildest dreams of accelerator builders. "The big bang," as Zeldovich had said and Schramm now repeated over and over, "is the poor man's particle accelerator."

He could be quite forceful. In the physics world Schramm seemed to be regarded as some kind of natural phenomenon. I caught his act one night at an astrophysics conference in a crowded workshop. The atmosphere in the room was thick with beer and garlic. A young scientist said several dubious things about supernovae. On one side of the room Schramm, surrounded by an entourage, rose immediately, like a mountain out of the sea, and gave a short, blunt critique.

The astronomer, on the defensive, offered a white flag. "We should talk about it," he muttered.

Schramm, still standing, smiled. "I thought we just did, heh, heh, heh," he said and sat down. Chuckles swept out from his corner of the room.

Schramm's connections continued to multiply. He became a member of the steering committee of a consortium of physicists who were proposing to string instruments underwater near the island of Maui and turn part of the Pacific Ocean into a detector of cosmic neutrinos from high-energy sources and supernova explosions, a sort of underwater telescope.

Aspen had had another effect on Schramm's life besides the hooking up with Steigman. Prior to then his experience while climbing had been confined to Hoyle's hikes walking up the 1,000-meter-high mountains called Munros in the Scottish Highlands. Schramm took one look at the Rockies, at the Maroon Bells—a craggy trio of 14,000-foot peaks rising shoulder to shoulder just outside Aspen—and decided to become a mountain climber and a skier. He bought a house in Aspen and began coming every summer.

Schramm's style of skiing was to point his skis downhill and crash through the moguls like a runaway tank. As a climber he was equally aggressive and ambitious. A helicopter had to pluck him and his companion off the famous Eiger in Switzerland when their clothes turned to ice as they were coming down its northwest flank. Another time he got caught in a whiteout while descending Mount McKinley in Alaska and had to spend four days sitting in a tent waiting to help carry a body from another team of climbers out of a crevasse. He climbed in the Andes and the Himalayas. I asked Schramm if he'd ever had any close calls and he answered smiling, "Not really. I always make the summit and then run into problems."

The title of chairman of a university department sounds impressive, but in reality it is often an oppressive load of administrative work that leaves no time for research or teaching. At the University of Chicago, as in many schools, the burden was rotated among department members. In 1977, a senior astrophysicist declined his turn and Schramm was appointed acting chairman of the astronomy department.

Schramm was not a complete neophyte; he had administered research projects under grants. "I tended to worry about people," he explained, describing his management style. "Some astronomers are not too concerned with humans, heh, heh, heh." He moved aggressively. During his tenure he increased the faculty from ten to sixteen by a variety of stratagems, claiming unfilled physics vacancies for his own, or getting new faculty slots created for Nobelists and then hiring somebody else when the stars fell through. "The university is run on a feudal system," said Schramm, sitting in his new sparsely furnished office in front of a clean blackboard. A "Mount Everest Flight Certificate" hung on the wall next to a row of group photographs of Hoyle's Cambridge summer institutes. "I managed to conquer this building for the astronomy department."

Schramm, an empire builder extraordinaire, was, of course, not about to be content with a single building. The first thing he did as chairman was chase after a big NASA plum in partnership with Leon Lederman, the leprechaunish director of the Fermi National Accelerator Laboratory out beyond the western suburbs. While at Brookhaven on Long Island, Lederman had been one of the discoverers of the muon neutrino, for which he ultimately received the Nobel Prize. In 1976 the space agency had approved plans for an orbiting 2.4-meter-diameter telescope. Above the troublesome atmosphere, equipped with spy satellite technology, tended and refurbished by spacewalking astronauts, the space telescope loomed as the biggest science project in the history of NASA and the biggest thing for astronomy since the 200-inch telescope. NASA invited proposals from universities to build and staff a Space Telescope Science Institute, from which the telescope would be run and its precious observing time and data doled out. Whoever won might be the center of astrophysics for the next twenty years. Lederman and Schramm wanted it. They launched a campaign to land the Space Telescope Science Institute for Chicago, planning to put it out at Fermilab, which had lots of open space, solidifying the links between cosmology and particle physics.

Schramm found the experience glamorous but disillusioning. "The political system works nothing like the high-school civics textbooks," he said, shaking his head. He went to Washington to lobby with the Illinois congressional delegation and got to know his way around the Chicago political machine. More important, he became close to Lederman and the other leaders of the particle physics community. Fermilab and the Chicago astronomers were bonding.

When they lost to Johns Hopkins, Schramm and Lederman switched gears. In the summer of 1981 Schramm took time off from a conference in Europe to go climbing with Lederman in the Italian Dolomites. "We agreed to meet at noon under the clock in the train station. Both of us showed up with our respective companions and headed for the mountains." While they were clambering up and down cliffs on ladders and cables, Schramm, in his usual fashion, was talking up the particle physics-cosmology connection. "Wouldn't it be nice," he asked, "to have a group of resident cosmologists at Fermilab?"

Lederman thought it was a good idea. He climbed down, went home, and called up NASA, which after all had jurisdiction over the universe, and basically dared Hans Mark, the associate administrator, to fund a cosmology group. They were invited to send in a proposal, which Schramm wrote, cannibalizing a research proposal that one of his Chicago colleagues had written to the DOE. A year later, in 1982, Schramm moved out to Fermilab for two years to get the cosmology group going. He bought land just outside the reservation and laid plans to build a big house of glass and stone, and began dating Lederman's assistant, Judy

Ward. They were eventually married in Aspen in an outdoor ceremony at Crater Lake. By then Aspen was well established as the summer camp of cosmology.

The placement of the cosmology group at Fermilab was an act of faith. At that time the most powerful particle accelerator in the world was at the European Centre for Nuclear Research (CERN) outside Geneva, in which protons and antiprotons were collided at combined energies of 200 billion electron volts. Fermilab, playing catch-up, was being equipped to raise that figure to almost a trillion electron volts, the energy that a typical elementary particle would have had when the universe was a millionth of a millionth of a second old.

What happened in these prearranged collisions of matter and antimatter under the prairie was a kind of modern miracle. If a particle was a hill on the manifold of dreams, its antiparticle was a hole—put them together and you got back pure energy. That energy, a little momentary fireball, would condense and reform into whatever forms were permitted by the laws of physics—always in pairs whose quantifiable, conserved properties always added to zero. Carlo Rubbia, a charismatic and hard-driving Italian physicist who was the lord of the CERN rings, described the whole process as colliding two cars. In ordinary life, you got a mess, but in particle physics you got twenty new cars out. The more energy you could marshal, the more massive particles you could create—particles not seen since the universe, the big bang, had been that hot. That was the whole principle behind those bigger and bigger, more and more violent particle accelerators. And that was the principle behind God's particle accelerator, the big bang.

When two particles—a proton and an antiproton, for example—collided, for an instant too short to be measured there was a fireball of pure possibility, a piece of the big bang circa a trillionth of a second. Whatever laws governed that far time, whatever forms matter and energy had been permitted to take back then, would recrystallize in a spit of fire under the Illinois prairie. Species of energy and matter now vanished more completely than the dinosaurs, but whose brief struts had shaped the universe, would return.

Awaiting and anticipating those revelations, Lederman and Schramm built on the third floor of Fermilab's high-rise a gang of cosmologists whose style—brash, athletic, fun loving, and irreverent—was as distinctive as their science. Steigman, who continued to collaborate with Schramm, was a sort of satellite of the Chicago crowd.

One of Schramm's first hires when he took over had been a Stanford graduate student named Michael Turner. The son of a southern California businessman, Turner grew up in the car and surf culture, and went to Caltech and then Stanford graduate school, where he played volleyball with the future astronaut Sally Ride. Turner had large, quizzical blue eyes under a high bony forehead and a wide, humorous mouth. His brown hair periodically sprouted to shoulder length while his temples thinned, and at various times he had beards and moustaches. Turner made a point of dressing outlandishly. On one visit to his office I found him in a Gold's Gym T-shirt, shorts, and hospital slippers. He used humor and feigned coolness to hide a sharklike seriousness about physics.

Schramm had met Turner, who was Wagoner's graduate student, during a visit to Stanford and was impressed with his intelligence. But Turner was in crisis. He kept oscillating between particle physics and astrophysics. It took him seven years to complete his Ph.D., with a thesis on general relativity. He wasn't sure whether to stay in research or to go and teach at a small school. Schramm brought him to Chicago as a postdoc, and Turner thrived, plunging into nucleosynthesis work. He transplanted his southern California lifestyle to the Midwest.

In 1981 a six-month workshop on particle physics and cosmology was held at the Institute for Theoretical Physics, which is on the campus of the University of California, Santa Barbara. Turner went, grew a long beard, and fell in with Edward "Rocky" Kolb, a physicist-cosmologist from Los Alamos. Kolb, who grew up in New Orleans, lived up to his nickname by being thrown out of Los Alamos basketball games for fighting. He had a deadpan sarcasm that matched Turner's; they hit it off. Schramm visited Santa Barbara for a week at the end of the workshop and became reacquainted with Kolb, who had been a graduate student doing supernova calculations when Schramm was at Texas. He convinced Lederman to hire Kolb and Turner as codirectors of the nascent Fermilab cosmology group for the first year. Turner, by now a professor, arranged to split his time between the university and Fermilab.

Teamed together, Kolb and Turner came into their own. They seemed to live in each other's offices and collaborated in humor as well as in science. One of their specialties was T-shirts, without which no cosmology conference was complete. When John Ellis, an English particle theorist who works at CERN and who (like Turner) is known for his hirsute looks, visited Fermilab, they grafted his face onto a picture of a buffalo and stuck it on a T-shirt proclaiming "John Ellis, FRS [Fellow of the Royal Society], World Tour" in the style of rock tour T-shirts. Among their peers Turner and Kolb had a reputation for being fast, if a little impatient. If a new theory appeared to court fashion or a new discovery appeared unexpectedly from the frontier, they would be among the first to dash off a paper on its cosmological consequences.

At conferences Turner gave witty lectures on the history of the universe, weaving all the physics and astronomy together and making it fun. "I can say I know what happened all the way back to a hundredth of a second after the universe began," he cracked, his face crinkling into a grin, "and the little men in white suits won't come and take me away."

A frustrated artist, he transformed into a high art the dull practice of viewgraphs—the eight and a half by eleven transparencies lecturers use to project diagrams and equations overhead. Many scientists just read their talks off them or project typewritten tables of data in letters too small to read. Turner's were wild cartoons featuring Darth Vader, executed in as many colors as you could buy grease pencils. One of them contained the entire history of the universe in cartoon form, from the "dread singularity" to the Milky Way. "Most people's viewgraphs are boring," he complained.

Turner's scientific style was intuitive, as if he were a hip version of Zeldovich. He was more interested in ideas than details, which meant he was often in and out of a particular subject so fast that he missed out on contributing to the definitive explanation and thus the lasting scientific credit. I asked him once what he thought his principal contribution was, or was it just a certain overall style? We were eating dinner in the Woody Creek Tavern, Hunter Thompson's hangout, outside Aspen, hoping for a glimpse of the gonzo writer himself. He snapped, "I'll take style."

One of Turner and Kolb's favorite pastimes was twitting their red-headed friend and colleague, implying that his ego was as big as the rest of him. They called him "Schrambo," and kept threatening to produce Schrambo T-shirts. "Actually," admitted Turner once, "Dave's been very good to me."

Schramm's lifestyle was, it was true, becoming the stuff of legend. It seemed as if he were still trying to prove that a scientist didn't have to be a nerd. He wrestled at the Y with members of the Chicago Bears football team. When his knees got too old to wrestle he took up bicycling; his idea of a ride was from Aspen (about 8,000 feet) up to Independence Pass at 12,500 feet, twenty miles away, and back down again. Gradually he abandoned plans for a big house out at Fermilab and instead bought a pair of condominiums on the fifty-second floor of a clover-leaf-shaped tower on the edge of Lake Michigan that had a flying saucer's 270-degree panorama of the lake and the city, combined them and furnished them in white shag. He kept up his political connections. When the University of Chicago hosted the biannual Texas Symposium, Schramm took a delegation of high-ranking astrophysicists down to the mayor's office to meet Harold Washington.

On a trip to Germany one winter, Schramm bought a Porsche 944, bright red, which he outfitted with license plates that said BIG BANG and used it to give reporters fast rides out to Fermilab. Not satisfied with speedy ground transportation, he celebrated his fortieth birthday by taking flying lessons and buying an airplane, a turbocharged single-engine Cessna. Schramm incorporated the airplane as Big Bang Aviation and rented it out for charter when he wasn't flying it to Aspen or other conferences around the country.

He was a publicity agent's dream. Once a year a fawning profile of Schramm would appear in the Chicago newspapers. He was named to the *Esquire* "Register" of prominent men under forty. Turner and Kolb wrote in a satirical column in the Fermilab newsletter that Schramm had been named to a list of "the 40 men most likely to be named to a list." Along with the University of Chicago public information officer, they spent an afternoon trying to see if they could get Schramm in a Dewar's Scotch ad. Often seen on the backs of magazine covers, Dewar's ads feature a black-and-white photograph of somebody with a glass of Scotch and a snappy itemized profile of the drinker.

QUOTE: "THE UNIVERSE IS THE POOR MAN'S PARTICLE ACCELERATOR."

Every Monday evening the entire Chicago and Fermilab cosmology contingent drove over to Turner's house next door to Fermilab to talk physics in his basement game room and broil hamburgers and fish or eat "primordial pizza." Was it just my imagination, or were the Chicago and Fermilab cosmologists and astrophysicists better looking and more athletic than the average run-of-the-mill scientific group? Was it my imagination, or did there seem to be more women among the graduate students at Chicago than elsewhere? I began to think of Chicago as the jock school of astrophysics.

Sometimes they acted that way; Chicago-style astrophysics could be tough physically as well as mentally. The Chicago Texas Symposium in 1986 was a case in point. Toward the end of the traditional banquet in the ballroom of the downtown Holiday Inn, I noticed that food was flying around the table occupied by the Chicago people. Soon hard rolls were careening on rainbow trajectories across the ornate spaces. The chief launcher, with a devil-may-care grin on his face, was Kolb. One of his missiles whizzed past my ear. Another hit Steigman, about a hundred feet away at another table, right on the nose, knocking off his glasses.

Shortly thereafter, outside in the lobby, Steigman accosted Kolb, grabbing him by the lapels and demanding an apology. Kolb, resisting, didn't seem to comprehend that Steigman was serious—he had been offended. Watching them, I thought of Steigman declining Bronx Science because he had wanted to be "normal" and wondered if he'd had his glasses knocked off a lot as a kid in the Bronx. Suddenly these two six-footers were swinging each other around, scattering astronomers like duckpins. They almost fell on top of me before they were separated, just as punches began to be coiled, and led away. A few minutes later Schramm strolled through the lobby. He listened to what had happened with his head cocked sympathetically and then he clucked bemusedly and strolled off.

Style, yes, but what else did this anarchic bunch have to contribute to science? Schramm explained that the role of the Fermilab cosmology group was to be a sort of border police between astronomers and particle physicists. "Our biggest achievement is getting things right," he said. "People venture into these areas and they blow it, heh, heh, heh. We've been quick to pick up on things, but not so quick [that] we do silly things."

Their most important role to date has remained being the defenders of the faith in the big bang nucleosynthesis calculations. The faith has two tenets: First, that ordinary matter cannot amount to more than about 10 percent of the critical mass density needed to close the universe; second, that cosmologists can dictate to the physicists how many kinds of neutrinos exist in the universe. "We've been pretty vocal about it, pretty high profile," conceded Schramm cheerily. "People are always trying to shoot us down. So far we're still afloat."

By the mid-eighties, continued observational refinements had lowered the observed primordial helium abundance in the universe to around 23 percent. According to the big bang models, that left room for only three—maybe four—species of neutrinos. If those models were right, physicists were getting close to a moment of truth. Three neutrinos— the electron, muon, and tau—were already vouched for, which left at most one to go. There might not be very many—or any—elementary particles left to discover.

Schramm said that he often heard rumors or gossip that he, Steigman, and Gunn had been nominated for the Nobel Prize.

Cosmology was prospering. A science that could accurately describe the evolution of the universe back to a hundredth of a second after the big bang and predict the burning out of the galaxies a trillion years hence was pretty impressive, "a triumph," as Peebles liked to say.

But it wasn't perfect. There were a couple of problems, almost paradoxes, about the expanding universe, that had been bothering him and Dicke for a long time. They were the kinds of things physicists talked about at lunch but never wrote down. Peebles had learned them from Dicke. "The ideas go back to my earliest memories," said Peebles. "It's hard to remember when I wasn't puzzled about these two problems, and when Bob Dicke wasn't." The more progress cosmology made, the more these conundrums gnawed. Something was being missed.

As it happened, 1979 was the 100th anniversary of the birth of

Albert Einstein—the year was packed with a round of celebratory seminars and volumes. Dicke and Peebles decided that if they didn't speak up now, they might never get the chance, and so wrote a paper called "The Big Bang Cosmology—Enigmas and Nostrums" that was published in a centenary volume edited by Hawking and Werner Israel, and also sent out to friends. It had a rather flip tone: "Most of the conundrums and nostrums discussed here trace back, in one form or another, to the lively discussions in the 1930's," they wrote, "before physical cosmology became encrusted with revealed truth."

The first puzzle was why the universe was so uniform. That it was, was without question. In every direction, as Hubble had discovered, if you looked deep enough, there were galaxies scattered like dust, made of the same atoms in the same proportions, following the same laws, blending on the largest scales into a homogeneous, smooth whole and expanding, according to Sandage, at the same rate. Even more striking, the microwave cosmic radiation, which represented the state of the original fireball, was the same temperature to within a ten-thousandth of a degree in every direction. Homogeneity was nice; without it, cosmology would be impossible. The problem was how the universe could have become so perfectly homogeneous. The odds against its being born that way were greater than astronomical: In nature, things usually started out messy and then came to equilibrium, the way coffee and cream start out hot and black, and cold and white, respectively, and wind up in the same cup warm and brown. The equilibrating process depends on an exchange of energy between heterogeneous bodies or regions—heat goes from the hottest to the coolest. But heat or energy, like everything else in the universe, cannot travel faster than the speed of light. Therein lay the problem.

The microwaves from opposite sides of the sky came from parts of the universe so far apart that there had not been time since the universe began for a signal to fly between them—even at the speed of light—to synchronize their properties. That meant that cosmologists had to assume that the universe was born perfectly uniform—God had done it.

The other conundrum involved the sacred number omega, the ratio of the mass-energy density of the universe to the critical density that would determine whether it was open or closed. Mathematically, Peebles and Dicke explained, the universe was balanced on a knife edge between collapse and endless expansion, between being curved one way or the other. In the middle was the ideal solution, with omega equal to 1.0 and a space that was geometrically flat. According to the Friedmann equations again, any slight deviation from perfect flatness at the beginning of time would have grown like a monster and curled the universe like a dried leaf by now. If omega was a little less than 1.0, it rapidly became infinitesimal; if it was a little more than 1.0, it became huge.

Observers like Sandage and Gunn had estimated that omega today was about 0.1, which sounded like a long way from a flat universe. But for omega to be even that close to 1.0 today, Peebles and Dicke calculated, the universe would have to have been born with an omega virtually indistinguishable from 1.0—precisely balanced between collapse and expansion. The difference between this and an exactly flat universe was so negligible at birth that many theorists (including Einstein) had concluded that the universe was *in fact* flat and omega had really always been 1.0 in spite of the admittedly rough astronomical results.

Dicke and Peebles pointed out that this fine-tuned balance had to apply not just to the overall universe, but to each of its separate parts. "Otherwise," they wrote, "the universe would run amuck. . . . It requires very careful regulation to ensure that we do not see such things as wholesale collapse of the part of the universe appearing in the Southern hemisphere, and general expansion in the other half." How did the universe acquire this exquisite, unlikely balance? Peebles and Dicke called it the flatness problem.

Peebles and Dicke were a little surprised at the attention their paper got. Consciously or not, they were making trouble for their fellow cosmologists. In science, the obvious questions are the hardest, both to recognize—to ask—and to answer. Why is the universe anything like it is? They were raising the ante, challenging their peers, reminding them not to take any feature of the universe, no matter how obvious, for granted. It was no longer good enough to describe what the universe was like, the mission of cosmology was to explain *why* it was the way it was.

Fermilab held a coming-out party for its cosmology group in the form of a week-long conference called Inner Space/Outer Space. Astronomers (including Tammann but not Sandage), cosmologists, and particle physicists attended. Turner, wearing sandals, a Betty Boop T-shirt, a flannel shirt, and a Chicago Cubs baseball cap, punctuated the proceedings by clanging a cowbell.

In the material sent out to attendees before the meeting, Kolb and Turner included a guide to Fermilab, describing it like a theme park called Fermiland. There was a map to such attractions as the (nonexistent) Space Telescope Science Institute, Schramm's house (which he never built), and the buffalo ranch (real). It occurred to me that they had a point.

Cosmology *was* a kind of fantasy theme park. Inner Space/Outer Space met within the physical confines of Fermilab, but it really happened somewhere else, in Fermiland. Fermilab was a place where physicists sought to discover the constituents of matter and the laws that govern them. Fermiland was a state of mind in which it was possible to find the answers to the largest questions about the universe—why there is matter and galaxies and the expanding night, or why there is a universe—in the relationships between quantum particles back when Hubble's entire gleaming realm had been the size of a grapefruit, its contents buzzing with hideous and irreproducible energies.

On the first Friday in May 1984, a week of cool rain ceased and spring blundered onto the Fermi environs. At noon 200 physicists and astronomers, the cream of the American cosmological community, thundered out of the high-rise, up a ramp, and onto the four-mile blacktop access road that runs along the top of the main ring berm, a river of noise and contention, banging knees and shins, thinning temples flushed, glasses fogged, dressed in state-of-the-art Nikes, loafers, jeans and slacks, nylon wisps, and Hawaiian baggies. They flooded up the ramp, turned left, and began trotting clockwise, the same way as the protons in the magnetic racetrack below, tanned graduate students with ascetic lifestyles and Scandinavian eyes striding like gazelles ahead of the pack, middle-aged duffers, white hairy legs pumping, bringing up the rear, one reporter in borrowed shorts bravely ready to go anywhere in search of action.

Schramm didn't run; his knees were shot from years of football, rugby, and wrestling. He looked on with an avuncular smile as this anarchic grumbling mass, slowly stretching like an Einsteinian space or a poorly bunched quantum wave packet, huffed and puffed past mock road signs giving protons the right of way and past the new building where tons of electronics were being assembled to record the trillion-volt collisions of protons and antiprotons. As the pack spread out, the noise level went down by the footfall. By the back quadrant it was mercifully quiet, only the painful thud of shoe on blacktop to challenge the birds and distant construction. The season seemed to have switched to summer. Suddenly my reverie was broken by the tortured panting of a pair of graduate students with California tans breaking from the pack to catch and kick past one more faltering runner before the finish line.

Then it began again, a wedge of din rumbling like a distant buffalo herd, the talk, the endless talk, the great argument with God, physicists pounding in, clustered, sweat stained in the sun, drinking juice as if life were one long movable endless coffee hour, rising like a roar as conversations left off in Princeton or Pasadena or Cambridge, the first second of time, yapping yapping yapping against the big sky, one mouth and a thousand tongues propelled faster and faster by the caffeine of—not exactly mystery, but the sense that—*we've almost got it,* one more twist of theory in the cosmological lock and the universe could spring open. It might be one of those leaping gazellelike postdocs who had the consummate idea, it might be tomorrow.

Turner clanged his cowbell and we all went back inside.

12

THE TRIUMPH OF BEAUTY

In 1979 a man came as close as anyone ever had to Figuring It All Out, to the magic idea that would cause the universe to unfold its wings from practically nothing and evolve smoothly and inevitably into the configuration of reality as astronomers could see it today.

His name was Alan Guth, at the time a thirty-two-year-old postdoctoral fellow at Stanford on leave from Cornell. He was a theoretical particle physicist who distinctly disdained cosmology. Up until the fall of that year his life had been a haphazard wandering pilgrim's progress across the landscape of physics. Square-jawed, short and stocky with piercing dark brown eyes, long wavy brown bangs, and a ready grin, Guth had a mathematical mind and an undergraduate's enthusiasm for physics that had survived the setbacks in his peripatetic career.

At thirty-two, when he was getting a little long in the tooth for a postdoc position, he was dragged into a series of calculations that led him into a collision with cosmology and in so doing pushed the limits of time and space to within a whisker of eternity. The idea he discovered was called inflation, a sort of hyperexplosive bubbling of the very earliest universe. Inflation explained where the universe had come from, how and why it had to be the way it was—flat, even temperatured, yet spangled with galaxies—even to what end it was ballooning away in the night and why that last question would never be part of astronomers' lives.

Twenty years in
the making, the 200-inch
Hale reflector is dedicated on
Palomar Mountain, June 3, 1948.

Edwin Hubble stares out of
the observer's cage of the 200-
inch telescope. Hubble, discoverer of the expansion of the
universe, had planned a long
campaign of observations to
answer questions about the size
and fate of the universe, but
he died soon after the telescope
went into operation.

Hubble's uncompleted quest fell to Allan Sandage, a young Mount Wilson astronomer and recent Caltech graduate who had been his observing assistant. In 1954, at the age of twenty-eight, he was proclaimed by *Fortune* as one of America's promising young scientists.

Fritz Zwicky, an irrascible Caltech astronomer, disparaged Hubble and Sandage. He invented his own system of "morphological astronomy" but made the prescient observation that 90 percent of the mass in clusters of galaxies seems to be invisible.

Caltech's Jesse Greenstein was teacher and "subfather" to a generation of Palomar astronomers.

Maarten Schmidt decoded the spectra of the enigmatic quasars and discovered they were brilliant beacons in the far depths of time and space.

Fred Hoyle pauses for a breather near his Lake District home. A fearless iconoclast, Hoyle, along with Thomas Gold and Hermann Bondi, championed the steady state theory of a universe which had no beginning or end, and helped elucidate the chain of nuclear reactions by which stars produced the elements.

As a professor at Princeton and the University of Texas, John Archibald Wheeler initiated several generations of physicists into the mysteries of curved space and invented the term *black hole*. In the union of quantum theory and general relativity he sought the principle that made the universe "fly."

Stephen Hawking and the author on the steps of the Royal Society in 1978. While his body fought a wasting neuromuscular disease, Hawking's mind roamed black holes and the beginning of time, and returned with insights that changed the landscape of physics.

Martin Rees, another of the brilliant young Cambridge theorists, explained how black holes could power quasars.

Jim and Alison Peebles relax during an astronomical conference in Munich. A low-keyed theorist who never left Princeton, the male Peebles invented most of the mathematical tools with which cosmologists evaluated and debated the large-scale structure of the universe.

Jim Gunn had three careers: ingenious instrument builder, prodigious observer of quasars and galaxies, and gifted theorist.

Beatrice Tinsley spearheaded a group of young astronomers, including Jim Gunn, J. Richard Gott, and David Schramm, who wrote a manifesto declaring that the universe would expand forever.

As aggressive in the seminar room as on the wrestling mat or a mountain slope, David Schramm argued that the fate of the universe could be read in the microphysics of its first second.

Yakov Zeldovich cavorts in Chicago with Judy Schramm. The peppery, self-educated Zeldovich left the Soviet H-bomb project for astrophysics and became the leading genius and dictator of Eastern cosmology.

Fermilab director Leon Lederman shares a ceremonial laugh with Michael Turner and Edward "Rocky" Kolb, leaders of the Fermilab cosmology group.

Mike Turner used wit and a sharklike intensity about physics to elucidate the connections between particle physics and the structure of the universe.

Gary Steigman realized that measurements of the amount of helium produced in the big bang could tell physicists the number of families of elementary particles in the universe.

A particle physicist and reluctant cosmologist, Alan Guth discovered that the splitting of the primordial grand unified force would cause a glitch in the big bang, making the universe "inflate" wildly and altering its structure and geometry.

In the theory of new inflation authored by Paul Steinhardt, among others, quantum fluctuations in the first instant of time were revealed to be the source of galaxies.

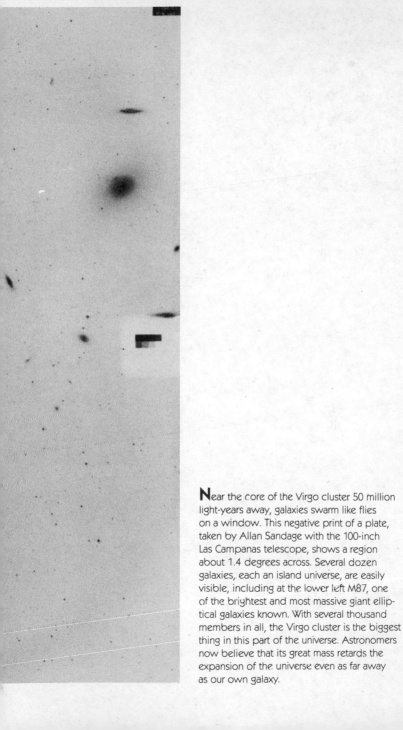

Near the core of the Virgo cluster 50 million light-years away, galaxies swarm like flies on a window. This negative print of a plate, taken by Allan Sandage with the 100-inch Las Campanas telescope, shows a region about 1.4 degrees across. Several dozen galaxies, each an island universe, are easily visible, including at the lower left M87, one of the brightest and most massive giant elliptical galaxies known. With several thousand members in all, the Virgo cluster is the biggest thing in this part of the universe. Astronomers now believe that its great mass retards the expansion of the universe even as far away as our own galaxy.

Marc Aaronson, shown above with his wife, Marianne Kun, and children, pioneered a new method of measuring distances to galaxies in partnership with Jeremy Mould (*below*) and John Huchra. Their distances suggested a Hubble constant higher than Sandage and Tammann's.

Frustrated by the lack of quantitative data in cosmology, Marc Davis launched a campaign at the Harvard-Smithsonian Center for Astrophysics to map the locations of 2400 galaxies in space, using their redshifts as a measure of relative distance.

From their studies of galactic rotations, Vera Rubin and Kent Ford concluded that spiral galaxies were surrounded by clouds of invisible dark matter ten times as massive as the visible galaxies themselves.

Margaret Geller and John Huchra continued and expanded the redshift survey when Davis left Harvard for Berkeley. Their first map (*below*) showed that galaxies seemed to be located on the walls of giant bubbles tens of millions of light-years across, with relatively few in the voids between. These results were a major challenge to galaxy theorists.

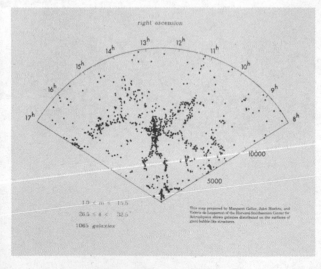

right ascension

14^h 13^h 12^h 11^h

15^h 10^h

16^h 9^h

17^h 8^h

10000

5000

$10 < m < 15.5$

$26.5 \leq \delta < 32.5$

1065 galaxies

This map prepared by Margaret Geller, John Huchra, and Valerie de Lapparent of the Harvard-Smithsonian Center for Astrophysics shows galaxies distributed on the surfaces of giant bubble-like structures.

Alex Szalay contemplates large-scale structure in the landscape outside Aspen. A disciple of the far-seeing Zeldovich, he mixed physics with rock and roll, and became an expert on the characteristics of universes dominated by dark-matter particles.

Joel Primack helped show how the gravitational influence of invisible subatomic particles known as cold dark matter could form galaxies as we know them, and served as sound man for his wife Nancy Abrams, whose ballads and techno-humor enlivened cosmological gatherings.

Alex Vilenkin proposed that the universe could emerge from Nothing, taking a quantum leap from eternity into time.

Murray Gell-Mann, inventor of quarks, and Yakov Zeldovich discuss quantum cosmology.

John Schwarz survived years of obscurity as a Caltech research associate while he pursued his mathematically elegant dream of a Theory of Everything based on superstrings.

Hawking, accompanied by a nurse, shows off his computer and voice synthesizer at a Fermilab meeting.

Sandage was fond of Mount Wilson's 100-inch Hooker telescope, which Hubble had used to discover the expansion of the universe.

The spiral galaxy M101, the Pinwheel, in Ursa Major was one of the linchpins in Sandage's and others' attempts to measure the rate of expansion of the universe. As astronomers fought over the distance to M101, the Hubble constant went up and down.

The participants of the Kona conference in January 1986, which was distinguished by reports of large-scale motions in the universe. Sandage and his archrival Gérard de Vaucouleurs (in dark glasses) are front and center. Conference organizer Brent Tully is at the left end of the front row in the broad-striped shirt; Marc Aaronson, another of Sandage's rivals, is at the other end of the row, holding a notebook.

Feed men and then ask of them wisdom. Maurice cooks for the cosmologists in Aspen.

Both Guth and cosmology were changed irrevocably in 1979. Not even Sandage, it turned out, would be immune from the impact of the theory of inflation. Guth's life took the kind of turn that an actor's takes when he becomes an overnight success after twenty years of bit parts and waiting tables. If Guth were not basically a cheerful guy, he would never have lasted long enough in physics to make his great breakthrough.

We were sitting in his big office at MIT, on a snowstorm day, nine years after his big breakthrough. Guth was leaning back, in baggy khaki pants and hiking boots with ferocious lug soles, trying manfully to reconstruct the past. His bicycle leaned against the wall. Guth has a Boy Scout eagerness to please when you talk to him. He scoots along on a nervous chuckle and a toothy grin, generously precise and patient about physics, maddeningly imprecise about his own self, as if he weren't important enough for it to have occurred to him to pay attention to himself. Like many bright physicists, his memory is atrocious, because he's always been able to figure things out on the spot.

The road to the edge of the universe began for Guth (a shortened version of a Russian-Jewish name) in New Jersey. He was born the middle child between two sisters in 1947 in New Brunswick, at the southern end of the New Jersey suburban sprawl attached like some great barnacle to the side of New York City. Both his parents had been raised in the same area; his grandparents died when he was young, so he never knew them. When he was three his family moved next door to Highland Park, to the house his parents still occupy. His father dabbled in a variety of businesses, ranging from a grocery to dry cleaning.

"From what I can recollect," Guth said, "until something like the seventh grade or so I was pretty much a reclusive sort of kid. I was at that time interested in science and math. How exactly I got interested I'm not sure. I know I watched 'Mr. Wizard,' and somewhere along the line I started reading *Scientific American*. And, uh, my recollection is that until about seventh grade I had a few friends, but not very many, and starting in seventh grade I became more sociable."

He recalled being terrible at baseball, among other things. He was terrific at math, however. While he was still in elementary school young Alan got hold of some advanced math books and began teaching himself square roots and algebra. When he wasn't doing math, he was drawing designs for model rocket ships that never got built. He was in fifth grade when the sputniks were launched and America was plunged into technological self-doubt. Highland Park, like many other schools, responded to the sputnik challenge by setting up a system of accelerated courses for smarter students. Guth was swept onto the escalator.

Later he would not be able to recall a time when he didn't like science and mathematics—the distinctions between math and the different branches of science were elusive to him. Instead there was simply what he calls a sharpening of focus as he went along. He had a short-lived fling with chemistry before deciding that it wasn't fundamental enough. "Even though I didn't know what science was about, I decided in high school that why I was interested was to try to understand what the fundamental laws of nature are. And soon I realized that physics was the branch of science about that.

"And I'm sure it was while I was in high school that I decided I would be a physicist, and probably while I was in college that I decided I would be an elementary particle physicist. I never decided to be a cosmologist, by the way."

Guth ran with a pack of similarly math- and science-inclined friends. They taught each other physics and math out of books and argued in their spare time about modern physics, a heady world prone to enigma. Guth was particularly fascinated by the paradoxical aspects of relativity and quantum theory. "It was weird but apparently true."

By the time they came to take physics, as juniors, Guth and his buddies already knew the course, so the teacher turned them loose in the backroom lab. They spent the year doing experiments and clowning around.

Near the end of that term, Guth's teachers suggested to him that he didn't have much to gain from sticking around Highland Park for another year; he should apply for early admission to college. The guidance counselors asked him where he wanted to go.

Harvard and MIT were the only colleges he knew anything about. Guth's older sister was attending Lesley, a teacher's college in Cambridge. Guth had been up to visit and liked the area. Harvard, he thought, was where you went if you wanted a good broad education; if you really wanted to learn science you went to MIT.

"So I said MIT," Guth reported. "They said it was after all the deadlines, but they said they'd call MIT and find out if they'd consider me. Within a day they told me they'd spoken to MIT and it sounded like they were prepared to accept me." Just on the basis of a phone call? "That's right. Well," he added modestly, "I guess I had a good record."

Supported by a full scholarship, Guth arrived in Cambridge in the fall of 1964, the same year as Schramm. They were both physics majors, future cosmologists, but it would be fifteen years before they would meet or interact.

"My image of MIT was that it was a very egghead place," recalled Guth, "so I was sure that when I got to MIT I would find myself somewhere in the middle of the class academically, but I figured I would still be the star of the debate team and the best broad jumper. Well, that turned out to be backward in both cases. In broad jumping I found there was some African student who could jump four feet farther than I could. In debating I discovered there was somebody who was part of the national championship high-school team the year before. So I was completely outclassed."

Guth spent a year broad jumping before giving it up, and four years on the debate team, having fun, but never making it to the first string. In pictures from that time he looks lean and intense and Kennedyesque in his debate suit, with heavy dark-rimmed glasses straddling a hawk nose and high cheekbones, and his longish hair combed carefully over his head.

Academically, MIT was actually easier for Guth than high school. The courses were all science and math; the institute didn't make him take so much English and history. He honed his focus to physics, particle physics, and then theoretical particle physics, the study of the laws that govern elementary particles. In the summers he went back to New Jersey where he had a succession of white-collar jobs. One year he reduced data for a Rutgers sociologist; another summer he spent working on lasers for Bell Labs.

He got his one dose of real science when he served an experimental apprenticeship as an undergraduate on the MIT cyclotron, an antique particle accelerator that used magnetic fields to whirl electrons to 99 percent of the speed of light, and then slammed them into a target. MIT's cyclotron, in contrast to the machines that would sprout in the next decade, could fit in one room. One man could run it. By studying how the electrons came ricocheting out of the target, Guth could analyze the structure of the nucleus that had deflected them and deduce something about the forces that welded it together. "It was very good for me. When you are an undergraduate just taking courses, you feel like a student, you don't feel like a scientist. That was the first time I got to feel like a scientist."

At the end of his junior year, Guth began to get worried about the draft and arranged to switch into a five-year program in which he would get both a bachelor's and a master's degree at the same time. It was tantamount to getting into MIT graduate school. After he received his master's in 1969 he simply stayed on. MIT gave him a fellowship of $4000 a year plus tuition, which enabled him to live pretty well by the standards of the time. He and another graduate student took an apartment in nearby Somerville, a working-class enclave of sturdy three-deckers.

Guth's decision to avoid the draft by extending his tenure at MIT was taken with a sort of straight-ahead practicality, but the passions of the era soon caught up with him. At the end of the sixties, MIT along with the rest of society went up in revolutionary flames over the war in Vietnam. At the institute, with its millions of defense research dollars, the soul-searching was particularly acute. For whom should the "tech tool"—as MIT students characterized themselves—be tooling? Guth remembered going to meetings in a basement at MIT. In the fall of 1970, after the Kent State shootings and the Cambodian invasion had closed the universities, Guth joined the antiwar election campaign of Father Robert Drinan, a Jesuit priest and law professor from Boston College who was running for Congress in the district that surrounded but did not include Cambridge (which was in Tip O'Neill's district).

Guth answered the phone in Drinan's Watertown office, leafleted door to door in the style made famous by the McCarthy campaigners in 1968, and gave at least one antiwar speech to a church group. "I was an MIT debater," he reminded me. "I'd had a long interest in public speaking. I was not hesitant to talk in public. I was never a leader."

Drinan won and served several terms before the Pope ordered all priests out of electoral politics. Meanwhile another revolution claimed Guth.

By the mid-sixties physics had achieved a kind of triumphant disarray. On one hand it was a great, even arrogant feat of reductionism. It had by now determined that all of the variety of nature—from sunsets and waterfalls and mosquitoes to the lordly galaxies—could be reduced to the interplay of four basic forces or, as the physicists liked to say, interactions.

Two of these forces had been part of the human intellectual landscape since before there was physics: gravity, which Einstein's theory of general relativity explained as curved space-time; and electromagnetism, which in all its manifestations was responsible for everything from static cling to the "CBS Evening News" broadcast. Einstein had spent the last half of his life trying to combine these two into one theory. While he was struggling with that, twentieth-century nuclear physics was adding two more forces: the so-called weak and strong nuclear forces. The weak force caused some kinds of radioactive decay. The strong force held the nuclei of atoms together.

Physicists like Turner often spoke as if these forces—or "interactions"—were languages that particles used to communicate with each other. Part of the richness and difficulty of physics (and cosmology) stemmed from the fact that not every particle responded to every force. Not everybody in the atomic Babylon understood everything that was being "said." The proton understood and responded to all the other forces. Others, like the elusive neutrino, interacted only by way of the weak force and gravity; they didn't "speak" to the other forces and were deaf to much of the ruckus around them. It was the job of the particle physicist to find out what the constituents of matter were and what words they were whispering. And perhaps, most ambitiously, find some common ground between these different tongues.

In the early sixties the notion of writing a few equations to explain all physics seemed very distant. Each of these four forces was a little kingdom of theory and experiment unto itself. The cobwebbed kingdom of gravity was the most removed; the effects of general relativity only manifested themselves on scales too grand to reproduce in the laboratory. The other three forces had something in common, in that their playing field was in the atom. And in the atom the house rules were something called quantum field theory.

However strange its metaphysical foundations seemed to the generation that had discovered it, the physics embodying quantum mechanics had become a slick mathematical machine that graduate students of Guth's generation learned to run. According to field theory, elementary particles were concentrations of energy in force fields that permeated the universe— one field for each kind of particle—filling the space between particles like an invisible ocean and transmitting forces between them by way of intermediary particles. On the quantum level, the physics of the four basic forces was sort of like a big unruly game of catch. Forces were transmitted by little wave-packet bundles of energy that got tossed back and forth like baseballs between particles. The world could be divided into particles of matter, called fermions, and force-carrying particles, called bosons.

This all went under the rubric of particle physics, and MIT was a factory for particle physicists. The undergraduates took courses on the basics of the design of accelerator machines. Guth and his friends learned to "turn the crank" and calculate wave functions.

Guth cut his research teeth on the strong force, which binds the nuclei of atoms together, which is a particularly daunting trick. An atom is composed of a cloud of negatively charged electrons surrounding a little ball of protons and neutrons that, although only one hundred-thousandth the size of the atom itself, contains essentially all of its mass (protons and neutrons each being 2000 times as massive as an electron). The strong force has to overcome the tremendous electrostatic repulsion that positively charged protons feel for one another inside the tiny, dense nucleus.

According to a new theory proposed the year that Guth entered MIT, protons, neutrons, and some other heavy particles known as mesons (whose numbers had been proliferating madly) were not actually elementary particles at all, but were ensembles of even tinier entities. One of the authors of this theory, Gell-Mann of Caltech, named them quarks after a line from *Finnegans Wake*.

In this scheme, protons and neutrons were each composed of

three quarks and mesons of two quarks. Quarks were a very powerful idea, mathematically, for only three types of quark—labeled whimsically "up," "down," and "strange"—and their corresponding antiquarks could explain a whole panoply of elementary particles. But the theory was problematic in that nobody had ever seen anything that looked like an individual quark loose in the world. Physicists questioned whether quarks were physically real or just a convenient fiction.

One possible explanation popular at MIT, and the one that Guth adopted for his Ph.D. thesis, was that quarks were very, very massive and the force binding them together was very, very strong—so strong that modern particle accelerators didn't have the power yet to bust a quark loose from a proton. But someday . . . Guth tried to calculate the structure of mesons. The equations could only be approximately solved numerically, which meant that Guth had to plow through them on a computer.

In the middle of all this, in 1971, he married his high-school sweetheart, Susan, who had studied French at Douglas College and was then teaching in a private school. In the fall, Guth's thesis still unfinished, they moved to Princeton, where he had obtained a three-year postdoctoral fellowship. Guth was graduating into the teeth of hard times. The preferred kinds of jobs—assistant professorships that lead to consideration for academic tenure—weren't around because universities had already staffed themselves to the gills to cope with the baby boomers. A postdoc, in which you got paid $12,000 a year and were given office space to do research, was a sort of holding pattern, a chance to establish yourself and advertise your wares.

Guth entered his scuffling days in comfort. To him $12,000 was a princely sum. The Guths had the makings of middle-class comfort. Guth later said it was the most financially secure time of his life. They lived in a two-bedroom apartment owned by the university. There was enough money left over to buy a new Buick Skylark.

Princeton, moreover, was full of great theorists. During his stay there he hooked up with another postdoc, and they continued working away on Guth's quark calculations. Guth kept to himself at first, a reclusive kid again, dividing his time between teaching and finishing his thesis and preparing it for publication. His reclusive habits proved to be his undoing.

"It was really as soon as I finished that work that a new theory of quarks called quantum chromodynamics became invented," Guth said, grinning ruefully. "It was a totally different way of looking at the interactions between quarks. So, my thesis sort of instantly became archaic." Quantum chromodynamics (QCD) vested quarks with a new quantum property called color. Quarks could come in three "colors"—red, green, or blue—which attracted each other with a funny force that grew stronger the farther apart the quarks were. The color force was transmitted by particles called gluons. So the proton became a busy place. Inside were three pointlike quarks arrayed like the colors of the rainbow, dancing amid a hail of gluons like third graders in a snowball fight.

The reason individual "free" quarks were never seen in this theory was because the color force between them was like a spring. The harder you pulled on a quark, the harder the gluons pulled back on it until you had built up so much energy that a new quark-antiquark pair appeared and the spring broke with the new quarks on its ends. When they were close, quarks felt no pressure from each other at all, like a pair of prisoners chained together in a cell whose dimensions are shorter than the chain. If neutrinos were the sly spooks of the cosmos, quarks were like the Communist conspirators in thirties literature, bound anonymously together with their brothers in three-man cells.

Guth kicked himself when he heard about QCD, because an important part of the work had been done by people he knew at Princeton. He hadn't paid attention, to him it was just another theory.

When his term at Princeton was up, Guth realized that his career wasn't in very good shape. The only thing he'd published so far was a thesis based on an obsolete theory of quarks. It was time to play musical chairs again, a ritual that involves reading notices on bulletin boards, writing letters, and calling friends in physics departments—a giant networking game. Guth had a hard time. Finally, late in the season, with some help from the head of the Princeton physics department, he got offered another three-year postdoc at Columbia. In 1974 the Guths sucked in their breath and moved to Manhattan. It was still the time when a couple of modest means could find an affordable (barely) one-bedroom apartment, a tenth of a brownstone with a closet for a kitchen, on West Seventy-sixth Street, now the kingdom of the yuppies. The Buick wound up in New Jersey. Susan was making less now, because the only job she could get was part-time in a private school. Still, New York was a lot of fun; they were big movie fans.

Once there, Guth realized he had some catching up to do. The success of QCD confirmed a revolution in particle physics. While it was conquering the strong force, theorists in Cambridge and elsewhere had been piecing together a new understanding of the weak force that unified it with electromagnetism. These were not just new theories—they belonged to a new *brand* of theories, called gauge theories, that had risen out of the anarchic welter of ideas in the fifties and sixties. Suddenly they were the wave of the future.

This revolution had more or less happened behind Guth's back. He hadn't been taught gauge theory when he was at MIT, and he hadn't picked it up at Princeton, even though, he admitted, "Princeton was the capital of gauge theory." At Columbia he had no teaching duties, and Guth decided he'd better teach himself the new physics.

The heart of the new physics was an ancient idea, the concept of symmetry. Anyone who has ever looked in the mirror or marveled at the form of a vase or appreciated the six-foldedness of a snowflake knows the beauty of symmetry. In art and nature something is symmetrical if it looks the same from different angles or vantages—like a snowflake rotated sixty degrees. It turns out that the concept of symmetry plays a deep role in mathematics and, therefore, in physics. In mathematics a symmetry is some aspect of a system that remains unchanged when the system is transformed. For example, the length of an arrow stays the same when the arrow is rotated or shot out of a bow.

Einstein realized that the search for a universal truth or law that is true under all conditions and times—whether you are here or on Mars, whether you look up or down, whether you are moving or standing still or spinning in a circle, whether time goes forward or backward—is also a search for a kind of symmetry. The dictate that the laws of physics should be the same for moving and nonmoving observers formed the basis of relativity. For physicists truth has always been beauty, and beauty has always been truth.

It turned out that nature and natural law abounded in symmetries. The statements that energy or electric charge was conserved were just expressions of symmetry: things that stayed the same during transformations in the world. Early in the century a woman named Emmy Noether at the University of Göttingen extended this insight. Noether showed that wherever there is a symmetry in nature, there is something conserved. The search for solutions to physics equations was, therefore, a search for the mathematical symmetries they embodied.

This line of thought reached its bravest expression in 1954 at the Brookhaven National Laboratory on Long Island. A pair of theorists there, Chen Ning Yang and Robert Mills, invented what would be called, for obscure historical reasons, gauge theories. They concluded that all force fields were just the result of nature trying to restore and maintain symmetries at every point in space, somewhat like water always leveling itself. If a rolling ball slowed down and lost momentum, it must mean there was a force—perhaps the gravity of an upper slope—operating on it. If you tried to arbitrarily twist and stretch a snowflake, nature would fight back with forces of compression and extension to maintain the integrity of its balance sheet of conserved quantities. In effect Yang and Mills said there was nothing in nature but symmetry. For every property or quantum number there was a force that kept it so, and every force could be related to symmetry in some mathematical space. Beauty would not be denied; it was the source of all physics.

For twenty years, the Yang-Mills theory was just one tool in the physicists' bag of tricks, until during the sixties and seventies it was used to unify two of the four forces of nature into one theory.

Three men working independently put that theory together. Guth had less than a worm's eye view of the process, in spite of the fact that two of the men had wound up in Cambridge at the same time as Guth—Sheldon Glashow at Harvard and Steven Weinberg at MIT. Weinberg and Glashow were one of the great stories in particle physics. They had grown up in New York and been friends and members of the science fiction club together at the Bronx School of Science and at Cornell before parting ways to different graduate schools. Ten years later they had both contributed to a Nobel Prize–winning theory—one of the landmarks of modern physics—and had both wound up at Harvard. They were a textbook case in opposites. Glashow is tall, silver haired, voluble, and outgoing, fond of fast cars and cigars that bounce up and down in his mouth as he talks, shedding ideas like sparks. Weinberg is red haired and introspective with a bullfrog voice and a somber mien. He was also an early fan of cosmology. Something of his frame of mind was revealed in his 1977 book about the big bang, *The First Three Minutes:* "The more the universe seems incomprehensible," he wrote, "the more it also seems pointless. . . . The effort to understand the universe is one of the very few things that lifts human life a little above the level of farce, and gives it some of the grace of tragedy."

The electroweak theory, as it came to be called, began with Glashow during a stay in Copenhagen at the Neils Bohr Institute in 1960. Prior to that time theorists had been stumped in trying to construct a theory of the weak force. Glashow suggested that it could be accomplished by combining the weak and electromagnetic forces in one theory. The theory of electromagnetism obeyed its own symmetry principle, but when juxtaposed with the weak force, a natural deeper symmetry emerged. The two forces complemented each other; they were both different manifestations of the same underlying force, or symmetry.

It was brilliant, but there was an obvious problem. Glashow's theory made brothers of the photon, the carrier of light, and the so-called W particles, the particles that transmitted the weak force bosons. The trouble was that photons were massless and traveled at the speed of light, while W particles would need to have large masses, that later turned out to be roughly 100 billion electron volts.[1] That was not a very symmetrical-

1. Particle physicists refer to the masses of particles in terms of their energy equivalents. An electron volt is the energy an electron would acquire by being accelerated through a 1-volt

looking relation. Glashow had no explanation for how his beautiful brotherhood of electroweak bosons had gotten skewed, some having mass and some not.

As a result, his theory languished for most of a decade until his old classmate Weinberg and a Pakistani theorist named Abdus Salam came in their separate ways to its rescue.

Weinberg and Salam each succeeded by adapting to particle physics a way to make beautiful theories come out ugly in reality that had been invented by Jeffrey Goldstone, now at MIT, and elaborated by a Scotch physicist, Peter Higgs, to help explain the phenomenon of superconductivity. The process is called spontaneous symmetry breaking. It is based on the principle that a symmetrical question can have an unsymmetrical answer. Consider, for example, a pencil balanced on its point. For an instant the pencil is the very model of symmetry and balance, but it won't last, for the pencil will quickly fall over. Which way will it fall? It can fall any way, but it will wind up falling only one way—lying on the table and pointing in some specific direction. An unsymmetrical outcome to a symmetrical question.

The pencil standing on its end corresponded to Glashow's model of perfect symmetry between the weak and electromagnetic forces—the electroweak force as God intended it, the Platonic ideal truth. In this model none of the bosons had mass. That was the way the world had been when it was born; at high energies and temperatures, the weak and electromagnetic forces were equal. The fallen pencil represented the expression of this ideal truth in today's world—some bosons with mass, others without. What we have today is broken symmetry. The world, said Weinberg and Salam, is based on beauty and symmetry, but at low energies the beauty and symmetry have been hidden, lost like some ancient Eden that can only be reconstructed mathematically from painful digging through the clues of physics.

What caused this fall from grace? How did symmetry break? To answer this question Weinberg and Salam made use of an important new cosmic entity called the Higgs field, which would allegedly permeate all space and interact with the other quantum fields to determine the properties of elementary particles.

To explain how symmetry breaking worked, Weinberg liked to compare the Higgs field to an ocean. The breaking of the electroweak symmetry was like the freezing of water. A water molecule of two hydrogen atoms attached to an oxygen atom is shaped like a shallow vee. Above the freezing point, 32°F, an individual water molecule can point

---

electrical field. The mass of an electron at rest is about 10^{-27} grams or 511,000 electron volts. Protons and neutrons both weigh about the same: 1.7×10^{-24} grams or 938 million electron volts.

in any direction it wants to—all directions are equal—space to the molecule is symmetric. Below the freezing point, the molecules hook up into the crystalline structure of ice; a molecule is no longer free—all directions are no longer the same.

The change from liquid to a solid, or gas to a liquid, is called a phase transition. Weinberg suggested that the universe underwent phase transitions as it went from high energy to low energy. Only in this case what "froze" was the Higgs field and what was lost was not a spatial directional symmetry, but the symmetry between forces and masses of elementary particles. The forces became unbalanced. Physics, which started off balanced, wound up off center, like a fallen pencil, corrupted by Higgs bosons. Some of the sweetness and beauty of physics becomes obscure.

The "freezing" and symmetry breaking of the electroweak force, Weinberg and Salam calculated, occurred at an energy of about 100 billion or "giga" electron volts. That meant that if physicists could produce that energy in their particle accelerators, they could briefly "remelt" the Higgs field in one microscopic explosion and recover the full symmetry of the electroweak force.

There were major cosmological implications in all this for anyone who believed in the big bang. An energy of 100 giga electron volts corresponded to when the universe was a trillionth of a second old. If Weinberg was right, the entire universe had undergone a phase transition back then, and that might not have been the only time it happened. There were other forces of nature that might be reconciled with the electroweak interaction by the application of still more powerful symmetry principles. Perhaps there was a whole hierarchy of symmetries that broke at higher and higher temperatures, earlier and earlier in history, accompanied by a whole series of phase changes—"freezings" of the Higgs field.

In short, during the early, hottest stages of the big bang the laws of physics themselves may have evolved as the nascent universe expanded and cooled, going in stages from being simple and beautiful to being complex and ugly—one force after another splintering away from primordial unity like ancient gods quarreling. Even physics itself—it seemed—was a relic of events in the big bang. If that was true, then the Higgs field was less a shadowy supporting actor than the main character in the evolution of physics. It was the very essence of what physicists call the vacuum—the naked, unadorned space-time in which the laws of physics are embodied.

At the time, however, in the late sixties, few physicists, Weinberg excepted, took the big bang seriously. Glashow himself often made disparaging remarks about cosmology. And his reaction to the notion of symmetry breaking was witty, but somewhat ungracious. He said it was an ugly—although necessary—thing to do to a beautiful theory. He compared it to a toilet, saying that while that fixture was a necessary part of the house, it was not something you show off. He made fun of Weinberg, calling him the Thomas Crapper of particle physics. By the time Weinberg and Glashow sat side by side with bottles of champagne at a party celebrating the announcement that they and Salam had won the Nobel Prize in 1979, their friendship had cooled considerably. Weinberg finally left Harvard for a lucrative job at the University of Texas.

Weinberg was something of a hero of Guth's when he was at MIT, although he didn't know anything about Weinberg's work. He frowned, trying to remember. "Weinberg arrived, yeah, I think, about the time I became a graduate student, 1967 or 1968, right around there. Did he know who I was? No. He did say hi to me when we passed when I was a graduate student; he recognized my face.

"My understanding of particle theory was sort of a *Scientific American* level still. Although I was aware that Weinberg was a great theoretical influence, I didn't understand what for. I think I had the impression nobody knew what for, that he was a brilliant person but he had not yet made contributions that people thought were important."

One of the by-products of Glashow's original theory was the prediction of the so-called neutral current weak interactions that had so excited Schramm. When these were discovered by physicists at CERN and Fermilab in 1976, the scientific community became convinced that the theory was right. A new force had appeared right where symmetry had decreed it—a triumph of beauty.

When Guth got to Columbia, he decided he had better modernize himself again and learn gauge theory. "Uh, yeah, it is neat," he says now, "and I think I appreciated it is neat. It wasn't quite like learning particle physics from scratch. It was an elaboration of the same kind of quantum field theory techniques that I did learn about when I was a graduate student." He paused, as if to see if the coast was clear. "I think it's probably fair to say that even in terms of traditional quantum field theory, my education coming out of MIT was somewhat lacking."

In any case, the Yang-Mills view of reality came quickly to Guth. He joined a small group of Columbia theorists who were trying to teach themselves gauge theory by posing and solving rather mathematical and technical problems related to symmetry breaking and Higgs fields. The model they were using for the electroweak force was not Glashow's, but a simpler variant, easier to calculate with, that they hoped would illustrate the basic principles, if not the detailed phenomenological results of the physics. Guth called it a "toy theory." In fact, he still did not know the Glashow-Weinberg-Salam model, based on a slightly fancier symmetry.

One thing they investigated was the strange idea that the symmetry-breaking process could inflict a scar on the universe. The Higgs field was not just a metaphysical abstraction, it turned out: According to calculations, the field could tie itself into knots of energy that wouldn't go away. This bizarre-sounding result was to transform cosmology. It was a consequence of the fact that the metaphorical pencil that represented the Higgs field could fall in any direction. In a whole forest of them, some in one part of the woods could fall one way and some somewhere else another, like windblown hair.

At each point in space, as the universe cooled, the symmetrical laws of physics could fall apart slightly differently, just as ice crystals could start to form with different orientations in different parts of the same freezing pond. Where these different domains grew into one another there would be flaws—discontinuities—in the ice or in the vacuum, the seat of physics itself. Depending on the symmetry being broken and the laws of particle physics, these discontinuities, scars, flaws, could turn out to be points, lines, or walls in space. Each would be thrumming with the mysterious Higgs energy.

In Moscow, Zeldovich became a little alarmed when he heard about this. Of these three possible discontinuities in the Higgs field, walls would be the most noticeable. They would stretch across the universe blazing with energy and thus mass; they would move at the speed of light, obliterating anything they hit. In the mid-seventies, physicists at CERN were laying plans for an accelerator that would smash together protons and antiprotons at energies high enough to remelt the Higgs field and re-create the original reunited symmetrical electroweak force—a rebirth of a primordial god. Afraid that one of these experiments could trigger the formation of a wall that would sweep out and destroy the universe, Zeldovich spent several months calculating feverishly before deciding that such walls probably didn't actually exist. If they did, the universe couldn't be as uniform as it appears; the galaxies would all be piled up somewhere and the microwave background would be lopsided.

The toy models that Guth and the others were working with produced pointlike discontinuities—little kernels of the unbroken symmetry, tiny regions where the old laws still held. These kernels would have the properties of long-sought hypothetical particles called magnetic monopoles. In nature as we know it, magnets always come with two poles—a north and a south—connected as indivisibly, it seemed, as the ends of a string or a pair of quarks. A so-called monopole would be a north or south magnetic pole unconnected to its opposite.

Monopoles were the unicorns of particle physics—fantastic to contemplate but scarcely in evidence. No law forbade their existence, however, so physicists had searched for magnetic monopoles in moon rocks, ocean sediment, and cosmic rays with no success. So monopoles, even though they were just a twist in an imaginary field, were a great lure. There was a Nobel Prize awaiting their discovery.

Even if it didn't have any direct relevance to the parade of particles out of particle accelerators, the monopole was the kind of problem that Guth realized he liked and was best suited for. He chuckles as he says this now. "I liked things to be precise. I really liked problems that were well-defined mathematically, with a well-defined answer and a limited amount of input. Where you could state what it was that you were trying to solve, write down all the equations that were relevant, and then it was a mathematical question to find out what the answer was. State all the input and go on." Cosmology of course was nothing like that.

Guth and his friends knew they weren't on any Nobel Prize trail—they weren't even using a realistic theory. They were just using these twists in space-time as a mathematical practice field. Guth was doing the calculations by a technique known as path integrals that was more rigorous than the methods gauge theorists had been using. "If I'd been bolder," he said, "I might have skipped that step and gone whole hog into the new physics. But then I wouldn't have become an expert on monopoles."

And the history of cosmology might have been different.

13

THE RELUCTANT COSMOLOGIST

During his three years in the Big Apple, Guth became an expert on symmetry breaking, adept at computing the mathematical contortions of the Higgs field as it froze and tried to smooth the wrinkles out of itself. When 1977 and the end of his Columbia appointment rolled around, he felt like he was in better shape. The group had published several papers. Rockefeller University offered him a postdoc, which he declined.

"I was considered sort of a finalist for an assistant professorship at Harvard, which is like a death sentence," he said wryly. It was well known that Harvard assistant professors hardly ever got offered tenure at Harvard. Luckily perhaps, Guth escaped. Harvard hired a couple of people from Cornell, which created a vacancy in that department. As a result Guth got offered another three-year postdoc, at Cornell.

They packed up again and moved north. Susan was pregnant. After Manhattan, Ithaca seemed cold and small. It was his third postdoc. The road was getting a little old. How was it, I asked Guth, that he remained so cheerful when his contemporaries were leaping into industry? "I did not get discouraged, as far as I can remember. It seems strange," he conceded. "I didn't worry about a permanent job. I was enjoying doing physics and not worrying about the future."

As usual Guth's instincts for steering away from the hot topics in physics were operating infallibly. Weinberg, Glashow, and Salam were then closing in on a Nobel Prize for the electroweak theory, and Guth, having just gotten his feet wet in a simpler version of the theory, was planning to drop the subject. Although he had brought some work from Columbia with him, Guth was planning on going back into quarks.

This time, however, he was saved from himself. In Guth's second year at Cornell Henry Tye came on the scene. Tye, born in Hong Kong, had been two years behind Guth at MIT. He had spent time on the trail as a postdoc at the Stanford Linear Accelerator (SLAC) and at Fermilab trying to predict the outcomes of high-speed elementary-particle collisions. Cornell needed him because it was building a machine to collide electrons and positrons, but the device wasn't yet ready when he arrived in 1978, and Tye was itching to get his fingers into something more fundamental. Like cosmology, and he needed a partner.

Tye remembered Guth as a bright guy, fun to bounce ideas off of. He was happy to see that Guth was still the same person and still had his boyish enthusiasm for physics after all the years as a postdoc. All around him people were getting depressed, worried that they'd never make it, bailing out into industry. It would be easy for Guth to be depressed, thought Tye, but instead he radiated cheer and competence. Part of the credit, he thought, had to go to Susan, who was unflaggingly supportive.

Tye was interested in magnetic monopoles, too, but on a much grander scale than Guth's Columbia buddies. He knew nothing about them, however; but he knew where to find out. One day he asked Guth whether monopoles would be produced in grand unified theories.

Guth asked innocently, "What's a grand unified theory?"

Of course, grand unified theories, or GUTs for short, were the hottest thing in physics. Tye explained that GUTs were an attempt to unite all three quantum forces—electromagnetic, weak, and strong—the way the electroweak theory had united the weak and electromagnetic. In GUTs there were two symmetry-breaking episodes: first the strong and electroweak forces became separate and distinct, and then later the electroweak force split apart into its modern components in the manner already described. So in GUTs there were at least two chances in the first fleeting moment of history for the Higgs field to tie itself in knots. Guth nodded. Tye began teaching GUTs to him—not the whole theory, just enough to let them get on with the monopole problem.

And so it was that Alan Guth finally came up to speed.

The first grand unified theory was the brainchild of Glashow and a Harvard colleague Howard Georgi. It was the next logical step after the success of the electroweak unification made physicists crazy for so-called gauge theories. The principle was the same. According to those theories, recall, each of the so-called fundamental forces corresponded to a symmetry. Glashow had been able to discern that the weak and electromagnetic symmetries were fragments of a more general symmetry. He and Georgi reasoned that the electroweak and strong force symmetries might themselves be parts of a still deeper and more general symmetry of nature.

This was physics' version of the long address: We live on West Fourth Street, on the island of Manhattan, the city of New York, in the state of New York, in the United States, on the continent of North America, on the earth, in the solar system, in the Milky Way galaxy, in the local group, in the Virgo supercluster, and so on. The weak force was part of the electroweak force, which was part of the grand unified force, which was part of the primordial supergrand unified force (if they could ever rope in gravity), and so on. In principle he and Georgi could always enfold more fundamental forces in their unification scheme by appealing to even more general and powerful principles (until they ran out of forces).

The French mathematician Elie Cartan had spent much of his life listing and classifying the different possible forms of symmetry. They ranged from simple mirror-image symmetries to rotational symmetries in 496 dimensions. Electromagnetism was related to one of the simpler cases—rotations of a circle. The weak and strong forces could be represented by certain kinds of two- and three-dimensional rotations.

To construct their theory, Georgi and Glashow sat down with Cartan's list of symmetries. The simplest one that could encompass the electromagnetic, weak, and strong symmetries was something called SU(5), which involves certain rotations in a five-dimensional mathematical space.

The theory that emerged also made brothers of the quarks, which constitute the heavy particles like protons and neutrons, and the lighter particles like electrons and neutrinos—so-called leptons. These hitherto separate classes of particles, quarks and leptons, could change into one another by exchanging a massive force-carrying particle called an X-boson. This implied that the number of kinds of neutrinos and electrons should be the same as the number of different quarks, which meant that Schramm, Gunn, and Steigman's work on the number of neutrino species was of fundamental importance.

This interchangeability of quarks and other lightweight particles had two important consequences—one for the past and one for the future of the universe. Inserted into the expanding universe it turned out to be the missing factor that could explain how the mismatch between matter and antimatter had evolved during the early universe. At the extremely high energies of grand unification, matter or antimatter could be destroyed or created independently: A proton didn't have to meet an antiproton to die—it could just fall apart. In the long run this was a death sentence for the universe of ordinary matter, the universe of atoms and stars and apples and BMWs. It meant that protons, the bricks of matter as we know and love it, were ultimately radioactively unstable; eventually, in some 10^{30} years, if the universe lasted that long, by emitting X-bosons they would all dissolve out from under us.[1] Matter would have only been a moment, a passing thought, in the eternal shifting fortunes of the universe; no artifact from it would remain, no work of Michelangelo or Bach or Trump would survive. The atoms in the ink in all the documents in the world would dissolve, as would those of the paper that held it; computer memories would evaporate.

According to the calculations, the grand unification of forces required a temperature of 10^{27} degrees K or energies of 10^{14} giga electron volts, for its shining symmetry to be restored and observed directly. That was 100 billion times the energy of the projected improvements to Fermilab. Wags like to point out that it would take a particle accelerator light-years long to achieve those energies. Physicists' only hope for testing GUTs was the poor man's accelerator—the universe.

A temperature of 10^{16} giga electron volts corresponded to a time in the big bang when the universe was only 10^{-35} seconds old and the entire cosmos observable today was the size of a grapefruit. Grand unified theories opened to cosmological speculation realms of time that made the electroweak and nucleosynthesis eras seem ancient. If they were any guide, it was now possible to run the film of the expanding universe backward to a fraction of the first second of time, when existence was just a hot spark, an elegant pumpkin seed of compressed possibility.

Quantum uncertainty itself determined how close to the putative beginning—the singularity—you could slice time. At a ten-millionth of a trillionth of a trillionth of a trillionth of a second—a number written as 10^{-43}—after the big bang, an era known as the Planck time, quantum uncertainties in the geometry of space-time were the same size as the universe, which meant that space and time considered in chunks shorter than that were simply incomprehensible. They were in the same limbo

1. This set off a great international search in the late seventies for signs of decaying protons. Physicists are still searching. The hunt was conducted far underground, to avoid interference from cosmic rays, where heaps of iron or tanks of ultrapure water were wired with detectors, counters, and computers to record the fragments of a proton exploding in their depths. None were seen, and as the lower limit on the proton's life crept up to 10^{32} years, theorists began to consider other unification schemes along the lines of SU(5), but fancier. The experimenters received another reward for what had been called "Zen physics"—sitting and waiting. In February 1987 at least two of the water tanks detected the passage of a burst of neutrinos at the same time that astronomers detected a supernova in the large Magellanic cloud, confirming the theory that copious neutrinos are produced in the collapsing core that triggers such an explosion and providing one of the first instances of neutrino astronomy.

land of probability as the electron before its place or motion had been measured.

With the advent of GUTs in the late seventies, gravity became the odd man out, the only force not explicable as a quantum game of catch. The law of laws that Wheeler called pregeometry, the principle that made the universe fly, was still not yet imagined, but it wasn't such a wild idea any more. It was possible to hope that some more general and powerful symmetry would enfold gravity in the mathematical embrace of unified theories. Some called it, only half jokingly, the Theory of Everything. In the Fermiland cosmic history the universe started in such a moment of blinding symmetry and went downhill from there. General relativity and quantum theory, which may have spoken the same lost language before the Planck time, spoke different languages afterward.

As the universe expanded and cooled, the energies of its contents—a dense inferno of radiation and particles—dropped; every so often the temperature fell below a critical value, and there was one of those phase transitions in which forces fissioned and fissioned again. The laws of physics got less symmetrical in steps.

Gravity presumably dropped out of the primordial unity when the universe was 10^{-43} second old and still as hot as 10^{32} degrees K. That was the first divorce, leaving the cosmos ruled by two rules: gravity and GUTs. Next the strong force, and finally the weak and electromagnetic forces fissioned away. By the time the universe was a billionth of a second old there were, according to standard physics, four forces and about two dozen kinds of elementary particles jangling away in the blossoming cosmos.

It was at the moment of the second divorce in this scheme of things, when the grand unified force fell apart into the strong and electroweak forces as the universe was 10^{-35} second old and as big as a grapefruit, that Tye wondered if the Higgs field would tie itself up into monopoles.

He and Guth talked every day, and then Guth went home to calculate on the kitchen table. Guth had already fallen into a pattern. He was a night person. He would work late, grinding through equations in a neat hand in his bound laboratory notebook. *Getting the answer*. In the office the next day he would compare notes with Tye. The phone would ring. Visitors would troop in, there were seminars to go to. Questions would arise, contradictions emerge. He was an eager student of unification.

"I guess when Henry showed me the SU(5) grand unified theory I was very impressed that one simple theory could incorporate so much physics."

In a few weeks he had learned enough to answer Tye's question.

Yes, monopoles would be formed in the SU(5) symmetry breaking process. But, Guth went on, they would be extraordinarily heavy for elementary particles, with a mass-energy equivalent to 10^{16} giga electron volts, so one monopole would weigh as much as 10,000 trillion protons, as much as a small bacterium—a living creature.

That meant that the monopoles were uninteresting, Guth explained, just like the other crazy heavy particles that grand unified theories predict. Normally, what was important in physics was what the theory predicted at the relatively low energies that particle accelerators could reach—where you could do experiments and test ideas.

Tye was undeterred. "Why don't we try to figure out how many of these magnetic monopoles were produced in the big bang?" he suggested.

Guth drew back. "At first, I was very leery about working on a project like that," he said, grinning wolfishly. "It was very different from anything I'd worked on before, because it was such an open-ended problem. And at the same time, although I don't think I at all questioned the existence of the big bang, I certainly believed we knew very, very little about the details of it, that it would be almost impossible to figure out something like how many monopoles would be produced." Tye kept nagging him on and off through the winter of 1978–1979, mostly getting nowhere.

One thing that helped change Guth's mind was a springtime visit to Cornell by Weinberg. Weinberg made an avocation out of cosmology and encouraged its practitioners. While he was there he gave a series of lectures on how grand unified theories could explain the imbalance between matter and antimatter in the universe. He was one of many, including Turner and Kolb, working on the problem of the origin of the elementary particles, the way Hoyle and Fowler had hammered out the origin and abundances of the elements. If reasonable assumptions were made about the details of GUTs physics and the early universe, the computations agreed roughly with what was known from the microwave background: There were about a billion photons in the universe for every proton. And no antiprotons—no antimatter at all. It was a great triumph for physics, one that impressed Tye. Once you decided that the prevalence of matter over antimatter in the universe belonged in the realm of physics and not of theology, he said, you had to have some kind of grand unified theory.

It even impressed Guth. "It was largely through Steve's influence that I decided maybe cosmology was indeed something that a respectable physicist could work on and draw reasonable conclusions about," he said.

During Weinberg's visit Guth at one point found himself in a discussion with him and a few others about symmetry breaking and the

Higgs field. What would happen to Higgs fields in cosmology? someone asked. Would they all fall in the same direction everywhere in space, or could there be domain walls separating different regions? The group concluded that it was an interesting question and it would be nice if somebody knew the answer.

Another oblique, almost premonitory, influence had been a visit to Cornell and lecture by Robert Dicke in 1978. Dicke described the conundrums and paradoxes that he and Peebles were writing about, particularly the so-called flatness problem. Guth listened carefully, but it would not mean much to him until later.

"This was before I was even thinking about working on cosmology," said Guth about Dicke's lecture. "It was just an odd fact to me, and had nothing to do with my research, but I was struck by it. The way he phrased the problem was that the expansion rate of the universe at one second after the big bang had to be fine-tuned to the accuracy of, I think the number he gave was one part in 10^{14}. And if the rate were one part in 10^{14} larger than it actually was, the universe would fly apart without galaxies ever forming. And if the universe were expanding at a rate of one part in 10^{14} slower, the universe would have long ago collapsed." It was just another odd fact about the so-called science of cosmology to be filed away in his brain.

Guth finally surrendered to Tye.

I asked Guth once why, in the long span of months, Tye simply didn't go ahead and do the calculation on his own. "Well, he was working on other things, too. I shouldn't make it sound as if the only thing he did was day after day nag me to try to get me to work on this," he said. "Most particle theorists prefer to work in groups, and Henry is certainly such a person. Almost all the papers he's written have been collaborations, and almost all the papers I've written have been collaborations, also."

He paused. "Part of it is loneliness. Part of it, I think, is just the idea that you can make more progress if you have more minds working against the problem. And I've always found that thought very useful in working with my collaborators. The only time it's sometimes a pain is when you get down to writing things out, and you need to pass drafts back and forth and compromise with everybody. That sometimes can be a pain."

Guth and Tye pushed and pulled each other through the calculations. By early summer they had refined their first rough estimates and had gradually become convinced that GUTs would flood the universe with as many monopoles as protons.

This was bad news. First of all monopoles were, on the experimental evidence, amazingly hard to detect for such an abundant species. Second, because each putative monopole was the mass equivalent of

10,000 trillion protons, their overall weight in such numbers would have crumpled the universe billions of years ago and dragged it back down to the big crunch. In fact, the estimated lifetime of a universe dense with monopoles was something like 6,000 years.

In fact it was important news—if it was true, it doomed GUTs. The theory predicted monopoles, and there were none. The poor man's particle accelerator had spoken.

Guth and Tye were still digesting the import of all this when they found out that a student of Weinberg's, John Preskill, had been doing the same calculation. They had been scooped. Sportingly they invited him to Cornell, and discovered Preskill had already gotten the same answer. He was convinced he had killed grand unified theories; populous monopoles were incompatible with a 15-billion-year-old universe. He went back home and published a paper that became an instant classic in cosmology.

Guth and Tye kicked themselves. Maybe they had been too cautious. "Why didn't we publish sooner? I didn't think we understood things well enough," Guth laughed. They didn't want to write a paper that just posed a dilemma. Rather than dance on SU(5)'s grave, they resolved to solve the monopole problem. "After his paper, I think both Henry and I thought we had to say something that he hadn't already said, so we leaned our subsequent work toward the question, What are the possible ways around the overproduction of magnetic monopoles?"

The gimmick they hit on exploited yet another similarity between the physics of the Higgs field and a glass of water. It was called supercooling. Physicists knew that if they were careful they could chill a glass of water below 32°F, and it would stay liquid as long as it wasn't disturbed. Rapping on the side of the glass or shaking it would break the spell, and the water would suddenly, belatedly, freeze. That was regular supercooling. Guth and Tye reasoned that there might be such a thing as cosmic supercooling of the Higgs field.

Suppose, they said, that as the universe cooled below the critical temperature of 10^{27} degrees the Higgs field did not immediately freeze, but stayed in its symmetrical state for a while longer. During that delay the universe would continue to enjoy the simplicity of grand unified physics while it got larger and cooler. If the delay were only a fraction of a trillionth of a second, when the Higgs field finally froze and tied itself in monopole knots, the universe would have had a longer time to align itself. As a result fewer knots would form—perhaps so few to account for the lack of monopoles in today's world.

"It was all a little strange to us," recalled Guth. "Phase transitions were certainly new to me. I felt somewhat ill at ease because I knew there were condensed-matter physicists who knew a lot more about them than I did."

They weren't completely alone. Other particle physicists, notably Sidney Coleman at Harvard, had investigated the possibility that there could be a "hang-up" in the symmetry-breaking process, a delayed reaction to the universe cooling. He called the temporary state of grace in which the universe continued to be imbued with symmetry the "false vacuum."

Guth knew Coleman from his Cambridge days and was familiar with the false vacuum notion. He and Tye figured that the false vacuum interlude could delay monopole production long enough to scatter them harmlessly. The preliminary calculations looked promising, and as fall approached Guth was getting more and more excited. He knew this was important physics. These calculations were not technical exercises on toy models or outdated theories; he and Tye were probing deep consequences of the most ambitious theories of our time. By following the trail of a nonexistent particle Guth had quietly steamed into uncharted waters. He was learning things nobody had ever known.

And if it was cosmology Guth found himself doing at the head of the line, well, it didn't seem so bad after all, or as sloppy. As Tye always said, when you really punched into it and did the calculations, cosmology was surprisingly restrictive; it was not easy to satisfy all the observations.

It was around then that Tye decided Guth needed still more seasoning. For a variety of reasons the coming year at Cornell—Guth's third—was going to be dull. Tye talked Guth into taking a leave of absence from Cornell. By now, he argued, if Guth had gone into an academic position when he first left MIT, he would have already qualified for a sabbatical. Guth called up SLAC and they agreed to let him go out West for a year—his fourth postdoc.

Guth was relieved because the leave prolonged his tenure at Cornell, delaying the next episode of job hunting. For his wife, Susan, it was a mixed blessing. She was happy to leave Ithaca but she was not happy about the prospect of having to move back in a year, and then having to move again yet a year later. Guth picked up a U-Haul and drove west. At least it was warm on the San Francisco peninsula. They settled into a rented ranch house near SLAC that, he says, was far beyond their means. But they were only going to be there a short time, after all. Guth was close enough to the lab to ride his bike to work. And it had a nice yard, the only one that he has ever had.

Stanford was full of experts on grand unified theories, including Coleman, who was visiting for the term. Guth spent a lot of time with Tye on the telephone. He was nervous. They puttered and refined and revised through the fall.

By early December they had nearly finished writing a paper that explained the lack of monopoles in the present-day universe by a split-second delay in the phase transition that marked the end of the grand unified era when the universe was 10^{-35} second old—an instant of false vacuum and prolonged symmetry that gave the Higgs field more time to align itself and thus resulted in fewer knots. Guth and Tye were going to be heroes; they had saved the grand unified theory from absurdity.

Having learned their lesson from having been beaten out by Preskill, Guth and Tye were eager to get their new paper done and published. As December approached they were frantic because Tye was going to China for six weeks over Christmas. Six weeks seemed like a cosmological epoch—the prospect of a delay that long was disheartening. Long coast-to-coast phone conversations ensued. It was during one of these phone calls that Tye mentioned a question that had been nagging him—whether the supercooling process itself would have any effect on the expansion of the universe. Guth should check it out, he suggested.

Tye's concern arose from yet another aspect of the analogy between water and the Higgs field. When water freezes and its molecules cease their restless jiggling and slip into still crystalline grooves, a surprising amount of energy is released and given off as heat. The energy freed by the freezing of a large swimming pool, if it could be captured, would heat a house for a couple of years. When water supercools this heat of condensation remains hidden, latent. When the Higgs field froze and symmetry broke, Tye and Guth knew, energy had to be released, the same way a pencil tipping over releases energy when it slaps the desk, or a ball rolling down a hill gains velocity. Under normal circumstances this energy went into beefing up the masses of particles like the weak force bosons that had been massless before. If the universe supercooled, however, all this energy would remain unreleased, latent, suffused through space for a while. The supercooled universe was a universe pregnant with energy.

Why should this make Tye queasy? Because according to Einstein, it was the density of matter and energy in the universe that determined the dynamics of space-time. The false vacuum energy was real energy; it could affect the universe and disrupt the big bang.

As he followed this chain of logic, suddenly Tye had realized they were flirting with something dangerous. He was dismayed. The issue of vacuum energy had been a tricky problem for physics ever since Einstein. According to quantum theory, even the ordinary "true" vacuum should be boiling with energy—infinite energy, in fact—due to the so-called vacuum fluctuations that produced the transient dense dance of virtual particles. This energy, according to cosmological equations, could exert a repulsive force on the cosmos just like the infamous cosmological constant that Einstein had invented back in 1918 to keep the universe from collapsing. Einstein had abandoned the constant when it was discovered that the universe was expanding quite nearly in accordance with his un-fudged equations. But the cosmological constant refused to die; quantum theories had reinvented it in the form of vacuum fluctuations. The orderly measured pace of the expansion of the universe suggested strongly that the cosmological constant was zero, yet quantum theory suggested it was infinite. Not even Hawking claimed to understand the cosmological constant problem. What had happened to the quantum energy of the vac-uum—how did it muffle itself? It was—and still is—a trapdoor deep at the heart of physics.

Tye was afraid that he and Guth were going to fall through that trapdoor. By imbuing the early universe, even for an instant, with the latent Higgs energy, they were meddling with the cosmological constant. Maybe they shouldn't publish. He couldn't sleep.

Guth wanted to go ahead with the paper; Tye recalls that his response was, "If you can't sleep, you might as well party." They didn't have to understand the cosmological constant, Guth said cheerily, to figure out what it would do. He agreed to check out the effect of the Higgs field on the expansion and evolution of the universe.

The stage was now set for Alan Guth's midnight creep through the meadows of cosmology.

As usual he put it off until after dinner. By eleven o'clock Larry and Susan were in bed. It was quiet, the time of day he worked best. In the spare bedroom where his office was set up, he flopped open his notebook and in a small, clear hand wrote across the top of the page.

> EVOLUTION OF THE UNIVERSE
> *I would like to consider the effects of*
> *(1) a cosmological constant, and*
> *(2) the freezing of degrees of freedom on the*
> *evolution of the universe*

Below that he wrote down the standard equations for an ex-panding universe. In the absence of any funny energy the early universe expanded at a steady rate, like the fragments of a grenade flying freely: Its size was proportional to the square root of its age—every time the age quadrupled the radius of the universe or the distance between any two points in it doubled, and the temperature went down accordingly. At the time of the breakdown of grand unified symmetry the region that we today call the observable universe—a sphere roughly 10 billion light-years in radius—was about the size of a grapefruit and sizzling at 10^{27} degrees K.

It took Guth three pages of neat algebraic calculations and two hours to put the stuck Higgs energy—the false vacuum—into this hot little grapefruit of a universe. When he did, he got a surprise.

The universe blew up.

The universe was already blowing up, of course, in the stately Hubble manner described above. If the universe were 1 foot across when it was 1 second old, when it was 4 seconds old it would be 2 feet across, and when it was 10 seconds old it would be 3 feet across. The universe with false vacuum in it didn't expand like that. It expanded exponentially. With every tick of the cosmic clock (a tick in this case being 10^{-34} second) the size of the universe doubled. If it started out at 1 foot, by the second tick it would be 4 feet, by the third tick 8 feet, by the fourth tick 16 feet. By the tenth tick it would be 1028 feet across.

It's easy to see that under such a pattern of growth the universe would swell like a monster. The bigger it got, the faster it would have to grow to keep doubling in each tick of time. In no time at all—a millionth of a trillionth of a trillionth of a second—the universe would double in size 100 times, becoming a trillion trillion times larger. Guth had a runaway universe on his hands. Disrupt the big bang? Letting the Higgs field supercool into a false vacuum state was like setting off an atomic bomb in the midst of a hand grenade explosion.

It was this runaway exponential doubling phenomenon that Guth was to call *inflation*—to distinguish it from mere Hubble *expansion*—but the name would not come to him for several months.

Once he had cleared the mathematical underbrush away from his calculations, it was easy for Guth to see what was happening to make the universe behave like this—although it was not so easy to explain. It sounded paradoxical, and whenever Guth has to explain it he breaks into a glib, toothy MIT grin. Sometimes doing cosmology and particle physics was like being the White Queen in *Through the Looking Glass,* who had to believe in six impossible things before breakfast.

Simple physics—believe it or not—predicted that the false vacuum would have negative pressure—put it in a balloon and the balloon would deflate. The universe acted differently, however—there was nothing to push against or be pushed by. This negative pressure, according to general relativity, generated negative gravity, which overcame all the other forces in the universe. As a result, the false vacuum is a cosmic bomb: It acts like a universal repulsion and blows the universe apart.

A kind of feedback effect would occur. As the universe ballooned it would increase the volume of supercooled symmetric space-time. The bigger the universe was, the more false vacuum there would be. Instead of being diluted by the expansion, the total amount of false vacuum energy would, therefore, increase. When the universe doubled in diameter, its volume went up eightfold, and there would be eight times as much false vacuum energy. The more false vacuum energy there was, of course, the faster the universe would push itself apart. The universe would go into what is called exponential expansion, in which there is increasing acceleration; the bigger it got the faster it would expand.

"I was definitely excited that night," Guth recalled. "I did realize that, uh, I mean, I was trepidatious about the whole thing. I wasn't *confident* that I had scored a big thing, but I *suspected* that I had scored a big thing, and was excited by that."

Well, that was an interesting evening's work. He had indeed scored a big thing. If Guth was right, the entire history of the universe as it had been worked out so far was wrong. Instead of expanding like an explosion that started out fast and was then gradually steadily slowing down, the universe apparently had expanded jerkily. Instead of slowing down, at least once and possibly other times the expansion of the universe had *accelerated*. The cosmological consequences of this speedup were to be immense, but they were slow in coming to Guth, partly because Guth knew so little about cosmology.

The major news was that if Guth was right, the universe was incredibly vaster than astronomers had dared to dream. The universe would have grown so fast during this exponential doubling that in only an eyeblink the entire realm visible to telescopes today could have burst forth from a piece of the big bang that was no larger than a proton. Henceforth a distinction would have to be made between the *observable* universe and the *real* universe, of which it was an insignificant and perhaps not even representative part. Who would have thought that this little detail of symmetry breaking that particle physicists invoked so glibly in their quest for the hidden beauty of physics could have such a disruptive effect on the cosmos? Well, it would certainly take care of the monopoles; after this burst of hyperexpansion they would be sprinkled billions of light-years apart.

Guth's mind was racing overtime. As he was getting ready for bed and brushing his teeth, his mind slid back to the curious lecture Dicke had given at Cornell in 1978, before Guth had become a cosmologist. Dicke's conundrum was that the density of the universe was too close to the magic value of 1.0 for comfort or coincidence. Geometrically it amounted to asking how space-time had managed to stay so flat when any deviation from balance would crumple or warp it. Guth, his head full of expanding universe equations, dimly realized that this outward whoosh back in microscopic time could explain it.

In the morning he got up and pedaled through the crisp peninsula air to his office in record time, flipped open his notebook, and on a fresh page wrote:

SPECTACULAR REALIZATION: *this kind of supercooling can explain why the universe today is so incredibly flat—and therefore resolve the fine-tuning paradox pointed out by Bob Dicke in his Einstein Day lectures.*

Then he drew a double box around it and started calculating again.

This was another instance in which mental pictures were not always helpful. Mathematically, omega gravitated rapidly to a perfectly balanced 1.0 during the accelerated expansion. According to Guth and most other cosmologists, the best way to think about it is to picture the universe as a balloon with galaxies dotted on its surface. The conceit of the Sandages and the Gunns was that we saw with our telescopes a sizable fraction of the balloon—remember the "balloon" can be concave as well as convex or any other shape—but that was wrong. During inflation, however, the balloon universe had been blown up to gargantuan proportions by the wild false vacuum energy. What we saw, Guth concluded, was really an infinitesimal patch of the universe, a chip that had been orders of magnitude smaller than an atom when the GUT symmetry broke or first tried to break. It *should* look flat because any curved surface looks flat if you examine a small enough piece of it close up. Football players need not take the curvature of the earth into account in figuring out a field goal kick or a downfield bomb. The universe looked the way it did, Guth concluded, because it was so big, blown out in some sense like a bald tire by the false vacuum bomb.

Of course, the Sandages and the Gunns didn't see the universe as flat. They had just collectively spent decades of telescope time establishing that the observable universe was apparently concave, with only 10 percent of the mass density needed to make it flat. On one hand, a factor of ten was so miraculously close to unity, as Peebles and Dicke had pointed out, implying a negligible difference during the big bang, that it wasn't worth quibbling about to the theorists. On the other hand, Guth's theory predicted that omega should be exactly 1.0 today, which made that missing factor of ten a glaring anomaly. Increasingly, from 1979 on, theorists were to see the task of the observers as closing the omega gap, finding the mass-energy they had missed before, and reconciling the universe with theory. Observers were to see the theorists as increasingly divorced from reality.

Excitedly Guth called up Tye and told him the news about this funny expansion. Tye was skeptical. He didn't want to delay their paper long enough to include a discussion of it. Guth asked if Tye minded if he went ahead and pursued the subject on his own.

Guth knew he was taking a risk putting out his and Tye's

supercooling paper without describing its effect on the expansion rate of the universe. Any experienced cosmologist who saw it could do the same calculation that he had just performed. Indeed, he and Tye knew of at least one other group that was doing the same work.

Guth next went to see Coleman, who seemed interested in his results. That was encouraging. Two days later he blew a calculation and became convinced that the whole thing was a mistake. It took him two weeks to recover.

By now he was devoting all his time to "inflation," as he dubbed his discovery. He doesn't remember when he coined the term. "Inflation was being talked about in the economic sense, and certainly what the universe did was inflate, so it seemed like a natural word." Guth's confidence and cosmological savvy were increasing.

Early in January, one day over lunch at the SLAC cafeteria, some physicists began discussing another cosmological conundrum. It was what astronomers had called the horizon problem: How had the universe, as exemplified by the evenness of the cosmic background radiation, become so uniform?

Guth got his friends to explain it to him, and it made him even more excited. By now he recognized that these paradoxes involving the scale of the universe might be resolvable by inflation theory. He went home to his study and convinced himself that horizons were indeed a problem. And he realized that inflation was indeed the answer.

The reason for that and other big bang paradoxes, Guth realized, was that in extrapolating the size of the universe back in time, conventional big bang theory assumed that it had always expanded at a steadily slowing rate. Inflation, however, would give the expansion a boost, making it possible for the universe to have sprung from a much smaller seed. If Guth's theory was right, it meant that the visible universe—what astronomers up to then had considered *the universe*—was descended from a much smaller speck of the original fireball than astronomers had imagined.

And from that, Guth saw, vast cosmological and astronomical consequences would flow. The original speck would be so small that any uneven spots would have already disappeared. It would already be at the same temperature, for example, and any local bumps or wiggles would be ironed out by the enormous expansion factor, like wrinkles on a fat person, so the resulting space-time would appear as featureless and even as a tennis court. Monopoles would be diluted to about one per observable universe. In short, almost no matter how the universe had begun, inflation would ensure that it turned out, in Guth's word, "dull."

Inflation was a heady triumph for first principles. In science the obvious questions were the hardest to see, to ask—and to answer. In a short month of theorizing on the outskirts of particle physics, Guth had vanquished in principle three questions that cosmologists had barely gathered the courage and insight to ask.

When Tye came back from China he found that Guth had turned into a "born-again cosmologist," charging around SLAC like a kid with a new toy. For some time Susan Guth had noticed that her husband was getting more and more excited as he and Tye plunged into their project, but she didn't know why. It was not until much later that she realized that there had been a significant change in December 1979. Alan Guth was not one to rush home, or into the next room, and say something terribly wonderful had happened. He never told his wife he had overturned the universe.

By the end of January, however, he was ready to tell the world.

14

THE FREE LUNCH

Particle physics and cosmology had been moving closer and closer ever since Gamow had tried to use nuclear reactions to explain the big bang. Astrophysicists had learned to explicate the cosmos in terms of successively smaller and stranger objects. Atoms, microwave photons, quarks and neutrinos all had their moment on center stage as the theorists worked their way back in time closer and closer to the beginning. The scientists who shuffled into a SLAC seminar room at the end of January in 1980 had little idea that they were about to witness the consummation of this curious slow marriage of ideas—the very large and very small—and that the unknown postdoc grinning wolfishly and nervously at the front of the room was about to reinvent the universe, not out of subatomic particles or forces, but out of the vacuum in which they rattled around. In short, out of nothing, itself.

Guth was nervous on his coming-out day and yet confident. He had spent the last few days reading astronomy books. He had the natural cockiness of the kid who knows he has the right answer, who has aced the hardest problem on the test. In this case, the problem of existence itself.

The scenario that Guth described began shortly after the Planck time. The universe, having begun in a fireball of perfect energy or complete chaos—it didn't matter—is now expanding and cooling. Gravity has already become a distinct force, but the strong, weak, and electromagnetic forces are still bound in their grand unity. Somewhere in the writhing chaos that could be primordial space-time the temperature slides below 10^{27} degrees K, at which point the grand unified symmetry should break. Instead the Higgs field gets stuck, and a little speck of space about the size of a proton supercools. Symmetry lingers like an ominous silence.

Within that tiny zone is the equivalent of about 20 pounds of false vacuum, the latent Higgs energy. It pushes out with superexplosive force and begins doubling like a mad amoeba from some fifties science fiction film. Every 10^{-34} second the bubble grows twice as big, with eight times as much energy in it. About the time that the original proton-size bubble has swollen to the size of a grapefruit—barely an instant later—someplace within the inflating bubble the symmetry finally breaks, the Higgs field freezes, and small pockets of "true vacuum" appear, like bubbles in boiling water. Phase transitions, once started, spread like chicken pox. The bubbles of a true vacuum sweep outward through the inflated universe and coalesce. Inside them, as the symmetry breaks and the false vacuum decays to the real one, the Higgs energy condenses into real matter and radiation. It is this matter and energy that will one day become hydrogen and helium and galaxies and flowers and people. "Inflation," Guth trumpeted, "offers what is apparently the first plausible scientific explanation for the creation of essentially all the matter and energy in the universe."

Where does all the Higgs energy come from to keep filling the universe as it inflates? This, as Guth explained it, was one of the marvelous features of general relativity. In any cosmological model there are two components of energy. One is the mass-energy associated with the matter and radiation in the universe; the other is the energy of the gravitational field by which all this stuff is attracted to each other. Mathematically, the gravitational field represents negative energy because you have to spend regular energy—like a rocket burning fuel to get off the earth—to overcome it.

"In many cosmological models," explained Guth, "this negative gravitational energy precisely cancels the positive energy that created the matter. In other words, the total energy of the universe is zero." As the universe inflates, positive energy pours into it in the form of the false vacuum Higgs energy. By the end of inflation, when it condenses into matter, the mass of the Higgs field, and thus the universe, has grown fantastically, by a hundred orders of magnitude from the original 20 pounds. But at the same time, as everything gets farther and farther apart the gravitational energy also gets more negative. So the universe remains balanced. The bottom line still reads nothing."

He paused, gathering himself and grinning triumphantly. "It's often said that you can't get something for nothing, but the universe may be the ultimate free lunch."

When inflation is ended the universe is around 10^{-32} second old. Normal sedate expansion takes over. What will someday be the observable universe is now a uniform expanding fireball. The Einsteinian space-time geometry of this fireball has been ironed flat by the vast inflationary process; in this early early universe omega is exactly 1.0. Now the entire "standard" history of the universe so painfully worked out by people like Schramm and Turner and Peebles could unfold.

Guth, it seemed, had succeeded in making a universe out of almost nothing more than first principles. Except that it was no ordinary universe—it was, in a way, guaranteed to be the *most* ordinary universe, ironed flat, and homogeneous.

Neither he nor Susan relished the prospect of going back to Cornell for the last year of his postdoc. Guth decided to go on the lecture circuit, partly to popularize inflation and partly to see if he could scare up a permanent position. Seminars are comparable to auditions for job-hunting scientists, a chance to strut their stuff before the home crowd. Guth started filling out job applications.

The memory of his boldness perks Guth up. "I did decide that what I was doing was important to the people who would be interested and that I should be gutsy about it," Guth explained, leaning back and putting his lug soles on the coffee table. "And I had never really been gutsy before. Once it got going I felt perfectly free to call up places and tell them I'd be going from here to there—would they be interested in hearing me speak?"

Meanwhile he kept working to fill in the details of inflation, and he discovered a problem. The runaway inflation did not end as gracefully as he had ideally hoped, a bubble of regular space-time should pop into being amid the false vacuum, like a bubble in a boiling pot, and sedately expand into galaxies and dust and stars and human beings. Unfortunately, when he did that part of the calculation, the bubbles that theory formed were too small, way too small, to evolve into any semblance of today's universe. Guth had to find a way to make them merge. He needed to combine something like 10^{80} separate bubbles to make a universe. Would they meet and merge in the boiling pot of the big bang?

The answer lay in a hairy branch of mathematics called percolation theory, and it was too hairy even for the mathematically gifted Guth. So while he was on tour he sought out mathematicians who might be able to tell him if the theory he was advertising was fatally flawed. Meanwhile he hedged. He was sure the theory would work out somehow; it was too neat not to.

In the spring, one barnstorming stint took him clear around the country. The number and names of all the institutions are vague to him now. It started at the University of Minnesota, then rolled on to Fermilab, Harvard, Princeton, Cornell, the University of Pennsylvania, the University of Maryland, and some place in New York City—he can't remember if it was Columbia or Rockefeller University.

The initial reaction to Guth's idea was surprisingly good. He had the advantage that Coleman had become a propagandist for him. "I think that was a big help," Guth said, "otherwise people might not have listened very much. By and large the particle physics community was sort of instantly enamored by this idea. It took longer, I think, for the astrophysicists to be willing to absorb it."

Finally, at Cornell, he found a mathematician, Harry Kesten, who showed him how to solve the percolation problem. It didn't work. The bubbles banging together would destroy the nice uniformity that inflation had created, leaving a nightmarish, lopsided universe. So it seemed he was stuck, with the universe endlessly inflating.

Luckily, in the meantime the universities of Minnesota and Pennsylvania as well as a few other schools, including Harvard, came through with job offers. Gutsiness had become a habit with Guth. He called up a friend at MIT and suggested that his old alma mater take him back. "I had the MIT offer before it became known that the model didn't work," Guth said.

By then, however, Guth was hooked on the power of inflation to create the known universe and to resolve so many conundrums. Although inflation went from being the solution to the problems of the universe to being a prescription for their solution, Guth was hopeful that some version of it would still work out, somehow. In August he finally sent his paper to the *Physical Review*. In it he described his bubble predicament and appealed for a solution.

He took up residence at MIT, but he kept going out to stump for inflation. For the next few years Guth continued to accept nearly every invitation that came his way to speak. "I stopped keeping count of them after a hundred and fifty," he says.

I first came across him at an evening workshop at the December 1980 edition of the Texas Symposium, in Baltimore. The evening workshops are definitely the second string. The talks are short, only fifteen minutes. The rooms are crowded, and the physicists, hurrying back from dinner and wine, are overheated. Somebody is always dozing and somebody else is wondering if he wouldn't rather be wandering up the street to Baltimore's infamous block of bars and strip joints.

Guth had been scheduled into a small room, but a large crowd had flooded in. "It was very exciting," he recalled. He was still selling hard, with the earnest mien of a high-school debator. He seemed like a daring, inventive figure, with more confidence maybe than he deserved, poking farther back into the raging singular past than cosmologists had gone with any confidence. He described his inflating bubbles and then admitted that he had no "graceful exit" from this state of affairs. Nobody knew quite what to make of it.

Guth had already claimed some early converts to his theory. One of the first was Schramm, who by 1978 was convinced that a flat universe was the only aesthetic solution to Einstein's equations. Inflation, of course, predicted, demanded, a flat universe, so he liked it. In the fall of 1979 Schramm had visited Cornell and talked to Tye. Schramm's encouraging words were passed on to Guth out in California. He knew who Schramm was (although not that he had been a college classmate) from an article in *Physics Today*.

Nevertheless, in 1980 and especially at the Texas Symposium, Guth walked in the shadow of other discoveries. His theory was half-baked and unfinished, and he was too busy advertising it to fix it.

Other pressures crowded his life. He and Susan bought a condominium in Brookline, a luxurious close-in suburb of Boston famous for its school system and favored by MIT and Harvard types, and found themselves in a protracted legal snarl regarding Brookline's restrictive laws on condo conversions, which resulted in two trials.

On top of that, Guth found out he was sick. During his first year at MIT, Guth's bowels had begun to act up. The condition persisted, and eventually his doctors found a tumor in his colon. It was a slow-growing tumor that apparently had been there for years. Guth said he never had a chance to freak out about it. "I learned about it over a period of time," he explains with a shrug. He adds, "There was no thought of not taking it out." In the summer of 1981 Guth was operated on and his colon was removed. As usual, Guth was lucky. There was no sign of any spread. He never had chemotherapy or radiation treatment. Guth was nonchalant about the whole thing. "The condo ordeal took almost as much time as my colon," he says.

So, as for Hawking, daily life for Guth became a little more difficult than for the average person. The disease hasn't slowed him down, he says. "In retrospect," he admitted slightly facetiously, "its main effect on my life was to prevent me from inventing new inflation."

New inflation, the answer to Guth's problems with inflation, came in the mail one day in December of 1981 from a Moscow theorist named Andrei Linde. At the time Guth and his old Columbia collaborator Erick Weinberg were working on a paper about the percolation problem, explaining why the bubbles that came out of inflation were too small to make a whole universe. When a preprint from Linde arrived Guth scanned its abstract, noticed that it mentioned a single-bubble universe—the very thing they were in the midst of disproving—and muttered to himself "what nonsense."

He almost put it away, but he read on. Then he got excited as he realized that Linde had solved the inflation problem.

The key to Linde's solution was in the way the Higgs field froze when symmetry broke. In the original theory used by Guth, Coleman, and the others, the phase transition, when it finally happened, was almost instantaneous: The Higgs field froze in a snap. Linde proposed using a different version of the Higgs field in which the freezing occurred more slowly and gently. Beginning when the universe passed into its super-cooled state, the Higgs energy would start to decay slowly, like a ball rolling down a long shallow hill. All the time that it was decaying, the universe would still be in a false vacuum state, full of energy, and would continue to inflate. In effect, the universe would slowly congeal instead of freezing.

Inflation would stop of its own accord when the falling energy level finally hit zero. At that time a wave of true zero-energy vacuum, unsymmetrical space-time, would spread out at the speed of light. By then, however, inflation would have blown a bubble more than big enough to build the universe.

It was a brilliant stroke, and Guth saw at once that it worked. Indeed, thought Guth, Linde probably would have discovered inflation on his own if Guth hadn't formulated it when he did. In fact, Linde had discovered inflation on his own but had discarded it, not realizing that it would solve the flatness and horizon problems, and discouraged by the percolation problem. When he heard about Guth, he was thus doubly motivated to construct a new version of the theory.

Meanwhile a pair of physicists at the University of Pennsylvania, Paul Steinhardt and Andreas Albrecht, had been exploring the same ideas. Their preprint crossed Guth's desk not long after Linde's. Steinhardt had met Guth a year earlier, as a junior fellow at Harvard. He thought inflation was too nice an idea to die, and so had put his energies toward salvaging the theory.

That was fine with Guth, but as it turned out, not so fine with Stephen Hawking. An intense but quiet brouhaha developed about who had invented "new inflation," with Hawking on one side and most of the American cosmologists on the other.

Hawking, who was close to the Moscow school, had visited Linde in Moscow in 1981 while Linde was working on new inflation. In the fall of that year, Hawking made a trip to the United States to receive a medal from the Franklin Institute in Philadelphia. While in town he had given a talk at Drexel University. Afterward he claimed that he had mentioned Linde's ideas in this lecture and suggested that Steinhardt, who was present, could have gotten the notion for his own paper there. Hawking was afraid he had spilled his friend Linde's beans.

Steinhardt was devastated when he heard about the charge. He was only a junior professor at Pennsylvania; Hawking was the Lucasian Professor of Mathematics at Cambridge, the same post Newton and Dirac had held. He didn't remember Hawking's discussing Linde's theory. Neither did other American physicists who heard Hawking speak during that trip. But Hawking, as always, was stubborn.

Steinhardt was upset again when he found out that I knew about the new inflation controversy. Publicizing it, he protested, would only hurt his career more. When Hawking repeated the insinuation, however, in his best-selling book *A Brief History of Time*, Steinhardt was energized. He hunted around and dug up a videotape of Hawking's Philadelphia lecture, which proved that he did not mention Linde's new inflation, and sent a copy to Hawking. Hawking announced that the offending paragraph would be stricken from future editions of the book.

Turner, who idolized Hawking and was a friend and collaborator of Steinhardt's as well, had tried fruitlessly to convince Hawking he was wrong about Steinhardt. He found himself more and more disillusioned and alienated by Hawking's intransigence. Finally, at a meeting in Santa Barbara in 1988, Hawking came up to Turner and asked, "Are you ever going to speak to me again?" Turner admitted he was still miffed. Hawking offered to write a letter to *Physics Today*. It appeared in the February 1990 issue. Hawking claimed he had never meant to imply that Steinhardt had plagiarized Linde. "I am quite sure the work of Steinhardt and Albrecht was independent of Linde's," he concluded. "I am very sorry if some people have gotten the wrong impression from what I wrote." Steinhardt was not completely satisfied, but the case was closed. It was an episode in which the legendary Hawking stubbornness showed its dark side.

As it happened, the Hawking-Steinhardt dispute had initially flared as most of these same scientists were racing each other to grab the last great prize that inflation theory offered to cosmology: the answer to the question of where galaxies come from. According to people like Peebles, galaxies had grown gravitationally from small seeds or density perturbations, tiny ripples in the otherwise smooth and even distribution of matter and energy in the primordial fireball. But there was no theory that explained why these ripples should exist at all, or which predicted their relative sizes and numbers. Cosmologists just plugged whatever numbers they liked into their calculations.

It dawned nearly simultaneously on several people early in 1982 that inflation, combined with a notion not normally applied to the expansion of the universe, namely, the famous and mysterious uncertainty principle, could solve this riddle of origins. According to the uncertainty principle the Higgs field would not be strictly uniform over space but would be subject to random quantum fluctuations. When inflation ended and the Higgs field condensed into matter and radiation, these fluctuations would produce tiny ripples in the density of matter and energy, perhaps the very ripples in fact that galaxy theorists like Peebles and Zeldovich had always assumed were there. Eons later those ripples would grow into galaxies and clusters.

That, perhaps, was the penultimate Fermiland dream—to ascribe an entire galaxy, a majestic pinwheel of a hundred billion suns, the biggest individual objects in the cosmos, to a quantum fluctuation in a semi-metaphysical force field nobody had ever seen when the universe was a trillionth of a trillionth of a trillionth of a second old. The ultimate dream might be to describe the universe itself as some quantum effect.

Hawking was one of the first to try to calculate the actual lumpiness that would result in the universe from inflation. Meanwhile Steinhardt and Turner had teamed up to do the same thing. Later that spring Hawking circulated a preprint of a paper in which he claimed that inflation would indeed produce lumps of the right size and number to grow today's galaxies; they were just what the cosmological doctor had ordered. But the calculation was fiendishly difficult. Turner and Steinhardt got a different answer, that the lumps that grew out of inflation were far too small to grow into galaxies. Who was right?

As it happened, Hawking had organized a summer workshop on the early universe in Cambridge for the summer of 1982. It was the second in what was supposed to have been a series of three, sponsored by the Nuffield Foundation, but Hawking, excited by inflation, decided to splurge now and cancel the last meeting. This was a good time for a showdown. He proved himself this time to be as good a politician as a physicist. To get the good young Russians to attend, he invited a senior Soviet academician with clout, and suggested that he might like to bring some of his younger colleagues. Linde made it to Cambridge. So did his old friend Starobinsky, who, it turned out, had also done the inflation density perturbation calculation. Unfortunately Starobinsky had a severe stutter and nobody could understand what he was saying. Steinhardt came and wondered why Hawking was being so cool to him.

Inflation was sweeping cosmology like a prairie fire. Guth and a half dozen others had arrived in Cambridge determined to find out who, if anybody, really was right, and whether the entire content, size, and structure of the universe were the consequence of quantum fluctuations.

Guth likened it to being back in school. The workshop lasted three weeks. The scientists stayed in college dormitories, with no phones. Each day they met in Hawking's lair, the department of applied mathematics and theoretical physics, to hear lectures. The rest of the time was for working and talking. "There wasn't much in Cambridge to do besides work," says Guth. One Sunday, he remembers, a group of them went down to London for the day and had a great time sightseeing. They got back after eleven, when the gates to the college closed, and had to climb over the fence, just as undergraduates had in the days of yore.

At night the cosmologists divided into teams to redo the calculations of inflationary density perturbations, and thus to confirm or deny that they had the universe at last in their hands. A lot was at stake. If inflation was right, the density perturbations should be observable, someday when techniques were fine enough, as faintly hot and cool spots in the cosmic microwave background. To their team Turner and Steinhardt recruited Jim Bardeen, a crack theorist from the University of Washington and son of double Nobel Prize winner physicist John Bardeen. Bardeen thrived on competition and was convinced that Hawking and Starobinsky were wrong. He came banging on Turner's door at eleven one night exclaiming, "I can show you guys are right."

Nevertheless after three weeks of furious calculating they ended up changing their answer: the lumps produced by inflation were not too small to make galaxies, they were too big. Hawking got the same answer, and so did Guth, who had taken up the chase. Everybody agreed, and it was both good news and bad news for inflation. The density fluctuations induced by the Higgs field during inflation did indeed have the form and distribution considered most desirable by physical cosmologists like Peebles and Zeldovich. But they were 10,000 times too dense. If the universe had grown according to this scenario, all the galaxies would be today nothing but black holes. Starobinsky said this was what he had been trying to say all along.

The first part of the answer was considered a great victory for inflation and a sign that cosmology was on the right path. The second part, the overbearing size of the density fluctuations, was the fault of the underlying particle physics, namely the SU(5) model invented by Glashow and Georgi. A supremely confident cosmologist could have taken advantage of the occasion to declare that SU(5) was not the correct grand unified theory because it did not make galaxies—such a declaration would have been an audacious application of the poor man's particle accelerator.

Luckily, Guth explained, that was not necessary. Within a year SU(5), the simplest GUT, would be on its way out of favor with physicists. The reason was its prediction that protons would decay in 10^{30} years. By 1983 underground experiments to look for this decay had not detected any sign of it. That meant the proton lifetime had to be at least

10^{32} years, which was beyond the limits of Glashow and Georgi's theory.

The simplest grand unified theory was dead, but more ambitious and complicated models were waiting in the wings. The most popular models all had a feature that was called supersymmetry; they posited yet another new brotherhood of particles, this time between the elementary building blocks of matter, such as quarks, electrons, and neutrinos, and the force-carrying particles known as bosons, such as photons, gluons, and W-bosons. This theory had two interesting consequences for cosmology. One, it meant there were whole new families of elementary particles to be discovered, with exotic names like photinos, gluinos, gravitinos, winos, squarks, and sneutrinos; if these existed in the universe and in the big bang they might create observable effects. Second, the "breaking" of supersymmetry added yet another occasion in the very early history of the universe in which the Higgs field underwent a phase transition and "froze," another chance, in other words, for inflation.

As the particle physicist Heinz Pagels said, inflation was the only game in town. Once it had been invented, the occasions for inflation seemed to be everywhere. You could hardly keep it out of the universe. There were all those symmetry-breaking episodes, for example. Every one had a Higgs field and a phase transition associated with it. Was there inflation when the electroweak broke, or the quarks confined themselves, or gravity split off as a separate force? Linde invented what he called chaotic inflation; random quantum fluctuations in the fields that pervaded the universe, he argued, could raise the vacuum energy density somewhere and cause that part of the universe to spurt madly into inflation. Turner and Silk invented models in which the universe inflated twice. All these theories had problems. Turner began to call inflation "a paradigm in search of a theory."

What these theories had in common was that the universe did inflate. In some sense, as far as sensible physics was concerned, inflation *was* the beginning. What happened before was almost immaterial to what came later. "One of the great virtues of the theory," said Guth, "is its forgetfulness. The universe erases its history."

New inflation was cosmological magic. On one side of inflation—before—was this unimaginable stuff—Gamow's *ylem* was as good a word as any—hot, disconnected points of supersymmetric mass-energy. On the other was a rationalized smooth expanding space-time—homogeneous, free of monopoles, barely granulated with the seeds of future galaxies. What would become the 10 billion light-years visible from an arm of the Milky Way galaxy in the twentieth century was a tiny patch, about 4 inches across on this bubble.

"When I started to learn about cosmology," Guth later told *Omni*, "I was amazed more than anything else by the very size of the universe.

Ours is only one of a hundred billion suns in the Milky Way galaxy, which in itself is probably only one in a hundred billion galaxies in the visible universe. In the context of the inflationary model, even the visible universe is only an infinitesimal fraction of the entire universe that probably exists. In light of all that, it's rather hard to understand why what goes on in our tiny planet in this tiny corner of the universe could ever be considered important."

Guth got his picture in *Newsweek* magazine as the author of the "big bubble." He remained himself, with his obliging smile and giddy laugh, always accompanied like a careful Boy Scout by a bulging briefcase stocked with everything from a flashlight to his meticulous notebooks. He was still a nice guy ready to talk physics with that MIT fire-hose enthusiasm at the drop of a hat and keep going until the least physics-literate person nodded some sort of understanding. He got tenure and then a full professorship with a big new office with a couch. The demands on his time increased and pushed creative work completely out of his day into the Brookline evenings. When I remarked on his schedule, he protested, groaning, "I did do something creative once—five years ago."

In December 1982, when the tenth Texas Symposium met in Austin, Guth received the signal honor a cosmologist can receive in Fermiland. He gave the leadoff lecture, right after the mayor of Austin declared that it was Relativity Week, to the entire ballroom of astrophysicists, astronomers, general relativists, and physicists.

Afterward, over lunch in the towering lobby of the Austin Hyatt-Regency, I asked Peebles what he thought about inflation. Peebles took a breath and looked out from the balcony over a sea of clattering astrophysicists. "The inflationary universe allows the imagination to roam free," he started out. "Perhaps you start matter roaring around, collapsing and expanding. Somewhere you enter an inflationary phase and it all happens. From this chaos can pop up this and other universes. It's a tremendous conceptual breakthrough.

"Let there be chaos," his voice ranged deeper. "Let there be Guths to straighten it out, to let the crooked places be made straight, and the rough places planed."

He turned and smiled. "How's that?"

Astronomers were used to thinking of the universe that they could see—a ball of galaxies and stars from quasars to quasars, maybe 10 billion light-years across—as *the* universe. But if Guth was right it was all only a speck in a bigger, wilder ensemble. Sometimes he called it a meta-universe. Our own inflated bubble was probably trillions, not billions, of light-years across. And beyond that could be other bubbles, other enormously inflated island universes undergoing their own sagas of expansion and evolution, separated from us by domain walls or false vacuum seas, in their own separate space-times.

When I contemplated this great vista, I despaired. Inflation seemed like a pyrrhic victory, almost a tease. It made the universe more savage and unknowable than I had ever dreamed. *There's a great universe next door—let's go.* What could we know about the greater universe? The same process that guaranteed that our own cozy little universe was smooth and safe also guaranteed that everything interesting and fundamental was swept hopelessly, permanently out of reach. The price you paid for magic was ignorance. Strings and walls and zones where there were five forces or sixteen dimensions could be coming right at us at the speed of light from trillions of light-years away. Maybe it was the price of our own existence that we live in an early era before all the wildness came over the horizon.

The origins of the universe were erased. You could, as Guth admitted, never know what came before inflation. The universe was re-born during that split second, its past erased. Were there relics of the original creation that survived the false vacuum explosion?

One of the few observable predictions that inflation made was that the mass density of the observable universe—that is to say, that part of our own inflated bubble that we could ever hope to see—should be indistinguishably close to the critical density. The fate of our particular inflated bubble depended on whether omega was a whisker of a decimal bigger or smaller than absolute 1.0, whether the universe was one part in 10^{60} slightly convex or concave—an impossible task to ask even of a Sandage. The curve of the cosmos on our scale was less than a rumor on the wind. We would have to wait a trillion trillion years to find out what it was saying.

Asking for the fate of the whole universe was an insensible question—maybe it had been all along.

Guth agreed with this gloomy assessment, but it didn't seem to bother him. "In principle it's possible to divine the fate of our own bubble," he allowed judiciously. And because that's the only place we can live, it would be useful information, but that was a far cry from knowing the fate of the universe—all the other bubbles sideslipping away from us in the stream of time and matrix of false vacuum or uninflated primordial crud. In fact Guth seemed to find this kind of cheerful.

"The universe is not homogeneous on incredibly large scales," he told me, grinning wolfishly. The more he told me, the more uncertain the universe seemed; there was kind of an uncertainty principle operating in cosmology itself. "The universe could be open somewhere and closed somewhere. You could have a local big crunch. It would look like a black hole.

"It's not clear that the entire region will undergo one fate together," he went on. "Quantum fluctuations could produce perturbations on larger scales than the observable universe. The observable universe could collapse without taking the whole region of inflation with it." The resultant black holes would cascade upward into bigger and bigger black holes.

Meanwhile, there was chaos on the home front to contend with, and it led Guth to the "basement universe." Inspired by the notion of the free lunch universe, Guth had become an aficionado of theories of nothing. Because it only took an investment of 25 pounds of false vacuum to make a universe, Guth wondered what would happen if you could squeeze a handful of matter down to such incredible densities and temperatures that it reached the temperature of grand unification again and hopped up into the false vacuum?

Would your sample explode outward with exponential force obliterating the local universe? Would the gravitational field of that dense conglomeration wrap a black hole around itself and hide from the universe?

The answer, Guth and another MIT professor, Ed Fahri, found to their amusement, was both. If you could compress 25 pounds of matter into 10^{-24} centimeters, making a mass roughly 10^{75} times the density of water, and certain other conditions were met, a bubble of false vacuum, or what Guth called a "child universe," would be formed. From the outside it would look like a black hole. From the inside it would look like an inflating universe. "It's a safe experiment," explained Guth. The child universe would slip sort of sideways to our own and expand into its own space-time.

Its black hole umbilical cord would disappear within 10^{-34} seconds, leaving the child universe completely disconnected from ours. Although from the outside it would look like an evaporating black dot, from the inside it would appear to be a full-fledged universe that could expand to billions of light-years and grow galaxies.

Besides the energy required to do the initial compression of matter to false vacuum densities, Guth and Fahri found only one other obstacle to the basement universe. To save it from becoming just a run-of-the-mill black hole, this dense little pinprick had to be threaded on an already existing singularity—something the average basement presumably lacked.

But there was a loophole, Guth explained. Space, according to quantum theory, would be dense with such singularities in the form of virtual black holes. Their writhing collapse and constant evaporation would unhinge continuous space and time at the smallest scales. And perhaps supplying seed stock for new universes. So quantum fluctuations could be breeding countless universes under our feet all the time. There could be an infinite chain of universes begetting universes, a prospect Linde called "eternal inflation." As for the energy requirements of starting a new universe, just because it was beyond our own present technology did not mean it would be beyond the technology of the future or of an advanced race that might exist out there somewhere in the metauniverse of universes.

"In fact," Guth concluded, "our own universe might have been started in somebody's basement." And he chuckled.

III

THE SHADOW UNIVERSE

But I would not support madness, so I joined the ranks of those who have corrected thy work.

—F. Dostoyevsky

> *Heaven*
> *Heaven is a place*
> *A place where nothing*
> *Nothing ever happens*
> *—Talking Heads*

15

THE HUBBLE WARS

Allan Sandage's stature in astronomy was almost beyond measurement. By the mid-seventies he had been Mr. Cosmology—"SuperHubble" some people called him—for more than twenty years. Generations of astronomers had grown up who had never known Hubble and who had never known a time when Sandage was not huffing and puffing his way to the 200-inch, churning out papers on the Hubble constant, the deceleration parameter, and a thousand other subjects. He was a force of nature, a natural landmark, like one of those giant elliptical galaxies hogging the center of a galaxy cluster.

In 1975 Jesse Greenstein told writer Timothy Ferris, in *The Red Limit,*

> Much of what Sandage is doing, he has been doing for so
> long that for anybody else just to catch up would take years,
> and nobody would consider retracing his work anyway, be-
> cause he is viewed as a man of absolute integrity. I don't
> know any other field in the world where you can say that
> about somebody, that he has absolute integrity.

A year later, the unthinkable happened. Not from the pages of an obscure journal, but from the podium of the triennial meeting of the International

Astronomical Union (IAU), the largest assembly of astronomers in the world, in Grenoble, France, it was charged that Sandage had bungled the measurement of the Hubble constant through a long chain of confusion, questionable assumptions, observational errors, circular reasoning, and perhaps wishful thinking. The real Hubble constant was not 50, it was 100, which meant that the universe, instead of being 20 billion years old, could not be more than 10 billion. Sandage reeled under this attack and the others that followed. His morale plummeted like a stone in an angry sea. It opened a decade in which Sandage found himself being pummeled by the practitioners of the field he had helped found, the very handful of people in the world who understood what he did.

His challenger was a Frenchman turned Texan named Gérard de Vaucouleurs, proud of his Gallic heritage, but given to wearing cowboy boots and a Stetson. Slight and dapper with slicked-back hair and a stiff manner, de Vaucouleurs was no stranger to Sandage; they had been bumping heads around the cosmological circuit for twenty years. Their first encounter had set the tone for everything that followed.

De Vaucouleurs, a dedicated amateur astronomer, had grown up in Paris and studied physics and astronomy at the Sorbonne, getting his Ph.D. in 1949 after a long interruption due to the war and a year and a half in the French army. He had been following Hubble's work in faraway California with great envy. De Vaucouleurs lusted after the galaxies, but there were few extragalactic astronomers in Europe and no telescopes that could compete with those in America. Nevertheless he persisted, and began measuring the diameters and brightnesses of galaxies, using small telescopes and even amateur photographs. De Vaucouleurs also did planet work and, among other things, he measured the rotation rate of Mars to an accuracy surpassed only when spacecraft began visiting the planet. After his degree, seeking opportunity he went to England, and from there he got a fellowship to do planetary research at the Mount Stromlo Observatory of the Australian National University.

Once there, de Vaucouleurs wasted no time in turning the telescope out across the southern skies in search of galaxies. His perusal of Hubble's work convinced him that Hubble had made technical errors and had been less than rigorous in his approach to classifying galaxies, so de Vaucouleurs set out to reobserve the 1300 galaxies on the Shapley-Ames list and produce his own catalog. The southern sky had received little detailed attention; most of it had never been surveyed to any great depth, nor had many of its galaxies been cataloged or classified. De Vaucouleurs started to try to classify the galaxies according to what he knew of Hubble's work, but when he couldn't make sense out of what he was observing, he sent an SOS to Sandage.

At the time, Sandage was trying to put together the *Hubble Atlas,* a photographic compendium of galaxies and their types, a sort of extra-galactic bestiary that had been one of Hubble's fondest projects. Sandage felt a personal responsibility, by then, to carry out all the unfinished Mount Wilson projects. With duty, almost unnoticed, came pride.

Classifying galaxies was halfway between an art and a science. Since Hubble had first published his classification scheme in 1936, he had made some technical corrections, and reclassified a lot of galaxies. The data in support and explanation of all this were on Sandage's desk where he was assembling it into the *Hubble Atlas* as a memorial to his mentor. The galaxies had been what Hubble most cared about; now they were Sandage's.

When de Vaucouleurs arrived in the United States in 1955, he went straight to Santa Barbara Street, where Sandage greeted him like a long-lost brother. "It's so lonely working on galaxies," Sandage said moaning and proceeded to give him access to Hubble's material, including the plates that had been taken between 1936 and 1950. "It was like going into the Natural History Museum and seeing the collection nobody else had ever seen, if you were a taxonomist," said Sandage, "and somebody had organized the whole collection that the classic taxonomist, Charles Lyell, had laid out."

In the presence of the mother source, de Vaucouleurs quickly comprehended the galaxy classification scheme. He went back south, classified the southern galaxies, and wrote a paper.

In the meantime, the German encyclopedia *Handbuch der Physik* had asked Sandage to write an article about galaxy classifications. Too busy, he suggested to the editors that they get de Vaucouleurs. The match was successful, and when in 1958 de Vaucouleurs wrote to Sandage in search of illustrations for the project, Sandage sent him proofs for the atlas with the idea that de Vaucouleurs would use them somehow as a basis for drawings or diagrams. De Vaucouleurs misunderstood and printed them in the *Handbuch.* Hubble and Sandage had been scooped with their own material by this little outsider.

"That made me quite angry," Sandage said, "because he didn't ask permission to do that. And that was when the enmity somehow began. That was two years before the *Hubble Atlas* appeared. I was very hurt by his actions."

Although conciliatory letters eventually flowed, a sense of rivalry had been established, a line drawn in the cosmic turf. A style and theme had been set: Sandage, the arrogant keeper of the tradition and hoarder of data, the beleaguered heir, and de Vaucouleurs, the outsider.

In the intervening years, de Vaucouleurs had carved out his own niche as a cosmologist and galaxy expert, winding up at the University

of Texas (Texas had 84-inch and 102-inch telescopes at its own observatories in the Davis Mountains in dark and wild Texas), leader of a small
group of graduate students and postdocs who called him "GDV." He
had written a few papers of his own on the Hubble constant. He and his
wife, who was also an astronomer, had compiled an influential and highly
authoritative catalog of the brightest-appearing galaxies, and he was still
the world's expert on the rotation of Mars.

Most important, de Vaucouleurs took exception to much of the
Mount Wilson-Palomar orthodoxy about the universe. He disagreed with
Sandage, and by extension Hubble, about the homogeneity of the universe
on large scales. The traditional view was that galaxies and clusters were
fairly evenly scattered through space. De Vaucouleurs was convinced that
the universe was, on the contrary, ugly and lopsided. The clusters were
not the smooth balls of galaxies envisioned by people like Sandage and
Peebles, he argued, but were ungainly, unsymmetrical conglomerations.
There were vast areas of the universe rich in galaxies and others that were
poor. In 1953 de Vaucouleurs had pointed out that most of the bright
galaxies in the sky—from local group neighbors to beyond the Virgo
cluster—were confined to a narrow belt perpendicular to the Milky Way.
He suggested that this belt was actually a disk seen edge-on and all those
thousands of galaxies, including the Milky Way, were part of one vast
agglomeration tens of millions of light-years across that he called the
Local Supercluster. In a way he was anticipating the top-down pancake
theories that Zeldovich would propose in the seventies, but de Vaucouleurs was not inclined to theory. The astronomical establishment had
responded to the Local Supercluster with resounding silence, if not out-
right rejection.

De Vaucouleurs had argued that the gravity of these immense
alleged superclusters would distort the cosmic expansion and complicate
the measurement of the Hubble constant. But Sandage and Tammann
had found no trace of so-called supercluster structure in their own investigations of the cosmos. Using an intricate ladder of distance indicators
beginning with Cepheid variable stars and ending with giant spiral galaxies
as so-called standard candles, they had concluded that the galaxies were
receding uniformly in every direction, at a rate consistent with a Hubble
constant of about 50 kilometers per second per million parsecs of distance.
When the pair began to discuss their preliminary results, de Vaucouleurs
started popping up and saying that they were being fooled, that the gravity
of the Virgo cluster was slowing things down in this region of space. If
Sandage and Tammann were to measure the expansion rate for galaxies
far outside the local supercluster, he insisted, the "Hubble value" would
go up.

Sandage remembered one of those occasions had been a retirement party for the astronomer Nick Mayall in 1972. "It was held in a resort south of Tucson," Sandage recalled. "I had to give one of the principal lectures and I outlined where we were in seven steps. It was the outline of six or seven years' work. Gérard was there, and I'm sure the distance-scale problem was what he had hoped to keep finally to do himself. When he saw the results as they came out of the 200-inch and this amount of work, I just knew when we were walking out that this was a challenge that was put up to him, and he had to answer. And he began to answer it four years later."

In the summer of 1976, as de Vaucouleurs told me later, he was on a sabbatical from Texas at the Edinburgh Observatory, in Scotland. The British and European astronomers were gearing up for a massive photographic survey of the relatively unexplored southern sky. One of the local astronomers, Malcolm Smith, was asked to review Sandage and Tammann's Hubble constant work, particularly as it applied to the Southern Hemisphere. De Vaucouleurs, who claimed that he had not really dug into Sandage and Tammann's papers, offered to help. He set about to reread the entire set of six, called "Steps to the Hubble Constant."

The deeper he read them and tried to reconstruct the arguments by which Sandage and Tammann had constructed the cosmic distance scale, he said, the less he had liked it. The papers left him confused and irritated. The arguments seemed shaky and circular. "In paper two it says 'See paper four'; in paper four it says 'See paper two,'" he complained years later in a thick French accent. Moreover, Sandage's assumptions of uniformity in nature and in various aspects of the galaxies were too generous. The data were unconvincing to him; he felt that Sandage and Tammann had found the answer that they wanted to find, rather than the answer that nature might have provided. "I didn't like what I was reading," he recalled. "I was wandering around Edinburgh in a state of shock, asking other people to read the papers."

De Vaucouleurs went back through "Steps to the Hubble Constant" to make a systematic critique and claimed that he found twelve "blunders," each of which served to increase the imputed size and age of the universe.

"I decided that I felt a responsibility to correct something that was already going into textbooks," said de Vaucouleurs gravely. As it happened the perfect occasion on which to "ring the alarm" was coming up. In September the International Astronomical Union, consisting of thousands of astronomers, was to gather in Grenoble, France, for its triennial meeting. De Vaucouleurs had been invited to give one of the prestigious "invited discourses" before the entire assembly. In Grenoble he announced to the audience, which did not include Sandage, that Sandage and Tammann were wrong, and he offered his own estimate of the

Hubble constant, based on the brightness of globular clusters in the galaxies of the Virgo, Fornax, and Hydra clusters—a method Sandage himself had used once in 1968. He concluded that the universe was only about half as big and old as Sandage had said.

Next, de Vaucouleurs wrote up his talk and tried to get it published in *Nature*. Sandage, he says, advised him not to, saying that only six people in the world could follow the arguments, anyway. In return de Vaucouleurs told Sandage that if he had made as many mistakes as Sandage had, he would retire.

In his own mind, de Vaucouleurs was a midget smiting a giant, and it took courage to challenge Sandage and the orthodoxy. A lot of astronomers were afraid. Senior colleagues of his at Texas, he says, urged him to be cool, worried about the repercussions for the department. De Vaucouleurs was steadfast. "We cannot take the wrong course or take the wrong conclusions of nature to keep friends. I'm not happy to say this," he said. The first response was ostracism. The referees for *Nature* quibbled and stalled. At one point the editors ran out of referees and they had to ask de Vaucouleurs to suggest some European astronomers to read the paper.

"What survived is solid. If you say the king is naked, you better be right," de Vaucouleurs stated with satisfaction. "Young astronomers said they were glad they didn't have to say it. They were so damned scared. That stuck in my memory."

De Vaucouleurs went on to invent an entirely new cosmic distance system in competition with Sandage's. He added a whole raft of novel distance indicators, like the diameters of so-called ring galaxies, the brightest star clusters, and something called the luminosity index, to supplement and supplant the classic tried-and-true methods that went back to Hubble and Shapley. Some of them he used to determine distances, others he used to check his distance indicators. The whole scheme made Sandage and Tammann's look simple. His diagram of it was designed to look like the Eiffel Tower: several ladders of measurements rising to a pinnacle of truth, intricately cross braced by cross-checks, *a posteriori* comparisons, calibrations, linearity checks. In one respect his method was simple: Trusting no intuition or principle above any other, de Vaucouleurs used every way he could think of—no matter how half-cocked—to measure distances and then averaged all the results. Wily old nature wouldn't be able to hide from such a wide assault. He called it "spreading the risks."

The inevitable result of this new campaign, undertaken mostly by reanalyzing the vast amounts of data already published, was that de Vaucouleurs's Hubble constant rose even higher—to around 100. Which meant his version of the universe was half as big and half as old as Sandage and Tammann's.

De Vaucouleurs went on the road in the late seventies, selling his new Hubble constant of 100 like a car dealer, enthusiastically comparing the virtues of his deluxe model with the lamentable stripped-down competition. Over here, one primary distance indicator; over here, five!

Few astronomers would admit comprehending de Vaucouleurs's system, yet his criticisms scored. The younger generation, whom Sandage had not befriended, was not there for him when he came under attack. One young astrophysicist told me after hearing de Vaucouleurs's treatment, "I had read Sandage and Tammann and believed every word. Gérard de Vaucouleurs's talk made me think for the first time that the Hubble constant might be 100 instead of 50."

In 1980 de Vaucouleurs did a star turn at the prestigious Texas Symposium, unanswered by Sandage or Tammann. Afterward he agreed to talk, over beer and without a tape recorder, about the Hubble constant controversy. He was neatly dressed, with gray straight hair and a tired face. His manner was grave, as if Sandage's purported transgressions were a blot on science too serious to be conveyed beyond the priesthood, or as if his own honor forbade the airing of criticism of his colleagues. He had the air of a man who has reluctantly slain his idols and doesn't want to speak about it—the science should speak for itself.

This Old World restraint tugged against the feelings of a man who felt he had been wounded and never quite received the respect he felt he deserved from the big guys. De Vaucouleurs admits to no other interests, with the exception of stamp collecting, besides science. In his lecture and our conversation he kept returning to his rather stern precepts on what was and was not acceptably scientific methodology for a distance scale—the need, for example, to avoid extrapolations. Finally the urge to vindicate himself took precedence. De Vaucouleurs proceeded to draw diagrams all over my notebook.

Sandage and Tammann, he began, were too trusting of nature. They assumed that the universe was uniform. De Vaucouleurs, on the other hand, assumed it might not be. They relied on only a few select distance indicators, like Cepheids or the size of gas clouds, in each zone of remoteness, which meant that any errors made there would ripple through the whole Hubble calculation.

Sandage, he protested, was directed by intuition too much. If Sandage saw an Indian standing in the distance, he would assume that the Indian was as tall as he was. De Vaucouleurs, on the other hand, would consider that it could be a pygmy. To de Vaucouleurs everything was a trick. He didn't really believe that consistency existed anywhere in nature. In his universe, nothing fit, nothing added up to anything greater than it was. In fact the sum of the universe seemed to be less than its parts. There was no overarching view, no theory. Cosmology was addled. He was anarchy, and he wasn't a gentleman of the Mount Wilson school.

De Vaucouleurs was particularly incensed that Sandage and Tammann had made no correction for the fact that they were looking at all these galaxies from inside a dusty galaxy, the Milky Way. As everyone knew, our own interstellar dust would dim the brightness of outside galaxies and the Cepheid stars in them, making them appear farther away than they were. Although the effect was small, accounting for only 10 percent of the factor-of-two discrepancy between the two Hubble constants, it seemed to symbolize to de Vaucouleurs the lengths to which Sandage and Tammann would go to bend the data—they were arrogant enough to assume that God would leave a little dust-free hole through which they could peek out of the galaxy.

Most damning and indefensible to de Vaucouleurs, however, was the way that Sandage and Tammann had used M101, the giant spiral galaxy that was the linchpin of their distance scale. Sandage and Tammann, he argued, had extrapolated the properties of dimmer and smaller galaxies—such as the brightness of their brightest stars and of their hydrogen gas clouds—up to the giant spiral class of M101. Then, assuming that the extrapolation was valid, they had used it to "prove" that M101 and the other spirals were standard candles, and thus derived their distances, and the Hubble constant. That was another intuitive assumption of uniformity in nature. To de Vaucouleurs, an extrapolation was just a guess. "After they made the choice of the extrapolation, it became a fact," he declared.

The Sandage-Tammann distance scale, he joked, should come with a manufacturer's disclaimer.

Sandage seethed at de Vaucouleurs's attack, but it was not his style to engage in a return critique, at least not in the literature. He disparaged de Vaucouleurs's distance indicators as "gimmicks." "He's an obscurist," laughed Sandage bitterly. "He statistics the data to death. That's all right, everyone has his style.

"You look at what his indicators are and they all come down to taking the mean value of something that has a great spread. Well, that's fine, but the trouble with anything that has a great spread, like the linear diameters of spirals, is that you get a mean distance, but you have enormous individual errors." These errors, explained Sandage, always skew the distance scale toward underestimating the true luminosities and distances of faraway objects. The reason is that the galaxies at the dim end of the distribution drop out of sight at great distance; the farther we look, the fewer of the below-average galaxies we see, and the more our attention is concentrated on the high end of the distribution. That means that when somebody like de Vaucouleurs compares "average" galaxies, average galaxies nearby are unwittingly being compared to above-average galaxies faraway. So the distant galaxies are systematically underrated, and their distances are underestimated. Astronomers call this the Malmquist bias, and it has always been a subtle plague on cosmology.

De Vaucouleurs, Sandage concluded, had Malmquist bias. "He doesn't understand that. He simply doesn't understand that. And it doesn't do any good to write rebuttals to those papers. The fact that we don't rebut something so silly is maybe bad, but you'd spend your entire time trying to educate them."

There was more than pride at stake in this feud. Sandage and de Vaucouleurs weren't taking any prisoners. Only a sissy, Tammann once remarked to me, would consider making the obvious compromise at 75. The bigger reason was cosmology, or what de Vaucouleurs called "The holy book of Friedmann cosmology."

In an expanding universe that does not decelerate, the age of the universe is just the inverse of the Hubble constant. Sandage and Tammann's Hubble constant, before they corrected for any slowing down due to gravity, gave an age of 20 billion years; de Vaucouleurs's age was about 10 billion. According to the stellar evolution experts (including Sandage), however, the oldest stars in globular clusters were now 17 billion years old—as of 1980. And the galaxy, according to Fowler and his compatriots, was between 8 and 25 billion years old; the universe had to be a billion or so older.

If Sandage was right, the ages of the stars, galaxy, and universe all roughly agreed, and the simple Friedmann equation worked. If de Vaucouleurs was right, on the other hand, the stars would apparently be older than the universe. Wasn't that a powerful argument for Sandage's answer? De Vaucouleurs thought this was religious nonsense, the religion in this case being the Friedmann expanding universe theory.

"I never worry about the Bible," he said. "If I had announced an age of ten billion years three hundred years ago, I'd have gone to the stake. Facts cannot contradict each other. Theories are the source of contradiction. The ages of globular clusters," he added, "are more uncertain than the books lead you to believe."

In fact, Sandage's whole Hubble constant project, de Vaucouleurs wrote in *Sky & Telescope,* had been undertaken when the new older ages of globular clusters came out, which had made the old accepted value of 100 uncomfortable for the Friedmann theologians. Sandage denied it vehemently. When I read to him from de Vaucouleurs's article he leaned into the microphone of my tape recorder and said, "Now, you know

that's wrong, Gérard, the whole point is to make an independent estimate."

The difference between them was this: Sandage saw his job as
testing an entire world model, passing the definitive judgment on a whole
paradigm of creation—the Friedmann expanding universe. De Vaucouleurs was just measuring a number. What was at stake was not some dusty
parameter or gee whiz factor, but the sort of abstract principle for which
wars are fought, religions formed, personalities twisted, dinner invitations
given, and friendships canceled. Is God on our side here or not? Is there
a prospect of a comprehensible answer or not?

Sandage, a religious man, had been entrusted when Hubble died
in 1953 with a cosmos as simple and sturdy as a Shaker table. When he
looked around all the pieces fit. The galaxies looked like they had been
manufactured by Henry Ford. The Cepheids blipped away consistently
and obediently to the grand design. The simplest model worked. There
was a creation event. The universe was not dense enough to ever go home
again. People were stardust. Nature was out there, in the Hubble diagram,
all those galaxies and quasars riding the universal expansion, infinitely
into the night. If Sandage stood for anything it was the strength and
clarity of that vision, of its power to witness the glory of creation and
whatever was behind it.

Alas, de Vaucouleurs was not the only critic, for another, stronger challenge to the Sandagean cosmos arose. It had begun in 1972 with a pair
of graduate students, Brent Tully and Richard Fisher, at the University
of Maryland. Brainstorming together not long before their graduations,
Tully and Fisher thought of a new way to measure the true luminosities
of spiral galaxies, and thus their distances—a new standard yardstick, in
a way. They speculated that there should be a correlation between the
rotation rate of a galaxy and its luminosity. The reasoning was simple
and Newtonian: The faster a galaxy is whirling around, the more mass
must be inside it to supply the requisite centripetal force; at the same
time, more mass meant more stars and thus more luminosity.

They knew radio astronomy and thus knew a convenient and
sensitive way to measure rotations. Hydrogen atoms in interstellar space
(and elsewhere) are prone to an instability that causes their nuclei to flip
over once in a great while and emit a blip of radio energy at the precise
wavelength of 21 centimeters, a frequency of about 14 megahertz. As a
result the clouds of space, and especially the spiral galaxies, are enlivened
by the roar of this so-called 21-centimeter radiation. If a galaxy were
rotating, Tully and Fisher realized, the wavelength of this roar as seen in
a radio telescope would be spread out, rather than concentrated at precisely 21 centimeters. Radiation from the gas on the side of the galaxy

turning away from us would have its wavelength lengthened by the Doppler effect—the same effect that redshifted the light of receding galaxies. On the other side, turning toward us, the wavelength would be shortened. The faster a galaxy rotated, the greater the overall spread in wavelength would be seen at the telescope. Following the jargon of nineteenth-century optical spectroscopists, radio astronomers call the 21-centimeter signal a "line." A wider line would indicate faster rotation and presumably a more massive galaxy.

If Tully and Fisher were right, more luminous galaxies should have wider 21-centimeter lines. The way to test it was to look at a bunch of galaxies all at the same distance, in the Virgo cluster, say, and see if the brighter ones were rotating faster. After graduation, Fisher left for the National Radio Astronomy Observatory in Green Bank, West Virginia, to do just that.

Tully, who would grow into yet another of the iconoclasts of cosmology, wanted to learn French, so he accepted a postdoctoral position at the University of Marseilles. Tall, with dirty blond hair, a sloucher in jeans and velour shirts, casual to the point of being lackadaisical, Tully seems all rounded corners and an aw-shucks manner. He traveled the long way around the world to France, taking a year to get there. By the time he did, in 1973, Fisher's "wonderful" radio data were waiting for him. When he compared the line widths to the blue magnitudes of the galaxies there was a correlation—their conjecture was confirmed. They had their new standard candle. A simple radio measurement could tell you the intrinsic luminosity and thus the distance of a galaxy.

Tully and Fisher wrote up their findings. Among their preliminary conclusions were distances shorter than Sandage's and thus a Hubble constant more like de Vaucouleurs's. It took them two years to get their paper published. It went back to the anonymous referees twice. "Some people say there was skullduggery," admits Tully, good-naturedly shrugging it off.

By the time Tully and Fisher's paper appeared, Sandage and Tammann had already published their own paper in response to it, and its implicit challenge to their Hubble constant. Sandage didn't trust the Tully-Fisher relation, and sometimes he referred to it as "Fishy-Tuller." He suspected that Tully and Fisher were also infected with the Malmquist bias, which caused them to underestimate galaxy distances with larger error the farther away they were. Moreover, the Tully-Fisher relation had to be calibrated with spiral galaxies of known distances and luminosity, and these were few and subject to the same old uncertainties that he and de Vaucouleurs and Madore had been fighting about for years.

In their new paper Sandage and Tammann junked almost their entire work of the previous twenty years and started over again to measure

the Hubble constant. They reanalyzed Fisher and Tully's data and rede-
rived their own version of the relation, plugging in their own calibrating
galaxies, the nearby spirals M81 and NGC 2403, whose distances they
had torturously divined from Cepheids and other indicators. They used
the Tully-Fisher relation on the spirals in Virgo to get the distance of
that cluster and to calibrate the brightness of its biggest galaxies, M87
and NGC 4472, the nearest of the giant elliptical galaxies. Sandage had
spent years showing that the giant ellipticals were standard candles and
using them in the Hubble diagram to search for traces of the deceleration
of the universe. Knowing at last the absolute luminosity of that candle,
he could use it to measure the Hubble constant to great distances.

The new answer was the same as the old one, about 50, they
announced in a seventh "Steps to the Hubble Constant"—the series that
would not die. That paper, said University of Washington astronomer
Paul Hodge writing in the *Annual Review of Astronomy and Astrophysics*
in 1981, "should be regarded as the major result of their program."
Sandage and Tammann's paper appeared a year before Tully and Fisher's.
Neither the answer nor its timing enhanced their credibility in the growing
community of Hubble-constant measurers.

Sandage's distrust of Tully-Fisher did have a sound basis. First
of all, the visible light astronomers measured in those galaxies came pri-
marily from bright blue massive stars, but most of the mass was in dim
little red stars. Second of all, even the blue unrepresentative stars were
hard to see. The reason was dust, which collects in the disklike planes of
spiral galaxies. The best galaxies for measuring rotation line widths were
spirals seen edge-on, where the gas came directly at you and away from
you. But these were the very galaxies in which you had to look through
the most dust. In photographs many of them had a wide black band right
across the brightest part of the galaxy.

"It's been stated," said Sandage, "that Gustav and I use only one
datum, that we put all our confidence in a single method. That's not true
in essence, but nothing's black and white. Is physics better understood?
Perhaps, perhaps not. Cepheids differ from galaxy to galaxy, but platitudes
are easy to come by, and how do you go about proving them? You prove
them by finding the distance by another method, if you can find one—
maybe you can find one. No scientific question is open forever; something
comes down the pike. So it's very good that these questions are raised,
but the people who raise these questions, if they have any guts at all,
ought to go out and try to solve them.

"Otherwise they're like guys who pour Molotov cocktails. They
ought to get down there and if the question is important for them they
ought to try to find the answer."

*   *   *

At the time that Tully-Fisher came out, the next important combatant in this story, Marc Aaronson, was a graduate student at Harvard. Born and raised in Los Angeles, Aaronson had the laid-back West Coast look. He wore blue jeans and sandals almost exclusively; he was short with dark curly hair and a serious manner belied by twinkling eyes and something mischievous in his manner. He had gotten interested in science by being dragged to science fiction movies by his parents. He was a movie buff; a subscription to *Variety* followed him from place to place. Aaronson had gone to Caltech, where he was a classmate of the future gonzo cosmologist Turner, and Gunn was his senior adviser. By the time he had moved on to Harvard he was a skilled and crafty observer, at home and at ease, on large telescopes.

In Cambridge he had become interested in infrared astronomy, a field burgeoning with the onslaught of new technology in the seventies. Infrared, sometimes called heat radiation, has longer wavelengths than visible radiation and occupies the part of the electromagnetic spectrum between it and the microwave-radio bands. Almost everything in the universe, from warm bodies to cool stars, radiates in the infrared, but infrared photons are too weak to record on photographic emulsions, and their frequencies are too high to capture with radio techniques. Infrared sensors were first developed by the military, for spotting enemy soldiers in the dark and for studying missile nose cones during the fiery heat of reentry. The gradual release of some of this technology into the public sector as the Vietnam War waned sparked a great interest in infrared astronomy.

Aaronson was doing his Ph.D. thesis on the infrared properties of galaxies. Due to a bureaucratic wrangle, his thesis adviser left town when Aaronson still had a year to go and was not replaced, and Aaronson became a kind of orphan. He was befriended by a young assistant professor named John Huchra, an optical astronomer recently arrived from Caltech. "Astronomy was the most important thing in Marc's life," said Huchra. They agreed to teach each other their specialties. Huchra matched Aaronson in confidence and observing zeal. He had a dry, cynical wit that verged on the brusque and a lack of fear of his elders.

As part of his thesis, Aaronson had built an infrared photometer of indium antimonide, a technology that had just dribbled out from the military. The detector had two special features: It was portable, meaning he could haul it from telescope to telescope easily, and it had a wide field of view, which made it ideal for studying galaxies. So when he left Harvard in 1977 he took it with him to the University of Arizona's Steward Observatory.

While at Caltech, Huchra had witnessed an intense colloquium clash between Sandage and Tully, which piqued his interest, both sociologically and astronomically, in the Hubble constant. Mindful of the infrared photometer he and Aaronson looked into the controversial Tully-Fisher relation. They decided, with the help of an Australian infrared astronomer named Jeremy Mould, who was across the street from Steward at the Kitt Peak headquarters, that a lot of the problems with the method could be eliminated if the galaxy luminosities were measured in the infrared instead of blue. For one thing, dust—both in our own and in the originating galaxy—was practically transparent to infrared, so there would be less guesswork about absorption. Another indisputable fact was that the vast majority of stars in a galaxy are low mass and very red. It didn't make much sense to measure the blue brightness of a fundamentally red object; infrared would measure the dog, not the tail.

Granted four nights on a 36-inch telescope at Kitt Peak, Aaronson, Mould, and Huchra took the little infrared detector and measured the infrared flux from twenty or so spiral galaxies in the Virgo and Ursa Major clusters, the same clusters that Tully and Fisher had used. When Aaronson left the telescope he went straight to the Kitt Peak library to check his results against the radio data. Sure enough, the Tully-Fisher relation stood out clearly. The hydrogen line widths marched along right in step with the infrared magnitudes. The more massive and luminous a galaxy was, the faster it rotated. The galaxy line width was a key to its *infrared* luminosity. Armed with that knowledge, Aaronson concluded that they could measure the distances to galaxies in a way that was free of astronomical mumbo jumbo—it was just pure physics. There was room in the world for one more group to pursue the Hubble constant. On the same day his first daughter was born.

The preliminary results of that calculation were disquieting, however. Aaronson, Mould, and Huchra's own measurements of Virgo and Ursa Major gave a Hubble constant of around 65. As usual the actual number depended on the calibration of a few close-in galaxies, like M31, for which they used Sandage's values. Huchra was alarmed. He had presumed all along that they would get the "right" answer, namely, the Sandage answer. A Hubble constant of 65 or 70 wasn't going to please anybody. Scared that they had made a mistake, they checked and rechecked their calculations until they were convinced it was safe to publish.

Meanwhile they went back to the telescope to firm up their new technique and extend it farther out into the universe. This meant going beyond the galaxies that Tully and Fisher had observed, but luckily similar radio observations of galaxies in distant clusters were being made by Gregory Bothun and his collaborators at the University of Washington. When Aaronson's group compared their infrared measurements to Bothun's data, they got their second big surprise. The spirals in these more distant clusters—Pegasus, Cancer, and Perseus—appeared closer than expected. The deep space Hubble constant was around 90.

How could that be? This, of course, was exactly the pattern of expansion that de Vaucouleurs had been predicting all along: The gravity of the mighty Virgo cluster and its neighbors slowed down the flight of the galaxies in their vicinity and made the local Hubble constant less than the "cosmic" Hubble constant. The two Hubble constants could be reconciled if, once the velocities of galaxies due to the expansion of the universe were subtracted, the Milky Way and the rest of the local group galaxies were actually falling into Virgo at a rate of about 350 kilometers per second.

The effect of the Virgo supercluster, according to their data, was real. "De Vaucouleurs had been claiming it for twenty years," said Aaronson. "Sandage and Tammann had been denying it for twenty years."

The group laid plans for a massive survey of the local supercluster. The subsequent observing campaigns were notable for their good luck, variety of telescopes used, and bureaucratic favor. "I would go to the mountain, and it would magically clear up," marveled Aaronson. "After our first paper, we requested seven nights on the 84-inch at Kitt Peak, and they gave us seventeen."

In 1979, while the trio was still sorting through their results, Huchra gave a seminar at the Harvard-Smithsonian Center for Astrophysics. A reporter from the Harvard *Crimson* was present and wrote it up. His story, in turn, was picked up by the wire services. Newspaper headlines bloomed with the news that a mistake in the dimensions of the universe had been corrected and that the universe was now only 10 billion years old. Walter Cronkite reported it on the "CBS Evening News."

Distressed to read of this work first in the newspapers instead of in the journals, and with no mention of his own contribution, de Vaucouleurs sent a sarcastic open letter around the astronomical establishment congratulating Aaronson, Mould, and Huchra on seeing the truth, and chiding them for publishing this important result in the *Crimson*. He then reminded them of all the occasions on which he had anticipated their findings.

Sandage had been becoming more isolated over the years. Now his gloom reached new depths. Tammann had never seen him so down. He disconnected his phone. His door stayed closed. His comments outside of scientific publications dropped off to virtually nothing.

"There are only a handful of people in this game, and they all hate each other," Huchra told me. "Nobody likes to be proved wrong." He allowed as how he ran into Sandage on the mountain sometimes. "We usually talk about the weather."

Every time a new Aaronson, Mould, and Huchra paper came out, Sandage got more depressed. It seemed to him that his rivals were more interested in proving him wrong than in getting the right answer. Didn't they know how hard it had been? Sandage's attitude had always been that he would rather do what he could and be wrong than retire to a bar and moan. He just didn't understand all the triumphant carping. He was always willing to look at new data, though it filled him with trepidation— *what now?*

The more miserable he grew in private, the stiffer and more defensive he became in public. Asked for a comment about the Hubble controversy by *Science,* he declined. "The answer will come when responsible people go to the telescope," he said.

He and Tammann suspected that the discrepancy between the Hubble constants again had to do somehow with the dread, elusive Malmquist bias. A Hubble constant that grew with distance was the signature of that effect, but he couldn't prove it. His and Tammann's periodic attempts to prove that the Aaronson group's data had a hidden bias provoked bitter howls from the other camp that their data were being misused.

Sandage went so far as to mention the challenge to their distance scale in the observatories' *Annual Report,* but then reassured his readers that he knew what was going on. The real answer, he was confident, was still 50.

Aaronson, Mould, and Huchra liked to portray themselves as bringing fresh, uncommitted air to a stale feud; they joked in talks about two old astronomers, each claiming to have measured the Hubble constant with an error of only 15 percent, and yet differing by a factor of two. Their error bars didn't even overlap. Who could take either claim seriously? Nobody paid much attention to the baroque schemes of de Vaucouleurs, and Sandage and Tammann's results depended too much on shaky notions. Tully-Fisher, on the other hand, was based on physics; it was clean.

That appealed to the younger cosmologists. Many of them were physicists who thought it odd that astronomers couldn't agree on the size of the universe. It was as if there were two camps on the mass of the electron. The roots of the controversy and of observational cosmology itself were obscure to them; Hubble and Mount Wilson were vague legends. These guys didn't know constellations. All they knew was that they had this damned floating variable in their equations. Ironically, they tended to favor a low value of the Hubble constant because it made the various age estimates of the universe agree with each other. "The young guys have the best technique," said Turner, "but everyone believes Sandage is right."

Once, at a European conference on cosmology and particle physics, one physicist put the question to Sandage directly. Could he compare and criticize the famous factor of two? Sandage replied stiffly, "You are asking us to criticize our own program, and this we cannot do. Simply put we cannot see where we have made any mistake. We believe we know where the other group went wrong."

At about the same time that he gave that answer, Sandage and Tammann junked their distance scale again and started over from scratch with yet another standard candle. This time it was supernova explosions. Good choice—a supernova can briefly outshine most quasars. Their brilliance and ability to be seen halfway across the universe had led Zwicky to propose these stellar obituary flares as the ultimate distance indicators.

But were they standard? Astronomers recognized two types of supernovae. Type II, the more violent and erratic, were the runaway explosions of very massive stars. Type I supernovae were a hair less violent, but they made up for it apparently by being very uniform. They had a different genesis. A Type I began in a binary star system with a white dwarf, the shrunken dense cinder of an ordinary star, circling another star. Over time, the second star will run out of gas and swell up into a red giant. Material from the expanded star will be attracted to the white dwarf and begin piling up on it, increasing that dense little cinder's mass. Nothing much will happen in such a situation until one day the white dwarf's growing mass exceeds 1.4 times the mass of the sun—a mass known as the Chandrasekhar limit—at which point the temperature and pressure on the surface become intolerable, triggering a runaway thermonuclear explosion that quickly spreads throughout the star. Death again lights the galaxy.

Tammann had long tended an interest in supernovae as a diversion from the Hubble work. He had discovered that, like giant elliptical galaxies, they fell on a straight line on the Hubble diagram. The reason for this uniformity was the Chandrasekhar trigger, which ensured every one of these bombs went off under identical conditions, with the same mass. "They're all peas in a pod," Sandage cracked.

In 1982 he and Tammann published "Steps to the Hubble Constant" number eight. The Hubble constant came out a little lower, but in the same range. "The answer is 42," he chirruped happily, echoing a line from the Douglas Adams book *Hitchhiker's Guide to the Galaxy*. In that novel nobody knew what the question was, but the answer was the secret to the universe.

"I still don't know why we keep getting 50," he confessed ingenuously. "That's a great mystery. It's a mystery to us, too, but now we're so convinced it's 50 that it was a miracle that the first paper said it was 50. You can read into that all you want as to prejudice and inviolacy of an initial position. I cannot explain it, the psychology is—first of all, astronomy is almost an impossible field."

In no time at all, the famous factor of two reasserted itself. De Vaucouleurs attacked again, claiming that supernovae were one-fourth the brightness that Sandage and Tammann had claimed, which would make them only half as far away.

The problem with supernovae was that they happened only infrequently, when and where nature wanted them to. Most of them were in so-called anonymous galaxies, far away. To calibrate the supernova standard candle, Sandage and Tammann had to wait for chance to strike in a galaxy whose distance was known independently by some impeccable means. As it happened, supernovae of the Type I variety had graced two galaxies close enough to the local group for the brightest stars in those galaxies to be identified. From the magnitudes of those stars, Sandage and Tammann had been able to read the distances of the galaxies and thus calibrate their supernovae.

However, it was in defending those distances, on which the supernova distance scale rested, that Sandage was drawn into the third stage of Hubble warfare.

These distances rested on the notion that red giant stars in any galaxy never got brighter than an absolute magnitude of -8.[1] That had been one of the main conclusions of Sandage and Tammann's 1974 paper in which they had derived the pivotal distance of M101. Now that was mired in controversy, and in the Hubble constant game, disputes about individual stars in obscure galaxies could ripple across the cosmos and the egos of those consumed by measuring it.

The idea that there was a maximum luminosity for stars had caught the fancy of Roberta Humphreys, an expert on stellar evolution at the University of Minnesota. She and her husband Kris Davidson had studied the properties of supermassive stars. She thought it reasonable and likely that physics would specify an upper limit to red supergiants: Beyond some mass—say fifty suns—they might become unstable and oscillate, belching their outer layers into space before they had a chance to settle into steady burning. She called it the "guillotine effect."

Attractive as the idea sounded, however, shamefully little was known about the stars even in the nearest galaxies. The Andromeda spiral M31 had never been surveyed for red supergiants, and the only inventory in its neighbor M33 had been done by Humason in the fifties. Hoping

---

1. Recall that the lower the magnitude, the brighter the star. A star at magnitude -8 is 2.5 times more luminous than a star at magnitude -7.

to investigate the reality of the guillotine effect, Humphreys asked Sandage if she could borrow the Humason data. Instead she wound up collaborating with him on a reanalysis of the stellar content of the galaxy.

Even in semiwithdrawal and isolation, Sandage still had the power and charisma to sweep new people into his projects. And at the pace at which he worked he was always in need of new partners. Humphreys is petite and delicate looking and talks fast in a voice so soft you can hardly hear it sitting next to her. Sandage is volatile and overpowering, nudging and gesturing and grabbing elbows. It might seem an unequal partnership, but beneath her softness Humphreys exudes a steely confidence in her astrophysics.

Initially she and Sandage confirmed that the brightest stars in M33 were at about magnitude -8. Everything was fine. Then they came to blows over the distance to M33. At the time, the accepted distance was slightly greater than 2 million light-years, but it was a shaky finding indeed. "The distance was based on Hubble's statement that the Cepheids in M33 were a tenth of a magnitude brighter than in M31," complained Humphreys. "Every time M31 changed, M33 did too."

In 1983 a schoolteacher named George Carlson walked into Santa Barbara Street looking for something constructive to do to while away the summer hours. Sandage put him to work blinking plates and reducing Cepheid data for M33. The result was a new distance of about 3.5 million light-years—and a correspondingly brighter absolute magnitude for M33's brightest stars. The red giants were no longer pegged at -8. Sandage changed his mind; there was no absolute limit to their luminosity. Now Sandage said the maximal luminosities were statistical—the bigger the galaxy, the brighter its brightest stars.

"That shook everybody up quite a bit," Humphreys said. She got on his case like a teacher chasing an aberrant schoolboy. She argued that he was ignoring the effects of dust in M33, which would redden and dim its Cepheids, making them look more distant. "He and I already knew that both the red and blue supergiants in M33 were reddened," she recited. "A few months before he had told me the Cepheids were reddened in M33. He and I had mapped the dust lanes." It was similar to the criticism that Madore had leveled against the Cepheid measurements of NGC 2403 back in 1969.

Sandage responded that the average reddening for the Cepheids and the other stars would be about the same, and that no correction had to be made for the dense dust lanes in which the Cepheids, but not necessarily the red giants, lay. Things got stormy. Sandage banged a bunch of papers down on the desk one day. "I've already done this. There's no need for you to reobserve these stars," he declared.

The next day he apologized. "Now you know my other side."

"Allan," Humphreys replied gently, "I always knew it was there."

Sandage and Humphreys each claim that they won the argument. The result of all this, Humphreys said, was that "M33 is back where it was and the calibration of the brightest red supergiants is back at -8." Sandage nonetheless continued to quote and use the distance he had obtained with Carlson, putting M33 and its calibrators brighter and farther away.

Meanwhile Humphreys had taken her quest outward to M101, a bigger and more luminous galaxy. M101 was the key to the Hubble constant work on both sides of the controversy. Sandage and Tammann had devoted an entire hotly contested paper to it. And, claimed Humphreys, six out of nine methods that Aaronson and his crew used to get the distance to the Virgo cluster depended on M101.

Humphreys had dreams of refounding that distance scale on the rock of stellar physics: The brightest stars. Would the guillotine operate to limit the brightness of stars in giant M101 to -8? If so, then the upcoming Space Telescope (the 2.4-meter telescope then scheduled to be put in orbit by NASA in 1986 from a shuttle) could be used to solve the Hubble controversy forever. It would be able to resolve the brightest individual stars in galaxies as faraway as Virgo.

Sandage and Tammann had searched M101 for red supergiants ten years before and not found any, reinforcing the conclusion that the galaxy was too far away for them to be visible. Humphreys teamed up with Stephen Strom, a young astronomer on the staff at Kitt Peak, to scour M101 for red supergiants. They took photographic plates and enlisted another piece of new technology that Sandage hated. They fed the plates through a machine known as PDS, for "plate density scanner," which measures the blackness of the image on the plate, point by point, and converts it into a bank of numbers, 5 million of them per plate. These numbers become fodder for the computer.

Gone, in this new system, was the patient astronomer staring through a microscope at a blinking pair of plates. Instead the computer could just compare plates electronically, as well as calibrate them, clean them of defects, remove the foreground stars from galaxy pictures or the galaxies from star fields. No intuition was needed or wanted here, everything was hard and cold, mathematically defined, untouched by human prejudice—surely a step forward the physicists would appreciate.

They announced that they had found red supergiant stars brighter than those Sandage and Tammann had failed to find earlier. That left them with a choice. If they accepted Sandage and Tammann's distance to M101, the supergiants would be half a magnitude brighter in that galaxy than in the others; the guillotine effect and the power of it as a distance indicator would be gone. Or they could adopt the shorter distance advocated by de Vaucouleurs and others, which would give the red supergiants the right magnitude.

Humphreys and Strom opted for physics over classic astronomy. Red supergiants, they declared, had the same luminosity cutoff in every galaxy. That had the effect of rippling and condensing the cosmos. It made M101 closer, which moved in the Virgo cluster. That, in turn, meant that the Hubble constant was really around 90, according to Humphreys. The universe was billowing in and out like an accordion.

"This was where my head-on collision with Allan began," she reported tartly. He tried, she says, to kill their paper, suggesting that they fight it out privately. Sandage protested that their sample was contaminated with foreground stars, of which there were so many that Strom and Humphreys had resorted to a way of eliminating them statistically.

Sandage went back to his own plates in a huff, and reanalyzed them looking for red supergiants. He didn't expect to find anything. "The Strom and Humphreys claim was a great motivation. I went back to discredit it, really," he explained. "I blinked the same plates and, lo and behold, in a big star association I find a very nice candidate, which I think they missed, by the way." He went on to find red supergiants that were half a magnitude brighter than the ones Humphreys and Strom had found. The memory of it took the steam out of him momentarily. "We were not careful enough in our first paper, I guess. My anger at Strom and Humphreys came back to haunt me at my own amateurishness."

Sandage, however, clung to his original distance for M101. The absolute magnitude of the brightest giants in that galaxy was not −8, he reported, it was −9. They were no longer a standard candle.

"How could he suddenly change it?" Humphreys asked incredulously. The general attitude of the community was that Sandage had changed his tune again in order to save the Hubble constant.

Sandage was now in the position of arguing something he'd said in the past (about the brightest stars) was wrong while everybody else said it was right. He took a sort of martyred satisfaction in that. He and Tammann were destroying their own position, in the name of science. "We cut our own throats on that one," he said.

"Everybody just wants to prove Gustav and I wrong," Sandage grumbled. He blamed the space telescope for the desire of astronomers to establish brightest stars as a reliable distance indicator; everybody was jockeying for position to use it. "So you're now getting down to the nitty-gritty details," he murmured, "which is what science is all about. And what are those details like? Where are the Cepheids in M81? Which are the red giants in M101? Which group has missed them? Which group has got the apparent magnitude right, and so forth?"

His voice got loud and steely. "I cannot explain it exactly, why

I was incensed. Except Strom and Humphreys didn't do their homework right. Well, they didn't do any homework. They relied on automatic methods without entering with the human mind, which is the best discriminator against flaws, plate defects, stars in galaxies." He claimed that 30 percent of their red supergiants were really foreground stars in our own galaxy.

"They got a closer distance. Then they went whole hog and jumped ahead four or five steps to the Hubble constant. Everyone jumps four or five steps ahead from a given observation and then they make the great leap: if if if, then."

He sounded tired. "But everyone is entitled once in their life to write a paper about the value of the Hubble constant."

Humphreys, feeling not in the least vanquished, felt that history and the Hubble constant were leaning away from Sandage. She didn't seem particularly mad at him. She saw Sandage as charismatic and complex, on the defensive after decades of unquestioned authority, hurt that people would even consider redoing his work, a prickly grandfather being eased out, tragically and not always so gently, by a younger, stronger generation.

"I imagine de Vaucouleurs must be feeling greatly vindicated," she said one night at dinner in Minneapolis with her husband Davidson. The paradox of a universe younger than its oldest stars did not strike her as relevant or insoluble. Sandage, after all, had dated the globular clusters himself. "If he had brought himself to admit that globular cluster ages were driving the whole thing . . . " she said, sighing.

"For a long time," said Davidson, "he was the most important astronomer in the world. He spoke from on high and issued edicts. 'We have done X.' With no explanation—just an edict."

He paused. "He didn't get q-nought, did he? And maybe not even h-nought."

Sandage grew an Old Testament bushy beard. His conversation, which periodically throughout his life had become sprinkled with biblical references, became so again. He failed to show up at conferences. Rumors went around that he had a religious conversion, that he was a born-again Baptist minister. Sandage encouraged the rumors and played on them. It suited the frantic flipping of masks that was his persona to let it go on, and besides, he was never around and too formidable to be asked directly about it, anyway.

He teased de Vaucouleurs that God had told him that the Hubble constant was 50. De Vaucouleurs repeated the story with a straight face.

16

THE Z-MACHINE

Some people say the Harvard College Observatory is haunted. The Harvard-Smithsonian Center for Astrophysics, of which it is now a part, is set among chestnuts and maples on a shallow hill a fifteen-minute walk west from Harvard Square past Tory mansions and the Radcliffe dormitories. The oldest part of the complex, including the original observatory, is a three-story U-shaped brick building on the crown of the hill. Butted against its back and running down the hillside behind it is a modern building of glass and stone with gray carpets. One wing of the old building houses an old library where colloquia are given every Thursday afternoon. The other side has a lobby hung with dusty astronomical photographs leading to a small planetarium; upstairs, in lonely, brassy splendor in an antique observing gallery, sits the historic Merz and Mahler 15-inch refractor, built in 1847. Its retirement survived a halfhearted and controversial attempt at modernization in the 1950s.

In a white dome the size of a small room on the observatory roof lives a 9-inch refractor made by the famed telescope makers Alvan Clark and Sons. This telescope gets used once a month on Thursday nights, when the public is let in, and anytime the junior staffers—college students and the local astronomy buffs who run tours and use the observatory for their amateur meetings—want to take a quick gander at something in the sky.

It's on dark nights like that, when the old creaking building is a labyrinth of shadows and drafts, and a countercultural air drifts over the hallways (and perhaps the acrid scent of marijuana as well), that someone is liable to wonder if they just saw a figure in nineteenth-century evening dress strolling out from the 15-inch or Was that a man in a trench coat over in the library stacks just a second ago?

One of the most popular ghosts, according to Stephen O'Meara, a *Sky & Telescope* staffer and observatory regular, was George Bond, son of the observatory's father, Henry Cranch Bond, and its second director, a lonely figure who died of consumption and pneumonia in 1865 after spending a winter in the unheated dome taking pictures of Orion. Bond was one of the fathers of celestial photography, but left no portraits or photographs behind of his own likeness. The aura of mystery around him was heightened by his rumored role in a suicide. According to the legend, Bond had been the object of the attentions of a gay classmate named Edward Bromfield Philips. When Bond rejected him, Philips killed himself. The disappointed lover left all his money, naturally, to Harvard.

Marc Davis appeared in these environs in 1974 with a fresh Ph.D. from Princeton and a mission from Jim Peebles. The cool, lanky one, Peebles taught that there was much more to the universe and for cosmologists to learn than just two numbers. There was, for example, the question of how the matter in the universe was really arranged and where its luminous denizens, the galaxies, came from. The mission of the new cosmologist was to enforce some sort of quantitative discipline on the patterns of the sky, to find some quantitative measurement to take the place of pictures and metaphors like "pancakes." At Harvard and a few other places in the late seventies a new style of observational cosmology was about to come of age.

Davis, imbued with Peebles's analytic zeal, was destined to launch the biggest and most successful of these projects, one that would not only become a paradigm of the new cosmology, but would also help usher Harvard astronomy into the late twentieth century. If ghosts were all Davis had to contend with, he would have had it easy.

Tall and ruddy with thinning curly brown hair, brown eyes that crinkled into slits when he smiled, and a smirky little moustache, Davis was another member of what might be called the automatic generation. He had never thought about being anything but a scientist. Growing up in Ohio, Davis credited his scientific leanings to a NASA exhibit in Cleveland, from which he brought home a pile of material, and reading books about relativity and the universe. "When I was in the sixth grade, a teacher gave me a high-school chemistry book," he recalled. To his chagrin he found he couldn't make anything explode, however. His interest gradually shifted to physics; during high school he spent a summer at an NSF-sponsored program at Ohio State working in a physics lab making vacuum tubes. He went to MIT and majored in physics, graduating in 1969, a year after Guth and Schramm, and a year before Huchra. After being intimidated initially by all the valedictorians around him, Davis did well enough to get into Princeton graduate school.

What had gotten Davis mixed up with Peebles was ambivalence. Davis was what Peebles called ambidextrous. One year he was a theorist, the next year he was an experimenter; he couldn't seem to make up his mind what to be. Technically, he was not even Peebles's student. He was getting his Ph.D. for work he'd done next door with Wilkinson, building an infrared detector and using it to search the sky for the heat from nascent galaxies.

But toward the end of his Princeton career he had begun to get the theoretical cosmology bug. He sat through Peebles's course twice, taking notes, and coming around to talk. Peebles's quest for quantitative measures of the universe struck a chord. Like Peebles, Davis had a relaxed but no-nonsense air about him and a distrust of qualitative statements. One day in the autumn of 1974 he dropped into Peebles's office. He was about to go off to Harvard as an assistant professor and he wanted, as it were, some homework to take with him.

At the time, Peebles was trying to figure out what to do with his correlation function, which measured how far galaxies were likely to be from one another. What was the physics behind the growth and correlation of lumps in the universe? He happened to have just read a paper about an obscure and complicated theory called the BBGKY hierarchy that involved the growth of perturbations and shock waves in gases such as the early universe was probably composed of. He suggested that Davis look into it. They decided to apply the theory to the early universe and see if they could reproduce the galaxy statistics. Davis took the work with him to Cambridge.

BBGKY stands for the names of five physicists: Born, Bogoliubov, Green, Kirkwood, and Yvone. Applying the theory was horrible work. The BBGKY hierarchy consists of an infinite number of equations. It was the worst math problem Peebles and Davis had ever been associated with. "We went through a frustrating period," said Peebles. "I spent many hours traveling to Boston, long nights of anxious analysis to try and figure out how to swing this."

With the help of a lot of suspiciously bold assumptions, Davis managed finally to struggle through to an answer that bore a statistical resemblance to Peebles's correlation function. Davis was not happy. They were stuck; the theory was never going to get any better as long as physicists were restricted to analyzing a two-dimensional universe with all its galaxies plastered together on the dome of the sky.

The basic idea of cosmic organization had not changed since Hubble's time: The universe was composed of clusters of galaxies, dominated by ellipticals, and "field" galaxies, mostly spirals, scattered about in between. The idea that there were even larger assemblages—de Vaucouleurs's superclusters—was tantamount to heresy. What physical cosmologists like Peebles and Davis needed to interpret the patterns of the sky was the third dimension, the depths of galaxies.

Fortunately, Hubble's law had left cosmologists with the perfect tool to do systematic cosmic geography. Because a galaxy's distance from earth was proportional to its redshift, if you found a number of galaxies in the sky, plotted their celestial coordinates, and measured their redshifts, you would have a three-dimensional scale model of the cosmos. Even if you did not know the exact value of the Hubble constant, you could still know the pattern that galaxies made in relation to one another. Redshifts were the key to distance and structure.

Despite all the interest in the expansion of the universe and the Hubble constant, however, the redshift key had turned very few locks. In the early seventies, nearly forty years after Hubble's breakthrough, the redshift velocities of only about 250 galaxies had been measured and published—the ones that had been compiled by Humason, Mayall, and Sandage in 1956. That was less than a drop in the bucket. Sandage had been working for years to get redshifts for a list of about 1300 of the brightest galaxies that had been compiled by Shapley and Ames. Nobody seemed to know when it would ever be published. That would still be only a drop.

The reason for so little work having been done in this area was that redshift knowledge came at too high a price in telescope time. To measure a redshift, you had to disperse the light of a galaxy into its component wavelengths and record its spectrum. Even on the 200-inch telescope, it could take hours or all night for enough photons to seep into a photographic emulsion to make a readable spectrum.

This enormous lack of data meant that astronomers had practically no idea, beyond prejudice, about how the matter of the universe was arranged. It left them free to argue, as de Vaucouleurs and Sandage had, about whether galaxies were ganged up in superclusters that would deflect the Hubble flow or were just lonely raisins in a uniform field.

An Estonian astronomer, Jaan Einasto, attempted to analyze the distribution of galaxies based on the few hundred redshifts already in the literature—mostly from de Vaucouleurs's catalog. He concluded that the large-scale organization of galaxies had a netlike or cellular structure:

interconnected strings of galaxies surrounding vast, empty regions of space. Einasto mentioned his results at the 1976 IAU meeting, the same one at which de Vaucouleurs blasted Sandage, but few people paid much attention to it.

Redshifts were about to become easier, however. In the seventies electronic light detectors that were faster and more sensitive than photographic emulsions, and had other advantages as well, made it possible to record spectra much more quickly. A few astronomers began to exploit this technology to measure redshifts and do cosmic mapping.

Among the early American pioneers in the area were Laird Thompson and Stephen Gregory, graduate students of William G. Tifft's at the University of Arizona in the early seventies. Thompson and Gregory started recording redshifts with the new technology and plotting the distributions of galaxies in narrow sectors of the sky. "We had the intuition to recognize it was an irregular distribution," said Thompson.

He and Gregory centered their survey fields on large clusters like Hercules and Coma, starting at their cores and then working outward. They found that the clusters were not isolated, dense regions in a homogeneous sea of galaxies, but were more like knots in a lace handkerchief. Thompson and Gregory traced bridges of galaxies from one cluster to the next. There was almost no such thing as an isolated galaxy in their charts; the clusters seemed to be all connected, part of some structure, chain, or filament. And in front of them and behind them, all around, were holes, desert stretches of space. Just what Einasto had said.

What this meant was not clear. Peebles didn't trust their findings, especially because they were giving the wrong answer. According to his bottom-up theory, in which galaxies formed first and then joined into clusters, such large conglomerations should not exist yet on the cosmic scale. It was easy, he thought, to get a misleading impression that they did. Take his "One Million Galaxies" map. In the smoosh of galaxies over the sky, it was easy for the eye to pick out knots and clumps—Coma was somewhat conspicuous—and chains and filaments writhing and tangling across the sky like a petri dish full of cilia. Peebles, for one, doubted that such structures were real. He didn't trust the subjective judgments of the eyeball. Cosmology needed numbers.

Davis agreed. He didn't like spinning his wheels on inadequate data. Surveying the situation in cosmology, and the paucity of facts, he was discouraged about the prospects for constructing a meaningful theory of large-scale structure. He decided he was too motivated to be a theorist. As an observer, he could help correct the dearth of cosmological data. Looking around at the facilities at Harvard, however, it didn't take much to figure out that the prospects for doing serious research were dire.

Since its founding under the restless Bond's father in 1837, Harvard Observatory had had its ups and downs. Under Edward Pickering and his women "computers" at the beginning of the century it was a pioneer in the classification of stars. Under Harlow Shapley, Hubble's old rival, who left Mount Wilson just before Hubble began his chain of discoveries to become Harvard's director, the observatory became a great center of astrophysical education. Shapley was director for more than thirty years; he once estimated that half the astronomers in the United States had gotten their Ph.D.s from him.

By the early seventies, however, there was a feeling in astronomical circles that Harvard had fallen from the first rank. Good graduate students were attracted by its name, but the faculty was mediocre. During his first year Davis and another Princeton graduate and Peebles student, Margaret Geller, organized a seminar on extragalactic astronomy, but they wound up having to give most of the lectures themselves.

What research vigor remained on Garden Street came from the observatory's other occupants, the Smithsonian Astrophysical Observatory. The SAO had been founded in Washington, D.C., at the turn of the century. In the fifties it moved to Harvard under the direction of Fred Whipple, a Harvard professor and comet expert who invested grants from the NSF and the National Academy of Sciences in satellite tracking stations just in time, it turned out, for sputnik. Later Whipple moved the North American tracking station from Organ Pass, New Mexico, to Mount Hopkins, a narrow peak outside of Tucson, and had a 60-inch telescope installed. After his right-hand man, Charles Tillinghast, died of cancer at age thirty-six, Whipple christened it the Tillinghast reflector. The best of the tracking crew moved to Arizona and became astronomical observing assistants.

In the early seventies, because they were in the same building in the same business, Harvard and the Smithsonian agreed to merge their operations. Like the Mount Wilson-Caltech merger, this one was partly cosmetic. The books in the library and the observing facility were shared, but each organization maintained its own payroll and controlled its own appointments.

The Center for Astrophysics was born with George Field, a lanky, mild-mannered theoretician as its first director. Field, friendly and un-assuming to a fault, started his tenure with a big coup when he recruited Riccardo Giacconi and his entire staff of physicists and astronomers away from the American Science and Engineering Corporation, where they had built *Uhuru*, the first X-ray astronomy satellite. Overnight in 1973 the Harvard-Smithsonian Center for Astrophysics (CFA) became the center of X-ray astronomy. Giacconi got a Harvard professorship, but he and his crew were on the payroll of the Smithsonian. It was a source of bitter amusement for the Smithsonian side when the X-ray crew was referred to in newspaper shorthand as "Harvard astronomers." That was typical: Smithsonian paid the money and Harvard got the credit.

Despite the Smithsonian wealth, when Davis arrived and began to cast his eye around for observing projects, optical astronomy was particularly moribund at the CFA. Its observing facilities consisted of a 60-inch telescope in a cylindrical tin shed on the outskirts of the Boston suburbs (Shapley had insisted on calling it a 61-inch so that it would be listed ahead of the Mount Wilson 60-inch in telescope lists) and the 60-inch reflector on Mount Hopkins. The Smithsonian had begun collaborating with the University of Arizona to build a radical new kind of telescope called the multiple mirror telescope (MMT), in which six one-meter mirrors would work in concert to provide the light-gathering power of a single 4.5-meter (140-inch) mirror. They had to dynamite the summit of Mount Hopkins flat to make room for the MMT, and it wasn't going to be ready before the end of the seventies.

Meanwhile a small concession to the needs of modern astronomy had been made on the Tillinghast. Its spectrograph was outfitted with an image tube, which amplified weak spectra electronically so that they could be more readily recorded on film. It didn't work, but the younger astronomers on the CFA staff had trouble convincing their superiors of that.

Davis and another astronomer had gone out to Mount Hopkins once and had a horrible time trying to look at some galaxies with the 60-inch. They returned shaking their heads. "It was pathetic. I vowed never to go back," Davis said. But he did, replacing the image tube with a television tube. The television tube was too fragile; it fell apart.

Davis moped around. Nothing was working out for him, as he admitted; "I was not too productive for a while."

In the course of following up his thesis work on primeval galaxies he struck up a friendship with David Latham, an astronomer on the Smithsonian side. Latham was a Harvard graduate student who had never left, as well as a world-class motorcycle racer. In 1971 he won a gold medal at the famous motorcycle races in Isle of Man. A poster of Latham on his bike in full racing leathers adorns the inside of his office door. Latham himself has conservative clean looks and is one of the few CFA people who always appears in a tie and jacket. Latham was interested in the spectroscopy of unusual stars, but he spent a lot of his time building telescope instruments because the observing facilities were so appallingly bad.

Between talking about galaxies, Davis and Latham daydreamed

together about big observing projects. What was crying out to be done? They came up with several ideas, but Davis kept coming back to the idea of measuring redshifts.

Once he got a redshift survey into his head, Davis became quite messianic. It was the logical step in cosmology. Already groups elsewhere were harvesting redshifts electronically, mostly to map the extent of big clusters. Their results seemed to suggest that the clusters, snaking through space, were bigger than had been thought, and that there appeared to be nothing—no galaxies—for millions of light-years between them.

Davis was skeptical of such conclusions drawn from such limited data. He liked to say that superclusters were not the universe. "Let's not be pikers and just measure ten redshifts and then write a paper," he told Latham. Davis had decided that he wanted to measure the whole sky—hundreds, or thousands, of redshifts, enough to understand the distribution of the galaxies. It would require building a whole new spectroscopic instrument for the hated 60-inch, but that was okay. "I had electronic and metal-cutting experience from my thesis," he explained.

In 1976 Davis sat down and figured out that they could build an electronic detector for the 60-inch to carry out a survey of galaxy redshifts on Mount Hopkins for about a quarter of a million dollars. Latham thought Davis was crazy. One day they stood for an hour out on the observatory's loading dock, across the driveway from the Radcliffe tennis court, arguing loudly about it.

Davis began looking around trying to find funds to undertake this project, and he encountered a problem: There were none. "Marc was on the Harvard side," explained Latham, "he had no access to the Smithsonian money." Davis wrote a proposal and sent it off to the National Science Foundation (NSF). Nothing happened. NSF was perhaps naturally reluctant to send vast sums to the CFA and the Smithsonian, the same way that Rockefeller had balked at giving money to Carnegie.

Finally, in the fall of 1976, Field and Herb Gursky, one of his associate directors, at Latham's behest, took Davis under their wings. Field intervened with NSF to get him a small grant. Gursky started steering Smithsonian resources—like technicians, computer programmers, and small but regular chunks of money—toward Davis and Latham, who finally caved in and signed onto the redshift project. Cosmology wasn't really his bag, but Latham figured the survey would be a way to develop first-class instruments for Mount Hopkins and lift Harvard-Smithsonian astronomy out of the dark ages.

Okay, we're going to build something, Davis said gleefully to himself. But what? His ambition was to measure the redshifts and thus the relative locations of the 2,000 brightest galaxies in Zwicky's catalog, all the galaxies brighter than magnitude 14.5 in the northern sky. That was a lot of data to obtain and preserve and manage.

Davis called the putative instrument that could pluck all this information from the sky and reveal the hidden order of the galaxies the z-machine, z being the astronomers' mathematical symbol for redshift. In the original proposal, the z-machine called for a special electronic tube called a digicon, which transforms light to digital electrical signals, installed on the back of the infamous spectrograph, but Davis's lead on a digicon dried up. "It was insane. The guy could never deliver."

Enter the third major player in the redshift area, John Huchra, a postdoc with a newly minted Ph.D. from Caltech and later to collaborate with Aaronson and Mould on the Hubble constant. He was an omnivorous observer who had a nose for large projects. "There are lots of things I like to look at," he says. Huchra had arrived in the fall of 1976 and gravitated to Davis and Latham like a Yellowstone bear to a newly filled garbage can.

Huchra is short and wiry with a wrestler's build to which Cambridge life has imparted a small, definite convexity at front and center. He has a round face, a short trim beard, and thick rectangular glasses. His hair is dirty blond and mostly gone; his wardrobe could be described as Bowery. The first time I saw him he was wearing an old fedora and a wrinkled, dirty trench coat.

Huchra talks about telescopes and instruments as if they were living things, smart pets or children he could wheedle or con into doing tricks for him. They usually did. Unlike Davis, he had been born to the telescope. He had grown up in New Jersey on science fiction and the popular cosmological writings of Gamow and Hoyle. Huchra was yet another veteran of the MIT physics factory, graduating in 1970, a year after Davis. He knew Schramm from the wrestling team and like Guth, he had gotten involved in antiwar politics, although of a slightly wilder and woollier nature. One night some friends of his went into a machine shop and welded together a battering ram, used it to bash in the door of the office of the president of MIT, and then occupied the office.

I asked Huchra why he thought so many MIT physics majors from the sixties had wound up in cosmology. "There were two things we were shying away from," Huchra answered. "The field of particle physics was getting too compressed, the groups were too big. And there was a lot of defense work in physics."

Huchra had gone to Caltech graduate school expecting to be a theorist—"A big E should light up," he laughs, referring to the "error" sign that lights up on a baseball scoreboard after a first baseman lets a ground ball go through his legs. "There were no theorists at Caltech in astrophysics, okay?" In Pasadena Huchra supplemented a meager graduate fellowship with a job at Palomar, photographing the sky with the 18-inch Schmidt telescope to look for supernovae. He liked to say he was the dumbest graduate student at Caltech, and also the highest paid.

His job required him to be on the mountain during the prime dark nights and rub elbows with the likes of Sandage, whom the other graduate students never met. "I thought we were pretty good friends," he said of Sandage, referring to the days before he joined with Aaronson and Mould to remeasure the universe, "I think we were." Another result was that he became an avid and skilled observer. "I learned how to make telescopes sit up and dance, how to align a Schmidt telescope and feed and water counterweights."

Huchra became good friends with the versatile Gunn, who is also a science fiction freak; he and Huchra used to leave notes for each other written in Elvish, the language invented by *Lord of the Rings* author J.R.R. Tolkien.

In his waning days at Caltech Huchra had gotten involved in a collaboration to measure redshifts of bright galaxies. He had already done some observing along those lines on the 60-inch telescope on Palomar and he planned to continue the work from Cambridge.

On his arrival, Huchra had made the obligatory trip out to Mount Hopkins, tried to use the 60-inch telescope there, and gave up. "The spectrograph," Huchra announced in his colorful fashion, "blew dead bears."

As a postdoc with a few years to do enough good research to obtain another job, Huchra didn't have time to get engaged in fund-raising or major construction projects. He joined the Davis collaboration once it began to show signs of actually happening. He was, after all, the only one of them who had ever taken the spectrum of a galaxy.

He also had excellent connections, which he put to use when Davis's digicon fell through. Huchra called up one of the pioneers in electronic astronomy, Steve Shectman, who was a good friend of his. Shectman had built a device for a University of Michigan telescope to do a job similar to what Davis wanted, and offered to let Huchra and his friends copy his instrument.

Shectman was now at Mount Wilson on famous Santa Barbara Street. He sent Davis a parts list. In the summer Davis boarded a plane for Los Angeles with a bag full of electronic parts and the empty circuit boards on which to string them. He spent a month in Pasadena, taking apart Shectman's electronics and duplicating them board by board.

By then things had really heated up back in Cambridge. John Tonry, a graduate student whom Davis had known at Princeton, arrived, and Gursky had organized everybody into "tiger teams," a NASA tradition for dealing with trouble or pressure. Latham, as the senior Smithsonian person on the redshift project, became its de facto manager. He and Huchra pulled apart the spectrograph and rebuilt it. They shopped around for parts while Tonry began writing computer programs for a Data General Nova that would run the z-machine, take data, and measure redshifts of galaxy spectra automatically.

The heart of Shectman and Davis's creation was a three-stage image tube and, in place of photographic film, a Reticon, which is just a line of diodes that accumulate electrical charge in response to light.

Latham explained the way it was supposed to work. Light from a galaxy, bunched from billions of suns, is gathered oh so gently by the 60-inch mirror and focused on a diffraction grating, which fans the light out into a rainbow. This rainbow, too faint for the eye or even photographic film to detect, falls on the end of the image intensifier tube, causing a special coating sensitive to light to spit electrons. Accelerated by high voltage, these electrons stream down the tube and smash into another screen at the other end. More electrons gush out of another sensitive coating and are accelerated through the second stage of the tube. The process is repeated three times. In the end, the pattern of the initial spectrum is reproduced a million times brighter on a phosphor screen at the rear of the image tube.

The amplified spectrum is focused in turn on the Reticon, which responds to light by storing an electrical charge—the more light, the more charge.

Because the average galaxy that Davis planned to observe was so faint when its light was dispersed and because the image tube and Reticon were so fast, the light that enters the spectrograph was recorded photon by photon. For every photon that pings onto the front end of the image intensifier, a million or so electrons pound onto the P^{20} phosphor at the rear, creating a splash of light. Half a dozen or so of the Reticon diodes charge up. The computer reads them, figures out where the center of splash is, and credits that diode (which corresponds to a particular wavelength) with a photon. A thousand times a second the Reticon reads out. The computer counts every photon that comes through the spectrograph and notes what its wavelength was, and gradually compiles a spectrum, photon by photon, into peaks and valleys, from the deepest blue to the warmest red. On a nearby oscilloscope an observer could watch the spectrum building and end the exposure when it had become detailed enough to measure a redshift.

For an ordinary star or galaxy the important features in the spectrum are the valleys—the dark holes in the rainbow. Each valley is the signature of an atomic element that absorbs light of that wavelength. In most stars, recall, including the sun, calcium makes a pair of close dark lines down at the blue end of the spectrum. If the galaxies weren't moving, being made of basically the same stuff, the valleys—absorption lines—in their spectra would all line up. In a receding galaxy, however, the Doppler effect slides the whole pattern over to the red—the faster the galaxy is going away, the farther redward go the calcium lines.

In their shopping spree, Huchra and Latham bought off the shelf an image intensifier developed for night vision for the military. The Reticon they picked up had an extra row of diodes on it, which meant that a reference spectrum could be beamed onto it. Then the computer could determine, without human intervention, at what wavelengths the valleys in the spectrum were showing up and calculate how far the lines were redshifted from normal, and thus how fast the galaxy was receding.

The spectra were recorded on 50-megabyte hard disks and analyzed automatically by software written by Tonry. No human eyes or hands were going to leave their subjective marks on this study. This was astronomy as a physicist would do it.

The program unfolded itself in Davis's mind. Including every northern galaxy brighter than magnitude 14.5, he estimated, might take the survey to a depth of 300 million light-years, which just might be far enough to include a fair sample of the universe. And if that were the case, the survey would provide the data base for a new statistical quantitative cosmology. With such a scale model of the universe, it might at last be possible to make a reliable estimate of omega, and discover what the clustering properties of the universe really were.

Whatever further impetus Davis might have needed in his quest to map the cosmos was supplied in the summer of 1977, when cosmologists assembled in Tallinn, Estonia, a medieval city on the edge of the Baltic Sea, under the auspices of the International Astronomical Union for a symposium on large-scale structure in the universe. Einasto, the Estonian astronomer, startled the meeting with his analysis of the distribution of galaxies. The universe, Einasto told them, has a cellular structure, with the galaxies and clusters concentrated along the cell walls enclosing dark empty voids. His analysis was a challenge to the theory of galaxy formation that Peebles had been working on for ten years. It sounded a lot like Zeldovich's pancake model in which matter collapsed into large sheets and then the sheets fragmented into galaxies. "Everybody was talking about Einasto," said Latham.

Both Peebles and Zeldovich were at the Tallinn meeting. It was the first time that the cool, lanky one and the little pepper pot, authors of competing worldviews, had met. Peebles recalled that their conversation was restricted to social banter, but he admitted that he watched Zeldovich carefully. Zeldovich had brought his whole group: Sunyaev,

Shandarin, Doroshkevich, and Novikov. "He was exceedingly vigorous, and operated in a very interesting way," Peebles said. "His sharpest questions were always directed toward his immediate colleagues. And his immediate colleagues all knew and expected this and endured it with great fortitude. He was at great pains to make people who were not in his group feel at home, to look after people.

"He never took a swipe at anyone. He never even particularly pressed on people who weren't in his immediate group, but he certainly drove those people very hard and fast."

Davis also met with Zeldovich, and told him of his plans for a comprehensive redshift survey. Zeldovich seemed mysteriously unenthused, as if he considered it low-level work, and couldn't understand why a first-rank scientist would devote himself to it. Davis concluded that the Soviets had no appreciation for experimental science.

In the fall of 1977 Davis was relieved of teaching duties. His career by now consisted of the z-machine. "This project only worked because I was willing to give up writing papers every two months," he said defiantly, having weathered criticism from Harvard faculty who thought he wasn't being productive. In reality he was about to bring home the bacon for Peebles and the quantitative cosmologists. They were about to learn how the universe really was. He was going flat out, he recalls. The group had scheduled the beginning of its monstrous bite of the Mount Hopkins dark time in March.

In February of 1978 a series of stupendous blizzards hit the northeastern United States. The second one buried New England in two feet of snow. The governor of Massachusetts closed the roads in the greater Boston area to everything but police and medical vehicles while the snow cleanup went on. In Cambridge the population took to cross-country skis; for two weeks Harvard Square resembled an Alpine village. Many businesses and schools shut down, but the CFA redshift team had no choice but to keep preparing for their March date on Mount Hopkins. They were having trouble making their borrowed electronics work.

Every day Davis skied to the observatory. Latham, who lived in Watertown, up the river from Cambridge, had a harder commute. He drove in with his wife's black doctor's bag on the seat beside him. When he was stopped by the local police, he said he was a doctor, and they let him go. On the last day before the z-machine was scheduled to be shipped to Arizona and the telescope for its maiden testing, Latham walked all the way from Watertown over the snow-packed empty streets.

Tonry, the software whiz, and Davis followed the instrument to Mount Hopkins. It was the early spring, when the winter Milky Way has slid from the top of the sky and the inky intergalactic spaces, the realm of more distant nebulae, are available for viewing.

"It was clear I had not been trained as an astronomer," Davis said, recalling their trials with the device. On the first night they got the spectrograph working and pointed it at various objects in the sky, among them the Orion nebula. Davis and Tonry looked at the oscilloscope and were completely baffled. It had emission lines, instead of absorption lines, at funny wavelengths. What they didn't recognize was in fact the spectrum of the Orion nebula—a hot gaseous cloud that forms the second "star" in Orion's sword and one of the landmarks of modern astrophysics.

Their astronomical skills improved rapidly after that. Latham had predicted that it would take them a month to get the z-machine working, but within three days and nights, Davis and Tonry were collecting data. Not bad, Davis thought, for a pair of greenhorns. Then Huchra flew out to take charge. The redshift survey was on the air.

It was about a year from when they had gotten the go-ahead. No other observatory, Latham thought afterward, could have done it so quickly, launched such a major project from scratch in one short year. And even they, he knew, would never do it again. For one thing, Harvard-Smithsonian would never be quite so bereft of other instruments and other projects to keep going. To Latham that was the most satisfying aspect of the project. They had brought the observatory back from the dark ages.

The 60-inch Tillinghast reflector became a workhorse. During the bright times of the month, when the galaxies were washed from the sky by the moon, astronomers began to use the same system to study individual stars. Mount Hopkins bloomed.

Davis spent all spring in Arizona, and he recalled it was a miserable observing season. They got rained and snowed on. Mount Hopkins was not Palomar or Kitt Peak with their dining rooms, libraries, paved roads, and private rooms. Half a mile down the road from the summit was a small dormitory building with bunk beds and a kitchen where astronomers could cook some food on hot plates. Life at Mount Hopkins was more like camping out than staying in an academic resort.

The road up the mountain from Amado, a small town outside Tucson, where Mount Hopkins's headquarters were located, was a special treat. The astronomers weren't allowed to drive their own cars; the observatory's four-wheel–drive vehicles took them up and down. During bad desert storms the bridge over the Amado River washed out, stranding astronomers on top of the mountain. One night Davis was walking back to the telescope from dinner. It was pitch black and the road ran along a cliff. Davis couldn't see anything but he smelled fresh dirt. They stopped. In the morning they found that they had just missed a big mud slide that buried a backhoe and blocked the road. It was not prime observing weather.

When the waxing of the moon brought their first two-week "dark" run to an end, Davis went to California, visited Berkeley, and skied at Squaw Valley. He made the rounds of the mountains. On other breaks he skied at Mount Lemon, north of Tucson (and home to an infrared telescope). All this spring skiing left him with a tan uncharacteristic of astronomers. When Latham came out to Arizona, he didn't even recognize Davis.

The depth of the redshift survey had been determined by practical considerations. Davis would be up for promotion soon; Huchra had only a limited appointment at the Smithsonian. Tonry needed material for a Ph.D. thesis. They needed a survey that could be done, then, in two or three years, which meant they couldn't do all 30,000 galaxies in the Zwicky catalog, only about 2,400.

The task of organizing the actual observing fell to Huchra. Before any galaxy could be observed, there had to be a finding chart—a road map through the stars via which the astronomer could point the telescope and line up the spectrograph. Huchra's office rapidly filled with white five-by-seven note cards, roughly one per galaxy. On one side was glued a Polaroid from a Palomar Sky Survey print with bright stars near the galaxy labeled; on the other side he listed the coordinates of the galaxy and other data from the catalogs, such as magnitudes measured in different systems and redshift velocity, if it happened to be one of the few hundred that had been previously observed. The cards went into green boxes. I asked Huchra if he farmed them out to students, and he shrugged and shook his head, "I can fill out about a hundred of these in an hour."

The observing process itself was painful and cumbersome to Huchra. The control room, where the astronomers could watch the z-machine build up a galaxy spectrum, was in the basement of the 60-inch telescope dome. There was no way from downstairs to tell where the telescope was pointing. Every time they wanted to find a new galaxy, they had to run upstairs, turn up the lights, rotate the dome by hand so that the slit was toward the right part of the sky, move the telescope until the "nixies"—little dials on its mount—indicated it was facing the right direction, turn the lights down again, and run back downstairs.

The average galaxy took half an hour to imprint its spectrum onto the Reticon. The CFA team could click off ten to fifteen redshifts per night, when everything was going right, running up and down the stairs inside the dome. Huchra claims that he took about 65 percent of the redshifts during the first year of the survey. "That's what I'm good at. The happiest—no, the second happiest—nights of my life were spent sitting under the 60-inch."

Huchra would go to Tucson for six weeks at a pop. When he

wasn't banging away with the z-machine on Mount Hopkins, he was often across the valley at the Kitt Peak National Observatory with Aaronson and Mould.

After a year there was another infusion of funds from Gursky and from NSF, enough to hire a pair of former satellite trackers to man the telescope and take data, with occasional help from the East. In 1979 the multiple mirror telescope was also dedicated, the first new idea in telescope design since Newton invented the reflector, and Mount Hopkins was renamed Whipple Observatory.

The CFA redshift observers ran up and down the Tillinghast stairs, chewing their way across the sky, soldier-anting the universe, fuzzy nebula by fuzzy nebula. A thousand times a second the Reticon counted photons that had died in the image tube after traveling 100 million light-years. On the oscilloscope the same old song slowly sang itself, slightly shifted in register. The computer measured that shift, assigned a redshift velocity to the galaxy, and stored it away. On reels of computer tape the three-dimensional structure of the universe assembled itself. The pattern of the galaxies waited only to be read and decoded.

But while that pattern was assembling itself, its very meaning was changing and the CFA redshift survey, having been organized to answer one question—how the galaxies are formed and distributed—would find itself dragged into a much deeper and fundamental, almost philosophical question.

Between the time that Davis and Peebles, complaining in the groves of Princeton, had planted the seed of the redshift survey, and the CFA computer began to spit out the results in the secret handwriting of the sky, an obscure, nagging question that had been lurking in the backs of the minds of a few astronomers for almost half a century pushed to the forefront of astronomy. It was an issue of such surprising, yet obvious, importance that all the old questions—is the universe endless or finite, where do galaxies and clusters come from, and which came first?—scurried under its skirts like toddlers running to their mother. Answer this other question first, the Schramms and Peebleses thought, and the others might fall open.

What is the universe made out of, anyway?

17

SPRINGTIME FOR NEUTRINOS

Next to the question of why there is anything at all, the identity of that anything might be the most fundamental issue that science could hope to confront—and one that astronomers might well have thought they had already answered. Since Hubble's time it was accepted that the universe was stars and galaxies and dust and gas and dirtball rocks of planets, orchids, algae, and human beings. It was a woman who played the leading role in convincing them that astronomers were wrong; they had no idea what most of the universe consisted of. Most of the universe was not stars and galaxies. Most of the universe was invisible.

Vera Rubin was on the run when she discovered dark matter. Trouble and controversy had dogged her career. Part of it was because she was a woman; part of it was because she was good. All she wanted to do was observe her galaxies in peace, but she seemed to have an unerring knack for stumbling on disquieting observations.

Rubin doesn't look like a troublemaker. Short, with a round face and a white-haired butch haircut, she is a grandmother and plain spoken. Growing up in Washington, D.C., she became entranced by astronomy from watching the stars go past her bedroom window. Rubin went to Vassar and was a member of the class of 1948, which was immortalized in Mary McCarthy's famous novel *The Group*. Rubin was attracted there by the fact that an earlier astronomer, Maria Mitchell, had been the first American to discover a comet, so there was a tradition of astronomy on the lawns of Poughkeepsie. Upon graduation she got married to a graduate student at Cornell, and so followed him to Ithaca.

It was in graduate school at Cornell that she first ran into trouble. For her master's thesis Rubin undertook an analysis of the redshifts and magnitudes of spiral galaxies to see if there was any systematic effect in addition to the Hubble expansion, say, due to the rotation of the universe. At the time such data only existed for a hundred or so galaxies, most of it garnered by Humason. Using magnitudes as a gauge of distance, she did indeed perceive an anomaly. Spirals with the same apparent brightness, and thus presumably at the same distance, seemed to be flying away faster in one direction of the sky than in other directions. She drove with her one-month-old baby daughter through the snow to Philadelphia to report this result at the 1950 meeting of the American Astronomical Society in a talk entitled "Rotation of the Universe." She was a stranger, knowing no one. Her paper, she said still wincing thirty-five years later, was "ill received." Strangers—really important astronomers she didn't recognize—stood up to humiliate and chastise her. She fled the city and the subject. Ironically, her master's thesis helped inspire de Vaucouleurs to invent the idea of the supercluster. What she had called the universe, he realized, was just the local supercluster.

Following her husband to Washington, D.C., and limping through a Ph.D. program at Georgetown while taking care of a baby, Rubin got a further taste of what it was like to be a woman in astronomy. The great George Gamow, whose colleague Alpher shared an office with Rubin's husband, called her up one day and asked her if she would come over to the Applied Physics Laboratory to talk about her research. She was delighted to accept the invitation. Then Gamow explained that they would have to talk downstairs in the lobby, because women weren't allowed in the offices. She went on to do her Ph.D. thesis—an early version of Peebles's galaxy correlation function—under his supervision.

Such experiences were to make her a keen champion of women's rights, especially in science. She once declined to join a prestigious delegation of astronomers to meet the pope because of what she felt was the Church's and the pontiff's inimical stance toward women on issues such as birth control.

With her degree Rubin joined the charmingly named Department of Terrestrial Magnetism, which was part of the Carnegie Institution, in Washington, where she became, in effect, a cross-continental cousin of Sandage and the Santa Barbara Street crew. In the 1960s she and Kent Ford, a Carnegie colleague, decided to join the electronic revolution. They built one of the first spectrographs to incorporate image tubes—it weighed 300 pounds—and dragged it to Kitt Peak to study quasars with the new telescopes being built there. They were just in time for the savage redshift debates. Sandage and Arp were shredding each other in the journals and not speaking in the Mount Wilson hallways. Rubin's brief taste of controversy over her graduate work left her with no stomach for religious warfare.

"I wanted a problem that nobody would bother me about," Rubin said. She and Ford looked around for something safe to do with their fancy spectrograph. They decided to study the dynamics of *normal* spiral galaxies. Usually when an astronomer took a spectrum of a galaxy, he got a single average redshift for the entire galaxy. Rubin and Ford realized that their sensitive spectrograph should be able to measure the relative redshifts and blueshifts of stars in each individual part of the galaxy as they revolved around its center. By measuring the rotation of each sector of the galaxy independently they could construct what astronomers called a rotation curve, and find out how fast stars were moving in different parts of the galaxy. In principle that meant they could measure how mass was distributed through the galaxy. According to Newton's laws, the faster the stars were going around at any distance out from the center, the more mass must lie interior to the orbit of those stars. Their spectrograph was a scalpel with which Rubin and Ford could literally dissect a galaxy.

"When you try to pick out anything by itself, you find out it's attached to everything else in the universe," says Rubin, borrowing a quote from the naturalist John Muir. By far the densest and brightest part of a galaxy is its nucleus. Perhaps naively, she and Ford expected to find that the mass in a galaxy was distributed like its light, with most of it at the center.

In that case the rotation curve should follow a classic pattern. Coming out from the center of the galaxy it would rise steeply, as the star orbits encompassed more and more densely packed mass. Outside the core, the curve would moderate its rise. In the suburban parts of the disk, where stars were not so tightly packed, the curve would rise very gradually or stay flat. Finally, at the edge of the galaxy, where the increased size of the orbits encompassed little additional star mass, the stars' velocities would fall. That was the case in the solar system, most of whose mass is contained in the sun. Mercury, the innermost planet, whips around its orbit in eighty-eight days, but faraway Pluto moves so slowly that it has barely budged from the constellation in which it was discovered fifty years ago.

The Andromeda nebula, the great historic pinwheel, the Milky Way's sister galaxy, is almost too close for observing comfort. It spans 5 degrees of sky, depending on how far out from its pearly nub of a center you can trace starlight. But Andromeda, M31, is where all cosmic expeditions begin, and it is where Rubin and Ford tried to plot their first rotation curve, in 1970. Andromeda was too big to cover in one exposure with the spectrograph; they had to tiptoe star field by star field through the galaxy over many nights of observing.

When they completed their survey, however, a curious thing happened. Rubin lost her first chance for immortality. After fitting all these observations methodically together into one long sinuous plot, she and Ford noticed that the star speeds did not slow down in the outer regions of the Andromeda galaxy. In astronomical jargon, the rotation curve was flat. It meant there was something going on next door in the nearest, most well-observed galaxy in the universe that nobody understood.

But there was no "eureka." "We weren't wise enough," said Rubin a little ruefully, shaking her head, at a loss to explain their lack of perception any further. The full meaning of rotation curves had not quite sunk in yet. It didn't occur to them to connect the flat curve to the distribution of matter. After puzzling about the possibility of some kind of feedback effect that would regulate the stars' speeds, they dropped the subject.

Rubin and Ford went back to using their splendid image tube spectrograph to measure redshifts of spiral galaxies. Of course, they bumbled right back into the lopsided expansion of the universe that Rubin had unhappily discovered in graduate school. This time it got a name: the Rubin-Ford effect. The simplest interpretation of this effect was that a sizable chunk of the local universe, in addition to expanding, was being pulled by something and sliding in the general direction of Pegasus. This discovery had few fans among senior astronomers. Sandage, having established the uniformity of the expansion of the universe to his own satisfaction, suspected that Rubin's sample of galaxies was biased.

With a storm cloud over her head, Rubin and Ford, along with Norbert Thonnard and David Burstein, returned to their spiral galaxy dynamics, where the puzzling rotation curve of M31 lay in wait like a ticking bomb. This time they decided to do a thorough job of investigation. They would go through the spiral galaxy types systematically, starting with the most luminous ones, take spectrographs, and analyze the star motions. For a small or distant galaxy, this was a one-step job. When the entrance slit of the spectrograph was lined up with the long axis of the galaxy, the blueshifting of light from stars at one end of the disk coming toward us and the redshifting of light from stars going away at the other end twisted all the spectral lines into S-curves. The galaxy's rotation curve could be read off a single spectrogram.

As soon as Rubin and Ford began eagerly holding up freshly developed spectrograms in the darkroom and squinting at the curving spectral lines, they realized that something strange was going on. The rotation curves did not trail off the way they were supposed to; instead, they were flat, like M31's. The stars were not slowing down at the edges of the galaxy, where there was presumably little additional mass to propel them. If anything, they speeded up, as if the farther out in the galaxy a star lived, the more and more mass was available to whip it around faster and faster. But out there was less and less light, so what *was* the additional mass that was propelling the distant stars? "This time," said Rubin, "we knew immediately we had something phenomenal."

There could be only two explanations for such behavior. Either Newton's laws were wrong, or there was something else in the galaxy—something that was not luminescent, and that was not concentrated at the center of a galaxy, as the stars were. Rubin calculated how much of this extra "dark" mass would be needed to whip the stars around at the speeds they were traveling and keep the rotation curves rising. The answer was astounding: two to ten times more mass than seemed to exist in the visible galaxy.

"Great astronomers told us it didn't mean anything," Rubin said. "They said it was an effect of looking at bright galaxies." Go look at dim galaxies, she was advised.

So she did. She was not backing down this time. Rubin suspected that what she had found was the missing mass.

"Missing mass" was a term that had been coined by the irascible Fritz Zwicky to describe the results of a disturbing observation that had been made back in the thirties, and it had been tugging anxiously at the minds of thoughtful astronomers ever since.

Zwicky had measured the redshifts of a bunch of galaxies in the Coma cluster, figured out how fast they were moving with respect to one another, and calculated how much gravitational force would be required to keep the cluster from flying apart. Then he compared this gravitational mass to the luminous mass of the cluster, which he obtained by simply adding up all the starlight. To his surprise, the gravitational mass outweighed the luminous mass ten to one. The conclusion was inescapable: unless the Coma cluster was just a temporary optical illusion, 90 percent of the Coma cluster—and, it turned out, of other clusters—was invisible.

Zwicky called the invisible 90 percent the missing mass. It would prove to be his most lasting and damaging swipe at the cosmological establishment. The bottom-line implication of Zwicky's work was that astronomers didn't know what the universe was made of.

Like most astronomers, Rubin had gone through the exercise of redoing Zwicky's calculations in graduate school and concluded there was something not ready to be understood there. Since Zwicky's time, astronomers had found other examples of collections of galaxies that seemed to weigh more than the sums of the individual galaxies inside. In general it seemed that the larger the system being studied, the bigger the discrepancy was between visible and invisible weight—the more missing mass there was. Some astronomers clung to the hope that, on the largest scales, there would turn out to be enough missing mass to close the universe.

Peebles had an influential hand in this interest in missing mass. In 1973, during the period when he was enamored of numerical simulations on computers, he and a Princeton colleague Jeremiah Ostriker had attempted to simulate the structure of a spiral galaxy on a computer. They tried and tried, but kept failing. The disk of the galaxy, it seemed, was unstable; the gravitational forces between the stars in the disk pulled it apart. How was it then, that galaxies existed? Eventually, Peebles and Ostriker found that the disk would be stable if it was surrounded by a spherical halo of other matter, like a hamburger patty sandwiched between two halves of a bulky roll. Such a halo would naturally have to be invisible, and it didn't have to be very massive to stabilize the galaxy. Peebles recognized that by supposing that the dark halo was arbitrarily large and massive, he could in effect add enough invisible mass to galaxies to close the universe, although there was no evidence that so much was there. He and Ostriker had to hold themselves back from imputing too much dark mass to the halos when they wrote their paper.

A year later Peebles and Ostriker addressed the question of dark matter, galaxy masses, and omega head on, with the help of Amos Yahil, a young Israeli particle theorist on a postdoc. They analyzed all the different ways in which galaxy masses had been measured—from visible starlight to the dynamics of double galaxies orbiting each other and large groups. They found that as astronomers looked on larger and larger scales, the masses of galaxies seemed to go up. They concluded that Zwicky had been right, that galaxies were probably ten times as big and massive as they looked. Outside their own department, this was not well received. Princeton theorists had the reputation of favoring a closed universe, both because of Wheeler's predilection and because the galaxy correlation statistics pointed to a dense universe.

As it turned out, a spherical halo of dark matter fit Rubin's observations very well. She and her crew spent the next few years shuffling back to telescopes in Arizona and Chile filling out their repertoire of galaxies: dim galaxies, bright galaxies, loose spirals with anemic cores and tight spirals with arms, barred spirals, spindle-shaped galaxies. They collected a great variety of rotation curves. Rubin became expert at reading their nuances. She bragged, "If you give me a rotation curve and the

Hubble type of a galaxy, I can tell you the mass, luminosity, and radius of the galaxy."

What they all had in common was the signature of missing mass or dark matter, the stars whirling faster than the gravity of the stars alone could make them; luminous matter was like a kind of foam on a dark mystery wave. "Nobody ever told us all matter radiated," she blurted in her blunt style, "we just assumed it did."

Rubin concluded that what astronomers call galaxies—the spidery curls of starlight and gas—were in fact only the luminous cores of much larger, darker, and more massive clouds. She developed a kind of stump speech about her work that she titled "What's the Matter in Spiral Galaxies?" Dark matter, she said, appeared to surround and permeate a visible galaxy. Within the visible boundaries of the typical spiral galaxy Rubin estimated that about half the mass was dark matter. But there was every indication that the dark cloud extended even beyond the visible edge of the galaxy; its ultimate dimensions could not be traced. Dark matter could, as Zwicky's much earlier findings suggested, and Ostriker, Peebles, and Yahil had repeated, outweigh luminous matter 10 or even 100 times to 1.

Rubin's main point was philosophical. For 300 years astronomers had been presuming that the universe was what they saw. What they spent their time doing was sorting lumps. Atoms formed into stars, stars formed into galaxies, galaxies into clusters, clusters into superclusters, maybe. Now this woman was claiming that the cosmos was what they did *not* see.

"When we view the sky with our eyes, with a telescope, or with a photographic plate, what we actually see is that the distribution of *luminosity* is clumpy," Rubin said, leaning heavily on the word "luminosity," as if it were suspect. "Is there valid evidence that the distribution of optical luminosity describes the distribution of matter? The answer to this question is a resounding no."

Up until Rubin began making the rounds with her rotation curves, the lords of astronomy could ignore the missing mass as an anomaly of poor or misunderstood data. Sandage grumbled, "I resisted Vera's rotation curves for so long, but you just can't beat them down." In the late seventies and early eighties, however, dark matter spread like a kind of astrophysical plague. It was in elliptical galaxies; it was in the plane of the Milky Way; it was in dwarf galaxies. It was in the space between galaxies.

What was in the dark matter halos? Peebles hadn't given it too much thought. He figured the halo consisted of low-mass lumps of matter too small to light up as stars, what astronomers sometimes call brown dwarfs or "basketballs." "Nature likes to make low-mass objects," he said once. There was a model right here in the solar system. The planets and asteroids revolving about the sun form a thin disk about eight billion miles across; around it, at the much vaster distance of 100 billion miles from the sun, is a spherical shell of cometary debris called the Oort cloud, its existence deduced from studying comet orbits, full of ice chunks too hopelessly small and dim to ever be seen from earth. That was a typical Peebles stratagem: Stick close to home and exploit classic physics. More adventurous souls speculated that the dark matter could be clouds of black holes—either regular black holes or the Hawking miniholes—or even some kind of exotic elementary particles from the big bang.

Still others hoped it would all go away. In a memoirish article, Greenstein, the patriarch of Pasadena, said bluntly, "I hope the missing matter isn't there."

Rather than rail about it, some astronomers were trying to figure out how to use dark matter to solve cosmological problems. One of the first was Simon White, a graduate student at Cambridge University who was to become an aficionado of dark matter astrophysics. White, a gangly Irishman with unruly blond hair, had done his Ph.D. thesis on numerical simulations of galaxy clusters under Rees, the fast-talking, faster-thinking ex-student of Sciama's. He was keen on dark matter. White prevailed on Rees in 1976 to help him write a paper on dark matter's role in the vexing problem of galaxy formation.

If dark matter was indeed most of the universe, White reasoned, then it would be dark matter that determined the nature and scale of structure in the universe. It would be dark matter that was carved into galaxies and superclusters. Luminous matter was just dust in the wind, along for the ride. Dark matter—what he and Rees called "black objects"—was the primordial ooze.

That might have seemed less than a promising revelation from the standpoint of making any further progress on the problem of the universe; some unknown invisible stuff was doing all the work. But Rees and White realized that they didn't have to know any details about the "black objects"—like their identity—to deduce how these objects would behave en masse. It didn't matter whether they were dim stars, rocks, black holes, or squads of exotic elementary particles left over from the big bang. Gravitationally they would all act the same.

During the early universe, according to Rees and White, the dark and bright matter would be mixed together. Once the universe had cooled, small primordial lumps of higher density could start to grow into large, diffuse clouds. Energized by the long fall together, the black objects would buzz around inside these clouds like atoms in a gas. These clouds would act as gravitational traps for the trace amounts of ordinary matter—at this stage of evolution mostly hydrogen and helium gas—mixed in with the primordial black stuff. The regular matter would just be dragged along.

Once clumped into a cloud, however, the ordinary and dark matter would separate, like a churn of cream. The ordinary, soon-to-be-luminous matter would cool off by radiating away its energy, like a cup of coffee left unattended. As it cooled, the hydrogen and helium atoms would slow down and sink to the center of the cloud, where they could condense into stars and galaxies. After eons there would be a puddle of luminous matter surrounded by a dark cloud.

The Rees-White scenario was sort of a dark twist on Peebles's original vision of building the big things in the universe from the bottom up, out of little things. The bigger the original clump, the longer it would take for the atoms within to cool, condense, and light up. Globular clusters would happen first, then galaxies. While they were cooling, the dark matter clouds would be attracting each other and merging to form even larger clouds, which in turn try to merge into even larger clouds, dragging their little light-filled cores along like children to Sunday dinner at the grandparents. As time went by the merger process would repeat itself, they predicted, on larger and larger scales. Their universe was organized with lumps within lumps within lumps, like William Blake's flea, which had its own tiny fleas which in turn were infested with even tinier fleas ad infinitum.

If Rees and White were right, then astronomers had spent the last centuries dancing in the dark. The visible universe was just a scrim, like snow on a mountaintop, controlled by influences beyond their vision. The history of the universe, of galaxies and clusters and superclusters, was a story being written in the dark.

Rees and White were ahead of their Western colleagues in ascribing such importance to dark stuff, but when it came to dark matter and the structure of the universe, they were playing catch-up to Zeldovich, the intuitive Russian. In the East, dark matter wore a different face, and according to Zeldovich it built a different universe, a top-down universe. The face it wore was of those spook particles, neutrinos.

When Zeldovich had conceived of pancakes, the top-down way of forming galaxies, he had assumed the universe was composed of regular matter. But in the back of his mind always was the possibility that there was more to the universe. According to the standard big bang calculations, for example, there were 115 cosmic neutrinos per cubic centimeter in the present day universe. According to the standard physics folklore, these neutrinos were massless—they weighed nothing and traveled at the speed of light. But that was just an assumption. Particle physics didn't have the tools sensitive enough to weigh particles lighter than electrons. Zeldovich knew there was no law that *required* neutrinos to be absolutely massless; in fact, some of the new unified particle theories suggested that they should have a slight mass. Suppose they did?

Back in the sixties Zeldovich had realized that he could see the whole universe as a scale on which to weigh neutrinos and thus improve on the laboratories' measurements of their masses. If neutrinos had even a slight mass, Zeldovich realized, the trillions upon uncounted trillions of them would add up to a sizable weight on the universe. Their collective gravity would add to the drag on the expansion of the cosmos. If neutrinos were too heavy, in fact, their weight would have already collapsed the universe.

The fact that they had not collapsed the universe yet, Zeldovich and a student named G. Gerstein calculated, meant that neutrinos couldn't be more massive than about 30 electron volts—less than a ten-thousandth of an electron.

In some ways it was one of the most expensive lopsided experiments in the history of science. Zeldovich had based the age of the universe on moon rocks retrieved by the *Apollo* astronauts. The entire American space program was being used to measure the teensiest thing in the universe. It was a real collision of outer and inner space. Zeldovich characteristically made arithmetic errors, and his cosmical data weren't any more reliable than cosmic data ever are, so his result was more of a ballpark estimate than an answer. Still, it was better than anything available from pure physics.

In the West Zeldovich's idea was regarded as a typical imaginative proposal hobbled by errors and no data to support it. In the East "massive neutrinos" became one of the contingencies of nature. What would be their effect on the other areas of astrophysics? Zeldovich always gave away his best ideas. Through Zeldovich a professor at Eötvös University in Budapest, Hungary, became interested in neutrino universes. He in turn had a bright young student named Sandor Alexander, "Alex," Szalay. In 1972 he suggested that Szalay look into the implications of massive neutrinos.

For Szalay, the encouragement to study massive neutrinos was like an invitation to join the family business. Szalay's grandfather had been a physicist who abandoned research for the security of teaching high school. His father had studied nuclear physics at Cambridge's Rutherford Laboratory, which was a breeding ground for Nobel Prizes. With his wife as his only assistant Szalay senior moved to Debrecen, a leafy university town surrounded by farmland near the Rumanian border, and founded a research institute devoted to nuclear physics, and was elected to the Hungarian Academy of Sciences.

Nuclear physics at that time was the frontier, and nuclear physicists were in pursuit of the neutrino. In 1940 Szalay figured out a way to detect the neutrino and confirm its existence, but the war tore his laboratory to shambles. Because he was isolated it took a long time to rebuild. He was not able to do the experiment until 1955; by then a pair of Americans had detected neutrinos coming out of a nuclear reactor.

Although Szalay came in second in the neutrino race, a photograph that he took of atomic tracks in a cloud chamber became the classic neutrino portrait, widely reproduced in textbooks. It shows the track of a helium nucleus executing a long slow curve and then abruptly changing direction as it recoils from the emission of a neutrino. The actual neutrino is invisible.

Alex, who was born in 1949, and his younger brother, Andrew, were raised in an apartment on the grounds of the research institute. Reddish blond with a flat open face, a slight build, and an unassuming wardrobe, Szalay seems to have grown up more a Westerner than an Easterner. As befits the sons of physicists who had to build their own instruments, he and his brother grew up with an appreciation and aptitude for gadgets, particularly electronic ones. Alex majored in physics at the University of Debrecen, and then moved on to Budapest and Eötvös University for graduate school.

The old twin cities of Buda and Pest, divided or joined by the thick Danube, have retained the historic flavor that German cities lost during the war. The merged city also has a gypsy hustler's hipster spirit. The clothes are bright, the shops on the cobblestoned pedestrian-only street carry Adidas. Eötvös University is a collection of threadbare buildings near the center of Budapest, which itself looks unchanged from the eighteenth century. The facilities are spare. As late as 1966 there were no computers on campus, and the entire physics department shared one telephone.

In Hungary, Szalay explained to me, the research money and equipment funds flowed to the official research institutes, like the one his father ran, which were separate from the universities. It was, he grumbled, a prescription for mediocrity, because it drew the best people out of the universities and away from contact with students, who would then grow up to be mediocre scientists. When physicists like Szalay wanted to do serious computing, they had to go all the way to the Central Research Institute out in the Budapest suburbs.

Eötvös University is named for József Eötvös, a patriarch of a powerful and rich Budapest family, whose bronze likeness stands on a quay overlooking the Danube. József's son, Lóránt Eötvös, the Baron Vásárosnemény performed one of the landmark experiments in physics at the turn of the century. He spent twenty years measuring the gravitational force exerted by the earth on objects that had identical masses but were made of different materials, and concluded that gravity was indifferent to the composition of the masses. The Eötvös experiment was one of the cornerstones of general relativity.[1] Szalay was amused to see that Eötvös's apparatus, looking rather like a tarnished brass coatrack, was still sitting in the physics department supply closet.

Szalay knew nothing about astrophysics when he was invited to become an expert on neutrino universes. He hit the books of Peebles, Weinberg, and Zeldovich, and got himself sent to conferences and special summer and winter schools at places like Erice, Sicily, where experts such as Sandage or Schramm gave lectures in return for two weeks of tennis and beachcombing. At a neutrino conference at Lake Balaton, a resort area in western Hungary, he met Zeldovich. Hungary was the first country Zeldovich had been able to visit when he started doing regular science, and he went there often.

Szalay's earliest memory of Zeldovich from that meeting is that when Zeldovich went down to the beach to swim early in the morning and found that the beach was not yet open, he climbed over t[...] rounding fence. When he came back out the beach was open[...] guards told him to climb back over the fence to get out. Which [...] did, laughing.

Szalay began his research by redoing Zeldovich's neutrino calculations. Then he tried to compute the effect of these neutrinos, this dark mass, on the formation of galaxies. He got a lot of equations that he didn't really know how to solve, and so worked his way through them intuitively.

Giving neutrinos a little weight, some 30 electron volts, made them the dominant mass in the universe. It also slowed them down from the speed of light, but, Szalay figured, not by any significant amount. Primordial neutrinos would fly around at *almost* the speed of light, and as if the rest of the universe were invisible. These two qualities combined to give neutrinos a kind of slippery nature; they could not be corralled in small groups. There was a cutoff in the sizes of lumps that could form and survive in the neutrino universe. That size, when Szalay calculated it, turned out to be about the mass of a thousand trillion suns, that is to

---

1. In 1985 the Eötvös experiment was once again at the center of physics, when University of Washington physicist Ephraim Fischbach reanalyzed Eötvös's data to a higher level of precision. Fischbach claimed that the data showed that atomic composition did have an effect on gravitational attraction over short ranges of a few meters. This was evidence, he suggested, of a fifth fundamental force of nature. Fischbach's work sparked hot debate and new experiments, whose results have been inconclusive and even contradictory. Many physicists concluded that the Fischbach effect is either nonexistent or can be explained by old-fashioned physics, such as air currents or geological anomalies. The fifth force is not yet part of standard physics, and cosmologists have not yet rushed to exploit it in their theories.

say, the mass of a large cluster of galaxies. That was the mass of the first objects that could condense out of the hot gas in the early expanding universe, and give rise to a Zeldovich pancake.

In other words, a neutrino universe would make galaxies the Zeldovich way, from the top down, by fragmenting cluster pancakes. Was Zeldovich excited? I wondered. His style was to juggle lots of ideas, explained Szalay noncommittally, hand them out to subordinates, and then criticize them fiercely. "He saw okay this was a good idea," recalled Szalay.

Encouraged by Zeldovich's response, Szalay kept working on neutrino-dominated universes for the next several years. The big event in their relationship took place in 1975 when Zeldovich went to Budapest for three weeks on a sort of state tour of research institutes. Szalay was assigned to be his guide and gofer. On the first day Zeldovich took a liking to the young man and gave him a physics problem, telling him to call at five-thirty the next morning with the answer. Szalay didn't take him seriously. The next morning Zeldovich yelled at him and told him he was lazy. Thus began Szalay's reeducation. Sighed Szalay wistfully, "He seemed to have the ability to see just how much you could understand and then give you 20 percent more."

Szalay had been struggling to reinvent astrophysics by himself out of isolated ideas and facts, as he put it, "from the bottom up." Zeldovich gave him problems that joined the fragments. As a result, Szalay began to see cosmology and astrophysics from a more integrated perspective. Suddenly the pieces all fit together. "It was really amazing, and a relief," he recalled gratefully. Meanwhile, Zeldovich and Szalay's father, united both by physics and generation, also became bosom buddies.

Zeldovich encouraged his new protégé to follow up the neutrino pancake work. He invited Szalay to come to Moscow and work with him, but Szalay was tired of neutrinos. "At that time, you know, the general feeling was that this is a ridiculous idea. So I really got fed up with being the oddball person and doing oddball things, so I put the whole neutrino thing away." When Szalay is unhappy his voice drops practically to the inaudible.

He tried to escape by applying for postdocs in the United States and Western Europe, but his applications were turned down. Apparently many scientists agreed with him that the neutrino stuff was ridiculous. He decided to become a particle physicist, a pursuit that lasted two years, until he realized he really wasn't interested in the subject. In 1977 he went to a conference in the Caucasus on underground neutrino detectors and ran into Zeldovich, and he asked if the invitation was still open.

It took a year for the creaky bureaucratic machinery to crank out the necessary permissions for Szalay to work at Soviet universities and research institutes. He missed the IAU symposium on large-scale structure in Tallinn, at which Davis explained the CFA redshift survey to Zeldovich and Einasto proposed that the pancakes of the universe interlocked to produce a cellular structure. Meanwhile something else reared its head: rock and roll.

Under Hungary's economic liberalization workers were allowed to moonlight in their own enterprises. For furniture makers that meant renting the state factories at night to turn out their own, better dining tables; for intellectuals like the Szalay brothers it meant software consulting. (Alex's brother, Andrew, had also by then earned a Ph.D. from Eötvös.) One day they were approached by a local rock idol in search of new sounds; they built him a synthesizer. One thing led to another. The Szalays had always been musical, and the physics legacy was pressing on their heads like a pile of neutron stars. They built synthesizers for themselves and, along with a mathematician friend, formed a rock-and-roll band called "Panta Rei." The name, Szalay explained, is Greek, from Heraclitus's famous saying, "You can't step in the same river twice."

The music, he explained, was somewhere between John Cage and Emerson, Lake, and Palmer: high-tech instrumental warblings and whispers. Panta Rei was popular enough to land a three-record contract from a Hungarian recording company. One of their concerts was broadcast live on Russian television. The stage, Szalay recalled, was ringed by soldiers carrying machine guns, just in case somebody felt inspired to leap up and make a political statement. Szalay told me once that his favorite movie was *This is Spinal Tap,* Rob Reiner's satire of the travails of a mediocre heavy metal band on the road. He sighed, "I lived that life."

Szalay lived rock and roll by night and cosmology by day. The songs and the universes both came out of computers. Both were music of a sort, a sifting of synthetic patterns in homage to and in quest of the original harmony, the secret chord of dark matter. "Maybe most of my energy went into music," Szalay admits sheepishly.

I asked Szalay what effect rock and roll had had on his career. He answered, "I learned that you always have to perform."

To Szalay the physics life was not that much different from the rock-and-roll life. The dress, the hours, the hair, the sense of fraternity, the brothers who knew what and how you played, were common elements. New ideas had to be taken on the road. You went from conference to conference to colloquium singing for your supper, dazzling with your insights, your sleight of hand with math. You were judged, as Wheeler said, by your ability to project a thought with power. You flew all night and ate strange food with people you'd never met before, but you plied the same integrals, you had the same heroes. Afterward you jammed with your colleagues.

No talk, no performance was ever quite the same. There were always new riffs to be tried out, tighter ways to say the same thing, new metaphors, simpler and more powerful theorems, and new angles on last week's idea. A good performance at occasions like the Texas Symposium or the IAU could change a career. How suavely or snidely could you put down the competition? How aggressively could you dissect the analysis of someone else's data? Who had the most elegant high-powered approach? The butterflies were in the stomach the day before the appointed hour on stage. Szalay learned to perform.

Meanwhile he was commuting to Moscow and to Zeldovich. The bouncing little soccer ball (as Szalay described him) had ascended to the heights of Soviet science. He was head of the theory division of the Institute for Physical Problems, and taught at Moscow University, presiding over the special Monday-morning seminars, as Landau and Lifschitz—heroic names in Soviet physics—had before him, at Sternberg State. His former students were dispersed through the institutes and universities that made up Moscow's physics establishment. They were still bound. They all owed their jobs to him, according to Szalay, an entire generation of astrophysicists. Szalay was sort of a caboose to that train of emulation and dependence; he had an inside look at Zeldovich's empire.

Szalay went to Moscow for a month at a time, sometimes staying with Zeldovich and his second wife Anjelica, an economist. (His first wife, a physicist, with whom he had two children, also physicists, had died of a heart attack swimming in the Black Sea.) They lived in a three-bedroom apartment in a section of Moscow known as the Lenin Hills, near the Institute for Physical Problems. The apartment, Szalay recalled, was full of medicine balls and weights. There was a big blackboard in the living room. Zeldovich would be up at five o'clock every morning on the phone.

"I decided not to go away for more than a month," Szalay explained. "And, I mean, a month without computers is okay. You talk to everyone. And then, after a while—I was always computer oriented, so there are certain limits to what you can do within a month's time without computers."

They had computers in Moscow, of course, but as a foreigner Szalay had no access to them. He passed the time going to Zeldovich's classes and seminars. Talking about those Moscow days transported Szalay back in time: "Sometimes, in those seminars, you would think that he was just about falling asleep. So his eyes are closed and he almost starts snoring, and then he opens half of one of his eyes and then asks the most embarrassing questions of the speaker. So it means he was not asleep. He was just thinking! And usually he just cuts through the whole problem."

Szalay's eyes get misty and his voice gets low when he recalls those days. Zeldovich became a second father to him. "Apparently he was doing it with all his students, but somehow what was interesting over the years was the way that, as I improved in science, he somehow managed to change his style, the way he set problems, the way he made me do things. He had such a fantastic natural instinct to drive people. And there are some people who can't switch. Once you are a very good student, but then once you become more grown-up and dependent they somehow can't find the proper ways of communicating with you.

"He is just so full of ideas, so sort of always sparkling—and he doesn't care, so he's willing to give out his best ideas for others to work out the details. And he can always guess the answer within a factor of two or three just at a snap of the fingers."

When Zeldovich wanted to talk about sensitive matters, he would suggest a walk around the block. It was on one of those walks that he admitted to Szalay that he had worked on a problem just like the one Teller and Ulam had solved—that of igniting a hydrogen bomb. Another time, Szalay recalled, Zeldovich had figured out the explosive yield of a Western nuclear test simply by looking at the picture of its mushroom cloud in the newspaper.

Despite whatever restlessness Zeldovich might have felt during his chauffeured high-security research days, he was not shy about using the privileges that had accrued to him. One of the relics of his classified life, apparently, was a special driver's license. One day he and Szalay were in a hurry to play tennis, and Zeldovich took a shortcut, the wrong way down a one-way street. They were stopped by a cop. Zeldovich flashed a card, what Szalay called his "hero's certificate." The cop waved them on, all smiles.

Szalay forgot the unhappy, oddball days of the sort-of-massive neutrino. Goaded by the dynamic little Zeldovich, he instead cracked into regular pancake theory—how to build the large-scale structure of the universe from the top down. He analyzed correlation functions, he simulated universes, he learned to see the cosmos as a physicist does: Galaxies are not spidery webs of light, but rather, density perturbations, lumps in a primordial gravy.

Most of the work was never published, but they were positioning themselves. History, unbeknownst to them, was on their side. In 1980, when all their wildest dreams came true, they were ready.

Szalay and Zeldovich's wildest dream was that neutrinos would ultimately prove to have some small but significant mass. If that was true, all their theoretical work on the cosmological effects of massive neutrinos would not be in vain.

Could massive neutrinos be Rubin's dark matter? A chain of

Western papers said no: Zeldovich and Szalay had decreed that neutrinos couldn't weigh more than 30 electron volts; otherwise the universe would already be collapsing. With such low masses, however, neutrinos would be too fast and slippery to be packed into halos small enough to fit around single galaxies, as the theory predicted.

At about this time Schramm and Steigman, who were old friends of neutrinos and the big bang, got into the act. Schramm had gradually been convinced that omega had to be 1.0—it was the only aesthetic solution to the cosmological equations. In the spring of 1980 Guth began to make his rounds on the seminar circuit looking for a job and selling inflation. Schramm was one of the first to hear about inflation and glommed onto it. He got Guth invited on short notice to a meeting of the Royal Society in London. Inflation, he realized, was a brilliant application of GUTs to cosmology, the closest thing anybody had yet come up with as an idea that *explained it all*.

Inflation made a troublesome experimental prediction—namely, that the mass density of the universe was equal to the critical density, that omega was, in fact, 1.0. Because the best observational efforts of the Sandages and the Gunns of the world had concluded that omega appeared to be less than 0.1, however, Guth's theory produced a real missing matter problem. Schramm and Steigman thought that the answer might lie in massive neutrinos spread in clouds the size of superclusters.

What would it take, they asked themselves, for neutrinos to "close" the universe? Not much. Essentially Schramm and Steigman repeated Zeldovich's reasoning of a decade earlier, except that now there were three families of neutrinos to consider—normal electron neutrinos, muon, and tau neutrinos—each of which could have a different mass. There were supposedly on average 115 big bang neutrinos per cubic centimeter throughout the universe. The average mass of a neutrino, they calculated, need only amount to 33 electron volts, give or take a factor of two (for the elusive Hubble scale factor) to close the universe. It was the same answer Zeldovich had gotten, although without modern methods.

But what about galaxy halos? Here Schramm and Steigman were guided by the delicate numerology of the big bang element production. Luminous matter—the galaxies—comprised about 1 percent of the amount of mass it would take to close the universe. The halos were ten times as massive as the luminous galaxies; that meant that even with Rubin's dark matter omega would be only 0.1—the universe would contain only 10 percent of the mass needed to halt the cosmic expansion some day. As it happened, an omega of 0.1 was consistent with the big bang models of nucleosynthesis. A fireball universe in which the density of ordinary, so-called baryonic matter amounted to 10 percent of the critical density produced about 24 percent helium, 75 percent hydrogen, 0.01 percent deuterium, and 1 percent everything else—just what the astronomers measured.

In other words, both the luminous galaxies and their dark halos could consist of ordinary matter—dim stars, sticks, and stones—without violating the constraints imposed by the deuterium and helium abundances. But if the rest of the putative universe—the supposed 90 percent not accounted for by galaxies and their halos—was also composed of ordinary matter, the deuterium and helium calculations would be thrown off irreparably. If omega really was 1.0, as required by Guth's theory, the other 90 percent had to be some exotic form of matter that did not interact in the primeval nuclear furnace. Schramm and Steigman argued that slightly massive neutrinos were perfect for that role.

Their recipe for the perfect cosmos was then: 1 percent visible ordinary matter, 9 percent dark ordinary matter, and 90 percent massive neutrinos. Dark ordinary matter in halos, with massive neutrino clouds to glue the superclusters of galaxies together and close the universe: that was the ticket to completing the inflationary picture. When Schramm and Steigman wrote their paper, it was just speculation driven by beauty. "Everything we said ten years ago is still true," Schramm told me. "Before inflation, the only reasons for an omega of 1.0 were philosophical and theological, heh, heh, heh. Now you have strong theoretical arguments. These tell me there is matter outside normal realms."

Schramm and Steigman sent their paper off to the Gravity Foundation, which runs an annual contest for the best article about gravitational research—in this case their work was about the gravitational content of the universe. While the paper was in the mail, events occurred that I like to call the Neutrino Spring.

In May of 1980, two groups from opposite sides of the world burst forth with evidence—difficult, painfully acquired, fragile, and hotly contested evidence—but evidence just the same that neutrinos were in reality *not* entirely the ghost particles that they had been made out to be; in fact, neutrinos had mass.

The first group, led by a physicist named V. A. Lubimov, came from the Soviet Institute for Theoretical and Experimental Physics in Moscow. Lubimov claimed to have measured the mass of the neutrino directly by means of a hellishly delicate experiment involving the radioactive decay of tritium. Lubimov reported that the best value for the mass was between 18 and 45 electron volts—right where Zeldovich had pegged it. The experiment took years to perform, and it was easy to make a mistake. Some Western physicists who looked at the data pointed out that they were also consistent with the conclusion of no neutrino mass at all.

The other group came from the University of California, Irvine, and was led by Fred Reines, the man who had discovered the neutrino back in 1957. Moving a portable neutrino detector, which was basically a bucket of cleaning fluid on wheels, back and forth in front of the Savannah River reactor (the same reactor he'd used to make the initial discovery), Reines claimed to have detected a phenomenon called neutrino oscillation. A certain class of theories predicted that neutrinos were unstable with respect to type. An electron neutrino could become a muon neutrino which could become a tau neutrino; it was one particle with three masks that it kept changing rhythmically.

The main point was that neutrino oscillations couldn't happen unless each type of neutrino had a different mass. Which meant that neutrinos had mass—the Reines experiment couldn't say how much, however.

Reines, aka Mr. Neutrino, chose to make his announcement at a meeting of the American Physical Society in May in Washington, D.C. He was decked out in turquoise and a string tie, but he looked nervous, as if his tan were being drained from within. When he wasn't tongue-tied, his speech was stilted, as if he couldn't find the words grand yet precise enough to explain what he thought he had known a few short weeks: that neutrinos, the multiplicative ghost riders of the sky, have mass. That they were, in fact, the masters of the universe.

During a sweaty news conference, Reines drew a picture of a cat that was changing into a dog on the blackboard, as a way of trying to explain the subtleties of neutrino oscillation. Then he went before 2000 assembled physicists and said, shaking, "If this is right, the universe is not the way we thought it was."

Meanwhile, Lubimov had brought his early results to a cosmic ray meeting in Budapest. Zeldovich jumped on them immediately. He redid Szalay's old thesis calculations using what Szalay calls "more professional" methods and got the same answers. A universe made of these new massive neutrinos would produce pancakes. The first hints of that kind of pancake structure in the arrangement of galaxies had shown up in Einasto's work and Gregory and Thompson's limited surveys. Zeldovich and Szalay's vision was coming true.

"Somehow," said Szalay wonderingly in a soft voice, "all these things managed to come at the right time, and it was unbelievable. It was an unbelievable feeling because one was really hoping that maybe there was some truth. Things started to click together so nicely. It was really very nice."

Suddenly Szalay was in demand at conferences and summer schools. In the spring at a cosmic ray meeting in Erice he renewed a casual acquaintance with Schramm, who was always a superb judge of talent. Meanwhile, at yet another summer meeting, he met Joe Silk, the British Berkeley theorist. Letters from Silk and Schramm arrived in Budapest the same week, inviting Szalay to America. Zeldovich whipped up some recommendation letters.

Szalay arrived in the United States in December of 1980, in time for the Texas Symposium, which was held in Baltimore that year. He was invited to give a feature talk about the cosmological consequences of massive neutrinos. By then pancakes were on everybody's lips. Szalay gave a lecture on the origin of structure in the universe, in which he described massive neutrinos as a band pass filter that filtered all the small-scale structure out of the early universe, leaving only the large protosupercluster lumps.

Then it was the peripatetic life of the cosmologist. From there he went to Berkeley for three months. A six-month cosmology workshop was just winding up at Santa Barbara, and Turner, who had grown a beard and lived by the beach, talked him into driving down the coast. By then it was summer; they all traveled—Turner, Alex, and Alex's wife, Kathy—from there to Aspen. During the summer they flew to a meeting in Hawaii. Szalay and Kathy finished off the trip with a long stay in Chicago. During that period he and Turner, especially, became fast friends.

"We had no money," Szalay laughed, "and we are seeing all the decadency of the United States. Then we hit the prairie."

The neutrino spring of 1980 was one of the those transcendent moments in science, when a whole new way of thought jelled. Rubin had been pushing the necessity of dark matter for half a decade. Then the physicists suddenly invented it. Guth was barnstorming with the theory of inflation, which implied that the universe must be 99 percent dark matter.

Inflation raised the stakes for dark matter and transformed it from an empirical curiosity to the linchpin at the holy nexus of cosmology and particle physics. The universe made more sense with dark matter and inflation than it did without dark matter and inflation. In the beginning cosmologists weighed the universe to find out if it would end; now they would weigh the universe to find out if it was beautiful.

Like drunks searching for their keys under the street lamp, astronomers had never had any choice about where to look for the universe. It was only by following the light that they had been led into the dark. And there was the dream that there was something there, in the lines between the stars, that all the accelerators in the physics-speaking world had not been able yet to find. Gangs of funny particles, for instance. In 1980 the funniest particle in physics or astronomy was the neutrino. *It was a real particle*. It was the dark matter front runner.

What was the dark matter? At the end of 1980 massive neutrinos, if not a good bet, were at least far from the worst bet, but there were other possibilities, including that it was all a mistake. The notion that the main constituent of the universe was still unknown after millennia of astronomy and centuries of physics was as bracing to Szalay's generation as it was discouraging to Sandage's. The identity of the dark matter was a richer question to the Fermiland generation than the arguments about q_0 and the H_0, those two tired numbers. The proof of the pudding might come when efforts like the CFA redshift survey finally did reveal how the visible trace of the universe was put together; it might testify to the forces shaping the invisible landscape behind the scrim. There was physics in dark matter, and perhaps, finally, at the end of the trail, those two numbers after all, written in a dark hand.

18

ZWICKY'S REVENGE

In May of 1981, in the hallway outside of Marc Davis's office in the new gray-carpeted wing of the Harvard-Smithsonian Center for Astrophysics, sat a Plexiglas cube on a pedestal. The cube was a little cosmic diorama, about a meter on a side and lit with a black light. Inside, strung on nylon thread were 2400 pith balls representing galaxies—red balls for ellipticals, blue for spirals. At the center of the cube a white bead marked our own Milky Way galaxy. Next to it, like a shark threatening a small fish, was a cloud of balls representing the Virgo cluster. Beyond Virgo there was an empty space and then an even larger cloud of beads—the Coma cluster.

Within these clouds of patiently strung beads was perhaps the answer to whether massive neutrinos or some other exotic matter was the invisible hand guiding the organization and structure of the universe. They were the most graphic representation of the results of two years of mapping the local neighborhood by measuring the redshifts of some 2400 galaxies. The first CFA redshift survey was finally complete.

By 1980 the end had been in sight. Davis's ski trips to the Southwest had been brutally ended. More than a 1000 redshift measurements had piled up, unanalyzed. Latham pulled him off the observing. "You understand the science," Latham told Davis. "It's time to start writing."

What do you do with 2400 galaxies? Davis asked himself as he set in motion the awesome churning of statistics, maps, and graphs that would explicate the three-dimensional arrangement of the universe. He and Peebles could measure the three-dimensional galaxy correlation function now at last, and see if the answer matched the elaborate calculations they had sweated through five years earlier. They could analyze the whole pattern of the motions of galaxies and clusters relative to one another and find out if superclusters were real and if their gravity was distorting the expansion of the universe, as de Vaucouleurs had claimed. They could weigh whole superclusters and measure omega from the dynamics of how clusters affected one another. And they could look for the signature of the dark matter in the arrangement of the luminous matter. Perhaps the whole debate between top-down pancake versus bottom-up hierarchical theories of galaxy formation would be resolved.

By the time he and Huchra and their assistants had finished measuring redshifts and they had been compiled and mapped, neutrino universes had squirmed near the top of the agenda of possibilities to be examined. What would a neutrino universe look like? Szalay and Zeldovich had given a rough description: giant slablike clouds of galaxies enclosing empty spaces, like the walls of an unfurnished house. There would be large clumps and large voids, correlations stretching over tens or hundreds of millions of light-years, order on a gigantic scale. The largest structures in the universe would be the oldest. Intersecting galactic sheets would form a sort of cellular structure. The galaxies would be on the walls and especially the corners of the cells; in between would be empty space.

Despite all his hairy integral calculus training and his analytical bent, Davis, looking relaxed in a sweater and tanned, admitted that more than anything else he liked to look at his galaxies. The Plexiglas diorama had been built by a local high-school student for a science fair. Davis hoped that it would eventually end up in the Smithsonian Air and Space Museum.

"This is the biggest scale model in the world," he said proudly. "One inch equals sixteen million light-years." He pointed out the landmarks with a pencil. The universe looked as if it had been whipped with an egg beater, full of air pockets. There were lumps and chains of galaxies, long dribbles ending in knots, delicate loops and fat thuggish fingers, swirling and knotting and carving space into bubbles of emptiness. He poked his pencil through the interstices of his model. "All these holes are real."

Reaching for a word to describe the universe he saw, he kept coming up with *froth*.

As he talked, he started to look slightly worried—and faintly amused that he was worried. He spoke in even distant tones, like a clinician describing the symptoms of a fascinating disease he didn't want to demean by merely naming. Dr. Davis, does the universe have dark matter? Does it have massive neutrinos? Does it have pancakes?

Davis didn't think it looked like that. He didn't think it looked like anything.

He ticked off what he considered to be the failures of the Zeldovich and Peebles's worldviews: The holes that pancake theory predicted were there, all right, but Davis didn't see anything that looked like pancakes. As for the traditional Peebles theory that galaxies formed and then gathered themselves into clusters, the holes, Davis said, were too big—one was 100 million light-years across—and the galaxies were moving too slowly to have cleared that much space since the beginning of time. "Of course," he smirked, "how do we know the holes are empty and not just dark?" They could actually be filled, he explained, with very dim galaxies, gas clouds that never condensed into stars, or just swarms of strange particles.

Davis pronounced his conclusion in a rather stately fashion that reminded me of the Hubble voice that all cosmologists resorted to when they had to answer the big question, the way pilots all became Chuck Yeager when God rattled the wings. "I'm not sure any theories in their present form pass the test," he said.

"Fine," he added almost to himself, biting off the words. "Make the theorists work for a living." It was hard to tell if he was smug or sad.

That was Davis's view, but froth apparently was somewhat in the eye of the beholder. When the first CFA survey data were published, Zeldovich looked at redshift maps of galaxy clots curling around empty spaces like the lips of waves and saw the cells of Einasto. He and his crew published a triumphant note in *Nature,* pointing out that, in their own analysis of the Harvard data, 90 percent of the galaxies were in strings and clumps. About 10 percent of all of space, they said, was occupied by superclusters. It was triumph for the top-down pancake theory, they concluded.

Much of the astrophysical community seemed to agree. The redshift survey results looked much more like a pancake universe than the hierarchical Peebles model. Sure, the topography was messy, not as cleanly cellular as Einasto had said, but the most important thing was the scale: The voids and clumps were so big—hundreds of millions of light-years. That was a point for massive neutrinos. Glashow, who was a neutrino enthusiast and had a witty quip for every occasion, said, "Not only are we living on some obscure corner of the universe, but the very matter we use is the minority constituent of the universe."

Massive neutrinos as the dark matter got another boost when other groups of astronomers started turning up examples of enormous cosmic structures or, in the most celebrated case—the Bootes void—a sort of cosmic antistructure. Four astronomers—Robert Kirshner, Augustus Oemler, Paul Schechter, and Steve Shectman (who had loaned Davis his own electronic expertise)—from four institutions had put together their own version of a redshift survey. Their technique complemented the CFA team's; instead of scanning the whole sky to a uniform depth, they picked a few selected regions and studied them intensively, trying to see galaxies and measure redshifts to as far away as they could, rather like geologists drilling core samples in the sky. One of their sites was in the constellation Bootes.

In 1981 they announced that they had hit an air pocket. Bootes was full of galaxies, but according to the redshifts, all of them were either relatively nearby or very very faraway. In between these two clumpings was a cavity 300 million light-years[1] deep apparently devoid of galaxies. The Bootes void, as it was called, was three times the size of the CFA voids. It was, by one estimate, half the volume of the entire CFA survey, but it had been imperfectly plumbed. Bootes was so huge that the telescope fields couldn't cover it all; the survey, Kirshner and the others admitted, was like sticking pencils through a watermelon. Maybe they were missing the seeds just by chance.

Through the work of other scientists, the clouds of galaxies, the large-scale structures, also got bigger. Martha Haynes, a Cornell radio astronomer, and her husband Riccardo Giovanelli, traced one chain of galaxies for 500 million light-years, from Pisces to Hydra-Centaurus.

Davis felt upstaged, a bit unfairly, by these kinds of discoveries. He thought the Bootes void was overplayed: The CFA survey had things *almost* as big. There was no way to gauge the statistical significance of one off-scale phenomenon like the Bootes void.

The Bootes void in particular fanned the flames of pancake neutrino fever. Only clouds of massive neutrinos, it seemed, could carve the universe into such large structures. The top-bottom picture of galaxy formation, the Zeldovichian theory, was conquering the universe of cosmologists.

For a while Peebles was nearly alone in his disdain for pancakes. "I remember a time when I used to go around to conferences," he recalled. "I remember in particular a meeting of the Texas Symposium in which there was a day of discussion on galaxy formation, followed in the evening by a workshop on galaxy formation. I was the only person there skeptical of the pancake picture."

1. For the sake of consistency all the distances referred to in this book, unless otherwise noted, are based on a Hubble constant of 50. If the reader wishes to adjust for the possibility that de Vaucouleurs and the other Hubble constant critics are correct and the Hubble constant is more like 100, then he or she can do what the rest of the astrophysicists do: divide all the distances in half.

Where, he kept asking, was the local pancake, the sheet from which the Milky Way and its neighbors had fragmented, and why hadn't we already dove through Virgo?

"As for voids," he went on, "the point I keep harping on is that if galaxies cluster, they have to leave regions outside the cluster that are void. If you pile material up into lumps, you must take the galaxies from somewhere, so you must have low-density regions. The surprise in the survey was the sheetlike character of the galaxy distribution. Sometimes it seems as though the edge of a void is bounded by a sheet of galaxies. That was the big startling thing, the startling discovery that could not, was not, anticipated from the statistics we had at the time. It was just something to which correlation functions are not sensitive. And that was certainly and rightfully pointed out as a good argument for a pancakelike picture.

"At the same time," his voice here got more confident and a little arch, "as one looked in a little more detail at the pancake picture, one saw that it had some problems—namely, that they tend to form such big clusters in the first generation, that you'd better form them very recently, or else, after they form, they will continue to eat material and get just too massive for what we see." That meant, Peebles explained in a long, densely mathematical paper, that if galaxies had to form after the clusters, then galaxies must have formed awfully recently. The supercluster looked young, however; it had not settled itself into a sphere, but was still an irregular jumble, as if it were still in the process of assembling itself. It was the galaxies—with their 17-billion-year-old globular clusters and quasars, baby galaxies shining at redshifts of three, which dated them in the first quarter of the life of the universe—that actually looked old. It still made more sense to Peebles that the universe was built from the bottom up.

Peebles kept holding out for dim stars as the dark matter, but he sounded irrelevant, as if history were running him over, as it had Chip Arp ten years before during the quasar controversy. Massive neutrinos and pancakes were the rage. Peebles sympathized with Arp.

The astronomers, Peebles thought, were beginning to operate like the particle physicists, in a pack fashion, ganging up on one problem at a time and talking it to death. "Yeah, isn't that a natural human tendency?" he asked leaning back.

"Particle physics is moving toward the cosmological mode, as theories outstrip the energies attainable in particle accelerators, which I think is not a good thing for particle physics. When the situation is that way you are reduced to probabilistic arguments, or arguments of reasonableness. 'Here is an interpretation that sounds sensible to me.' But of course those are fighting words!" He laughed. "Someone else is sure to say, what about this other possibility? There is a tendency to follow a fad, but there is no shortage of iconoclasts who will fight a fad no matter how weird or well motivated. So at any time in astronomy you can find voices, lots of them, crying in the wilderness."

In 1981, Peebles, one of the founders of the discipline of physical cosmology, was among them.

In May of that year, in conjunction with the publication of their first results from the redshift survey, Harvard and the Smithsonian hosted a two-day conference on dark matter in Cambridge. Cosmologists from the Northeast attended, and the tone of the meeting was dark. People seemed to be saying apocalyptic things out loud: None of the theories worked. Or that light was not a reliable indicator of anything in the universe.

I found myself thinking about Zwicky and his feelings of persecution by the astronomical establishment. Would he feel vindicated to see that a conjecture he had made fifty years ago had become the central issue in cosmology, undermining all the certainties of his persecutors? Would he be amused by the sight of that same establishment meeting to consider the idea that what astronomers saw in the sky was not as important as what they *didn't* see? Would he pull himself erect at the end of a long day of arguing in Cambridge, look around, and ask, "So, vot's new?"

Rubin was there, looking quite vindicated. "Why should luminosity be a prerequisite for matter?" she asked. "Nobody ever told us all matter radiates, we just assumed it did. I myself question whether there's something funny going on, and we're just being dumb."

Being dumb—that was a classic Rubin phrase. She talks like your grandmother, not a physicist. She told the meeting about her work on a galaxy called NGC 3067, in which light from background quasars could be used to limn the motions of gas in the halo far beyond the outermost stars. In NGC 3067, she reported, 95 percent of the mass was dark.

The next morning we had breakfast. I asked her, if the visible part of a galaxy was so inconsequential, why did it form at all?

"Good question," she blurted between bites. She went on to volunteer that the NGC 3067 data suggested to her that the dark halos and missing mass were gas. I reminded her that that much gas would violate the nucleosynthesis constraints on the amount of ordinary matter in the universe.

She snapped, "If we find the halos, the theory will change," meaning that theorists would find a way to justify however much ordinary matter really existed.

"We know very little about the universe," she went on. "I personally don't believe it's uniform and the same everywhere. That's like saying the earth is flat."

Uniformity of the universe, after all these years, was still a big issue. Davis believed that the universe was uniform. It pained him when critics charged that the CFA had not gone sufficiently deep to verify that uniformity. He was convinced that the redshift survey had sampled enough of the universe to average the local craziness of voids and the superclusters. He sifted the statistics, and they seemed to bear him out; in any direction the average density of galaxies turned out to be the same to within 10 percent.

Aside from the Swiss-cheese topology—which attracted a lot of newspaper space from reporters who had never heard of Einasto's or Gregory and Thompson's work—the CFA survey, Davis knew, had nothing sexy to offer. What it did have were statistics, the definitive measurement of the correlation function, velocity fields, and densities. It was a physicists' survey. It was a statistical bonanza for physical cosmology.

And so one of the first things Davis and his cohorts did with their findings was the first thing that every cosmologist seems to do with a new set of data, which is to go for the big question, the Sandage question, and find out if they can measure omega and recertify the fate of the universe.

The technique they used on their boxful of galaxies was what Sandage had called "backyard cosmology." The belated discovery that the Milky Way and local group of galaxies are under the sway of the Virgo supercluster had quickly turned into another tool for determining the fate of the universe. It was a way, in essence, of putting a chunk of the universe some 100 million light-years across on a scale and weighing it. If the universe was lumpy, with big concentrations of mass like superclusters, then the lumps should attract one another. The denser the universe and the larger the lumps, the more this attraction, and the subsequent motions of the galaxies, should perturb the otherwise smooth expansion of the universe.

In 1972, when Sandage could find no evidence that the Hubble flow was distorted by so-called peculiar velocities, he concluded that gravity seemed cosmically impotent and could never recollapse the universe. But Aaronson and Mould and Huchra had gotten a different answer in the late seventies. They concluded that the local group of galaxies was indeed falling into Virgo (after the expansion of the universe was accounted for) at some 350 kilometers per second. In other words, given its distance from us, the Virgo cluster was not receding as fast as it should be according to the Hubble law. Aaronson calculated that in 100 billion years the Milky Way would plunge right through Virgo's center.

The Aaronson group's results started a stampede of observational cosmologists toward the Virgo cluster. Were we really falling in, and how fast? How big did that make omega? Sandage and Tammann joined the Virgo-centric chase in the company of Amos Yahil, now a theoretical physicist at the State University of New York at Stony Brook.

Yahil had written to Sandage criticizing some of the statistical work in the "Steps to the Hubble Constant." Perhaps because Yahil, son of the Israeli ambassador to Sweden, went to Sandage instead of the journals first, they became friends, and Yahil wound up joining a collaboration to remap the entire Virgo cluster and its surroundings. Yahil found Sandage fascinating and irritating, tough and opinionated, but fair.

Yahil, tall with a high-domed forehead, is a formal man, given to technical rigor, addicted to long equations filled with integral signs, which eventually drove Sandage crazy. "He doesn't have an ounce of intuition," Sandage complained. They had planned six papers together, but Sandage dropped Yahil from the collaboration after only three. According to Tammann, it was because he couldn't keep up with Sandage's pace. They remained friendly.

This time Sandage and his group found that the Virgo cluster did have an effect on the expansion rate of the local universe and was, in essence, pulling the Milky Way into itself. Of all the groups competing to measure the Milky Way's Virgo infall velocity, however, Sandage's got the lowest value, about 175 kilometers per second. Davis and Tonry, sifting their redshift statistics, got the highest answer of all, that we were falling into Virgo at the rate of 470 kilometers per second.

Part of the confusion and the reason for the multiplicity of answers, Davis explained, was that astronomers couldn't agree on what the redshift velocity of the Virgo cluster was. Was it the redshift of M87, the giant galaxy near the cluster's center? Was it the average of the galaxies around the core? Was it the average of all the galaxies in the cluster? And if the latter, how do you decide which galaxies are in the cluster and which are not? Does Virgo contain 1000 galaxies or 5000? This was the kind of dilemma that would cause Sandage to trudge dutifully back to the telescope for another ten years. And in fact he did, to make a complete inventory of the Virgo galaxies.

A different sort of ambiguity, Davis pointed out, infected the attempt to turn the Virgo infall velocity into a measurement of omega, the mass density of the universe. The value of omega that resulted from the calculations depended not only on how fast we were falling in—the faster we were falling, the bigger was omega—but also on how much denser in mass the Virgo cluster was than the rest of space. There were, say, 2000 galaxies in the Virgo cluster—a volume of space 100 million light-years across. Did that amount to three times the normal amount of mass in a volume that big? Ten times the cosmic average? Or were the galaxies overwhelmed by so much dark matter spread so uniformly that 2000 extra galaxies only amounted to a 10 percent increase in the total mass?

Davis was still frustrated. They had worked for five years, spent hundreds of thousands of dollars, and scanned the sky as systematically and robotically—as purely and objectively—as possible with the most advanced detectors and computers that did everything but move the telescope and compose the final reports. They had statistics on more than 2000 galaxies, an entire 300-million-light-year chunk of the universe in a computer along with masterful algorithms to sift its properties. And yet in a way they were still just looking at pictures.

When it came to the questions that really counted—what is omega, is the universe open or closed, does the universe have a pancake structure—all this rigor was strangely irrelevant. The answers always turned out to depend on the same old belief structures, assumptions, and prejudices that hadn't changed since humans were staring out from campfires. Was the universe uniform? Did the same laws apply everywhere? Does light trace the distribution of matter?

Davis chose to answer yes to all those questions and let the computational machinery operate on the CFA universe. According to their own galaxy counts, the Virgo supercluster was only about twice as dense with galaxies as average space, a surprisingly low contrast. That such a relatively shallow lump could induce such a high infall velocity gave the answer that omega was 0.5—an amazingly high result.

If that was right, the universe would have half of the critical mass density predicted by inflation and needed to keep space-time flat and precisely balanced between expansion and contraction. That was the largest value that anyone had ever actually *measured,* and Davis thought it was no coincidence that this measurement was based on the largest volume of the cosmos ever weighed. The larger a hunk of the universe you considered, the more dark matter you seemed to find. This conclusion extended the work of Ostriker, Peebles, and Yahil six years before. If you weighed enough of the universe, he thought, you might find that omega was 1.0, and the universe was closed.

"Note that the overall trend is unmistakable," he, Tonry, Huchra, and Latham wrote in 1980 in a paper that was published in the *Astrophysical Journal.*

> The mass-to-light ratio increases as the measuring scale increases, and as the density contrast decreases. . . . What is the dark matter that dominates the dynamics of systems larger than 100 kiloparsecs? Apparently it does not cluster as

strongly as does the light emitting component of the Universe. If the unseen matter so heavily outweighs the visible galaxies, there is no reason why mass-to-light ratios should be constant, and there may well exist massive systems which emit essentially no light.

That the background bulk could be neutrinos was the first thought on everybody's mind. "It's only the ability of ordinary matter to shed energy, radiate heat, and shake down that can separate it from neutrinos," Davis explained. Evocative as these results might be for the neutrino hypothesis, Davis, aware of Peebles's analysis, was nevertheless still wary. He agreed that they didn't seem to make individual galaxies very well. He was looking for ways to make a more definitive cosmological test.

Just as the redshift survey was blooming into paradigmatic status, relations between Davis and Harvard reached the breaking point. During the redshift survey Davis had come up for promotion, and he then had to pay the price for his devotion to the survey. With hardly any publications and little interaction with the regular faculty, the recommendations from his own department were weak. All his support came from outside Harvard, from people like Peebles and Gunn. "I was so pissed I was ready to kill the chairman," Davis complained with a mixture of amusement and disgust.

Finally Davis was promoted, but without tenure. It rankled him, and he began to look around for another job. When he heard that the University of California, Berkeley, had openings, he and another young Harvard professor applied. Davis had visited that campus on his first observing trip West back in 1979. One of the attractions at Berkeley was Simon White, the Irish theorist and specialist in computer simulations who with Martin Rees had invented dark matter galaxy formation. Davis and White had met at a summer workshop, liked each other, and thought they could work together. Computer simulations of the universe (another tool that Peebles, natch, had helped invent) might help clear up the confusion about neutrino universes.

In the fall of 1980 Davis told Harvard that he had a good chance of getting a new job. Harvard, unfortunately, wasn't paying attention. As part of his effort to rebuild the observatory, Field had decided it was necessary to hire a notable observational astronomer, someone of the stature of a Sandage or a Gunn. The traditional way to fill important academic slots is to appoint a search committee of both insiders and outsiders, which examines candidates and recommends their appointment to the full faculty. Field and his search committee went courting but they weren't having much luck. Harvard's name was no longer enough of a draw. It had little to offer in the way of observing facilities; the innovative multiple mirror telescope—third largest in the world in light-gathering power—to which Harvard had access by virtue of its Smithsonian side, was still largely an untested instrument.

One of the candidates on the short list for this senior professorship turned out to be Vera Rubin. It was a typically painful experience for her. She came to Cambridge, met the search committee, and was approved. The search committee, as search committees do, recommended her to the faculty. The faculty voted her down. Field was aghast. The reason, several people told Rubin (and me), was that she was a woman.

Rubin had no trouble believing that. She recalled another visit to Cambridge and a dinner at MIT. During dinner she was reminiscing with Irwin Shapiro, an MIT professor who is now head of the CFA, about their days in graduate school together at Cornell and was trying to remember the date of a particularly important discovery that had happened around them. Suddenly from the table behind them boomed a voice she recognized, a Harvard astronomer. "That's why I don't like women astronomers. They don't pay any attention to details."

Field gave up trying to attract an outside star, and took up Davis's long-delayed case, voting him tenure. By then, however, Berkeley had come through with a tenured offer, and Davis accepted. For Field it was a double disaster.

"It was nice to be offered tenure; I used it to get a raise at Berkeley," Davis said wryly. "They had screwed around so bad."

So by its first moment of glory, the redshift team was already breaking up. Davis felt they didn't need him any more. The CFA redshift group could not agree on what to do next. The survey had to be extended beyond the original 2400 galaxies, they decided, but how? Huchra had undergone a cornea transplant after the first survey was completed and had been confined to his desk for six months reducing data. Now he was raring to go. Both Huchra and Davis wanted to extend the survey to fainter magnitudes, thus to see fainter and more distant galaxies. Huchra, however, preferred to start in the cores of the rich clusters and then slowly fill in the spaces around them. Davis wanted to range over the whole sky, going deeper—to even fainter magnitudes—but only measuring, say, every fifth galaxy; he had his eye on statistics, as usual. But since he was leaving he lost the tug of war. Just as he was leaving Margaret Geller, another former graduate student of Peebles's with whom Davis had briefly taught the Harvard cosmology seminar, returned to Harvard from two years in Cambridge, England. Although Davis and Huchra continued to collaborate on other observing projects, he left the CFA redshift survey to Huchra and Geller.

"I wanted to work with Simon, I wanted to get back to theory," he said succinctly. "I had fun. I learned modern electronics. I learned observing, learned some astrophysics. It only worked because I was willing to stop writing papers every two months." He shook his head and said he still didn't know the constellations.

In Berkeley, Davis and White, along with Carlos Frenk, a former graduate student of White's who was there on a postdoc, happily began to set about re-creating the dynamics of neutrino universes, but one more twist of tenure fate remained. White had only a temporary research appointment. When his time was up, he lost out to a noncosmologist in a bid for a tenured slot. "It was a big fight and I lost," said Davis. In the mid-eighties he moved to the University of Arizona. The collaboration continued and grew to include George Estafiou, a talented Berkeley graduate student, when they decided they needed a better computer code.

In principle the game of computer universes was the same as when Peebles had started doing it ten years earlier. In practice, however, the astrophysics had gotten more complicated; there was inflation and dark matter to account for (in fact, it was the dark matter, in the case of the neutrino universes, that came to be represented by all the little dots on the computer maps, not luminous matter). Computer simulations were bedeviled with subtle problems. What do you do about the particles on the edges of your "box," for example? The real universe has no edges.

The first step in making a neutrino universe was to assign a mass to the neutrinos—say, 30 electron volts. That would determine how fast they were moving during the crucial stages of galaxy formation and how slippery they were, and thus what were the smallest clumps into which they could be corralled. That, in turn, would determine the spectrum of lumps and bumps the universe would have when it came roaring out of its periods of explosion, inflation, symmetry breaking, and cooking.

Next, a few thousand or a few million points of mass—the number depending on the size and speed of the computer—were scattered in some volume of an imaginary expanding universe in a pattern that mimicked the primordial clumping properties of the neutrinos. The computer keeps track of the motions and gravitational forces on each mass point and produces periodic snapshots of the evolving pattern. It could, of course, also calculate correlation coefficients; velocity dispersions of clumps; and average speeds and densities of the mass points as they raced through speeded-up cosmic time and clumped, swirled, knotted, clashed, and wandered like stray bullets across little voids. By now Davis had a full arsenal of measurements of the real universe with which to compare those numbers.

Some of the first computer neutrino universes were made by Adrian Mellot of the University of Chicago and Joan Centrella from Texas on a Cray 1, the Cadillac of high-powered computers. They turned the results into imaginative computer graphics that showed clouds of galaxies that looked like modernistic impressions of rhinoceroses. There were holes and chains impressively like the free-form froth of the large-scale Harvard data.

All Davis, White, and Frenk had was a VAX, a minicomputer made by the Digital Equipment Corporation that is the workhorse of science. The comparable calculations, with a mere 32,000 mass points, took forty hours of computer time. Davis and White did them again and again, under varying conditions, neutrino masses, expansion rates. One of the most crucial choices was how dense to make their imaginary universe. They did each calculation twice, one for a so-called low-density universe with an omega of 0.2—the highest value that was supported by most responsible observations—and one for a universe in which omega was 1.0, the critical value, the flat space-time prescribed and predicted by inflation.

Davis's desk was soon overflowing with computer charts of these synthetic neutrino universes. The maps were fan-shaped, to resemble the CFA maps of the real universe—pie charts, they were called—with the earth at the vertex of a wedge of sky spreading outward.

"My partners thought it worked," said Davis ruefully of the simulated neutrino universes. "I didn't." At first glance the pattern of dots, dense in some places, connected in chains, sparse in other areas, looked a lot like the maps of filaments and voids that had been showing up on redshift maps of the universe. Moreover the correlation functions of the real and synthetic neutrino universes matched. But, said Davis, there was a problem. The neutrino universes just didn't seem to make galaxies fast enough, soon enough in the speeded-up cosmic computer time, for galaxies to be as old as they look today.

The result was a stand-off. They wrote what he called a "waffling" paper: Maybe neutrinos worked; maybe they didn't.

More calculations and computations followed. The battleground of the universe had shifted from the dome on Mount Hopkins or Palomar to the innards of VAXs and Crays. Gradually it became clear that neutrinos did not work. The problem was that, as Davis put it, in the scenario predicted by the neutrino universe, galaxies, or at least the knots of points that represented them, "formed yesterday." The observational evidence, however, as Peebles had pointed out, seemed to suggest that galaxies had at least formed the day before yesterday.

"We see neutrinos, we don't see galaxies in these simulations," Davis explained. A clump of neutrinos with the right mass for a galaxy might appear in the last frame of the simulation, but the luminous galaxy associated with that clump would not have the properties of the galaxies astronomers had been studying for the last sixty years—galaxies with 15-billion-year-old stars and interstellar dust enriched in heavy elements from generations of supernova explosions. It was possible to jack up the rate of galaxy formation in the simulations by increasing the density of the universe from the low omega of 0.2 to the perfect omega of 1.0—the denser the universe, the faster it evolved. But in that case the clustering process proved to be too efficient. In the pie charts of that universe, all the clumps of matter would have clustered so thoroughly that they ended in tight knots; the galaxies would all be black holes. The voids became gigantic and even more barren than in the real universe. There were ways to patch the picture, but they were ugly. Neutrinos just didn't fit.

Even as Peebles was wandering lonesomely around the Texas Symposium in 1982, the neutrino euphoria was already beginning to die. So far, all the efforts to duplicate the Soviet experiment and Reines's experiment, which had provided the scant evidence for massive neutrinos in the first place, had failed. The Russian experiments, Western physicists agreed, were not precise enough. No experiments, they conceded, might ever be precise enough to measure a mass as tiny as 20 or 30 electron volts.

For some astronomers the final blow was struck in a dwarfish little galaxy called Draco, an amorphous smudge of stars in the constellation of the same name. Draco is one of about half a dozen dwarf satellites of the Milky Way, less than a thousandth its luminosity and orbiting about a quarter of a million light-years out—an unlikely player in the cosmological quest.

In fact these pale clouds made excellent stellar laboratories. Aaronson and Mould were using them as windows into the chemistry of another galaxy. Here was Aaronson, with a genius for trouble, or promising problems, taking spectra of carbon stars in Draco and other dwarf galaxies with the multiple mirror telescope and a photon-counting spectrograph similar in principle to Davis's. Just being there was a thrill for Aaronson. The stars were so faint that he could watch the signal build up in his spectrograph literally photon by photon. "I sit there rooting for each photon," he told me.

And when he was done, he found that the stars in Draco were moving too fast—just as the galaxies in Coma had been too fast, and the outer stars in spiral galaxies revolved too fast. Dark matter had struck again. Even Draco, a tiny little dwarf galaxy, had its own halo of dark matter. Draco, he concluded, lumbers about the Milky Way neighborhood encumbered like Marley's ghost with an invisible yoke . . . of what?

Not neutrinos. The reason, in this case, was their packing properties. Neutrinos of any cosmologically realistic mass were clearly, obviously, uncontroversially, too fast and slippery—"hot" in the astrophysical jargon—to be squeezed and bound around something as unprepossessing as a dwarf galaxy.

If Aaronson was right, argued the critics of massive neutrinos and its associated evil, pancake theory, then massive neutrinos could not be the dark matter.

The reader at this point might be confused. The reader might recall from the previous chapter that Schramm and Steigman, when they proposed neutrinos as the dark matter that could close the universe, had excluded the necessity that dark galaxy halos had to be composed of neutrinos. There could be enough ordinary matter in the universe to account for the halos, they argued.

The fact is that, in cosmology, the confusion was not confined to the sidelines. Astronomers and physicists themselves often didn't seem to know what their colleagues were saying.[2]

Few people claimed or were given credit for grasping all the nuances of inflation, nucleosynthesis, and dark matter. Schramm, a neutrino advocate, was one who claimed to know the big picture. He liked neutrinos because they could "close" the universe and make omega a perfect 1.0. That was the answer required by inflation. If you understood anything at all about grand unified theories and inflation, you would realize that omega had to be 1.0. A true physicist could think no other way; it was the paradigm. The cosmologist's job was to reconcile the observations with that number.

We were outside his office talking to one of his graduate students, a young woman who was doing numerical simulations of the universe. She was doing great, he said, but why was she doing these runs with omega equal to 0.2?

That was what Davis and White did, she answered.

Schramm told her to get rid of those runs. "You're thinking like an astronomer instead of like a physicist," he snorted, adding, "Simon White never understood inflation."

The result was an impasse over what constituted grounds for accepting or rejecting massive neutrinos—or anything else—as the dark

2. The phrase *close the universe* was a perfect example. According to the equations of general relativity, a closed universe was one that would eventually recollapse, and it required omega to be greater than 1.0. An open universe, with omega less than 1.0, would expand forever. A universe with just the critical density, at omega of 1.0, was balanced between the two possibilities; the galaxies had just enough energy to sail to infinity—if anything it was more open than closed. The universe would last forever, but not one day longer. Nevertheless some time in the seventies astrophysicists began to refer to a universe with a 1.0 omega as "closed," perhaps out of some sort of sense of emotional closure of having put the universe back in mathematical balance. Perhaps because in such a perfect delicate balance maybe just the wish that the universe would someday fall back together would have enough gravity to tip the scales of fate.

matter. Despite all the efforts at garnering statistics, quantitative measures, cosmology kept coming down to a judgment call. In the eyes of most cosmologists, the failure of massive neutrinos to make galaxies soon enough ruled them out as the dark matter.

Schramm remained enthusiastic about massive neutrinos as the dominant component of the cosmos, but most of his colleagues did not. Many astronomers thought it ugly or unseemly to have two explanations for dark matter, even if one of them was ordinary stuff. There is a philosophical principle of simplicity called Occam's razor, which holds that the least complicated answer is always the best one. Turner had his own version of the principle, which Schramm quoted in a paper. "Invoking Turner's rule," he wrote, "you can't invoke the tooth fairy twice."

In fact, the confusion in physical cosmology was only beginning. Neutrinos were not the tooth fairy. But if they weren't, what was?

19

THE ASTROLOGER'S CURSE

The failure of neutrinos had a liberating effect on the imaginations of cosmologists. Having been lured out into the wonderful world of ghost particles and dark matter that could make the universe whole, beautiful, and rational, having taken the initial step, they were now lost, and consequently were almost forced to be more adventurous. For the next step they were ready to consider that dark matter might consist of particles that didn't exist, yet, but might. This time Peebles led the charge.

Massive neutrinos, he had long believed, would just not make proper galaxies, and he began to speculate on other possibilities for the dark matter. "I am not a particle theorist. I mean all I knew was that these particle theorists were coming out with all sorts of funny particles, and that many of them had the very convenient properties that they characterized simply as having negligible pressure, of acting like dust. In the wind there was a lot of speculation about different classes of weakly interacting particles. The first, as we saw, was massive neutrinos, but then people recognized there could be particles with other masses.

"I did not have any motivation for particle physics before that," he explained. "It was strictly simplicity, and a blind faith that if the particles were useful, particle theorists would come up with some candidates. I didn't anticipate how many."

The answer was a lot. "We theorists," said Glashow at the Cambridge dark matter meeting, "can invent all sorts of garbage to fill the universe." Many of the new particles that were proposed as the basis of dark matter were the predicted consequences of the rash of unification theories sweeping physics, especially the supersymmetry and supergravity theories. These latter, recall, postulated that fermions (the particles that constitute matter) and bosons (the particles that transmit forces) were interchangeable, which means that for every known particle there was an as-yet undiscovered supersymmetric "twin"—a new boson for each fermion and vice versa. The fact that none of these supersymmetry particles had been discovered yet suggested that there was still a bonanza in store for experimental particle physicists. It also meant that cosmologists had a lot of freedom to let their imaginations roam.

Like neutrinos, the supersymmetry particles would have been created in the big bang. Like neutrinos, the only forces they would feel would be gravity and the weak force. Like neutrinos, primordial photinos or gluinos or squarks—to name only a few putative species—would continually pierce us, in Updike's words, "Like tall and painless guillotines."

Unlike neutrinos, however, supersymmetry particles could have substantial masses. They could be as heavy as, or even much heavier than electrons or protons. The theories did not specify masses for the particles; in fact one conjectured reason they hadn't been discovered yet was that their masses might be beyond the energy range of existing particle accelerators. The more massive they were, the more energy it took to create them. (Another reason was that supersymmetry might be wrong, but that was no fun.)

The idea that these particles could have big masses, Peebles realized, would have enormous consequences for galaxy formation, if they were indeed the dark matter. Neutrinos failed the galaxy test because they were so light—a few tens of electron volts at most. That meant neutrinos moved so fast that they couldn't clump tightly enough to make galaxies; they were too "hot," in astrophysical jargon. In contrast, supersymmetric particles—photinos, say, a thousand or a million times heavier—would emerge from the big bang moving much more slowly. They would be "cold," and they would clump like wet snow. A universe dominated by such particles would be dimpled, dappled, and dumpled on increasingly fine scales. Making galaxies in such a universe would be no problem.

Peebles began giving talks about what would come to be called cold dark matter and its wondrous properties during workshops in the summer of 1982. He was not the first to realize the cosmological potential of these new particles. There was a sort of three-way traffic jam in astrophysical circles. A trio of physicists associated with the Stanford Linear

Accelerator in California (SLAC), Joel Primack, Heinz Pagels,[1] and George Blumenthal, wrote about a universe based on a hypothetical supersymmetry particle known as the gravitino, which was predicted to have a mass of about 1000 electron volts. Being heavier than neutrinos, gravitinos would be slower—not "hot" so much as "warm"—and clump somewhat more easily. Inspired by their work, Turner, Szalay, and Dick Bond, a young Canadian theorist at Stanford, wrote a paper in which they compared the clustering properties of particles over a whole range of masses, light to heavy, from hot dark matter to warm to cold.

Of the three early papers, Peebles's took the idea of cold dark matter most enthusiastically. In a universe made of these new, wonderfully compactible particles, the smallest cosmic structures, like globular clusters, would form first and then, dragged by their massive dark halos, grow together to form larger and larger units. He had reinvented his old idea of hierarchical clustering, building from the bottom up, but in the sexy new language of gauge theory particle physics.

Peebles could scarcely conceal his delight in making the universe out of this new magical panacea. To him it was a return to the old primeval globular clusters game. "One of the consequences again would be, under a pretty broad variety of conditions, of formation of these gas clouds, but this time with the difference that they'd have a halo of dark matter around them," he explained. "So I was led to write a paper on globular star clusters with dark halos of mass around them. That again annoyed some of my astronomical friends a bit, I think. Well, finding dark halos under every bed seemed too much for them, and to find dark halos around globular clusters seems strange. In fact there's very little evidence for globular clusters to have dark halos, but it's very difficult, also, to rule out the possibility that they have them."

A dwarf galaxy is not very much larger or more massive than a globular cluster. It was at about this time that Aaronson produced evidence that the dwarf galaxy in Draco seemed to be about ten times more massive than it looked. Could there be dark halos around dwarfs?

What Peebles and the others were suggesting was not very different in principle from the dark matter scenario that Rees and White had invented back in 1976. In their scheme clouds of so-called dark objects acted as gravitational traps for ordinary matter.

The key to this scheme, recall, was that the dark objects acted as if they were points of mass with no other properties. It didn't matter

---

1. Pagels, a Rockefeller University professor, was also a gifted physics popularizer, author of *The Cosmic Code, Perfect Symmetry,* and *Dreams of Reason*. He died from a fall while mountain climbing near Aspen in 1988. Ironically, or perhaps prophetically, Pagels had written about the possibility of being killed on a mountain climb, imagining himself falling but at peace because he knew that life would always go on even if he himself did not.

what the dark objects were—black holes, rocks, or funny particles—as long as they were massive and slow compared to neutrinos. Which meant that astronomers didn't need to wait for physicists to discover any of these particles to begin calculating their effects on the universe. In fact, astronomers, if they were clever, might discover them first. "It doesn't matter," said Primack, who moved to the University of California, Santa Cruz, and took up dark matter cosmology, "which figment of a physicist's imagination comes true."

Cosmologists came up with any number of names for these wonderful mystery particles: darkons, darkinos, cold dark matter, the missing mass, cosmions. The one that stuck was one coined by Turner: *wimps*—short for weakly interacting massive particles.

For a time, in 1984 and 1985, it appeared that wimps might be more than mere figments. Reports began circulating from CERN, the giant European particle collider, that Rubbia, the hard-driving Italian physicist whose 157-member team had finally detected the *W*- and *Z*-bosons, clinching both the Nobel Prize and electroweak unification theory, had found evidence in higher energy collisions for a new particle. It turned out to be a statistical fluke, but for about a year the possibility that it could be one of the sought-after supersymmetric particles (the photino being a popular choice) had physicists and cosmologists in a fever. "If we have discovered the photino," said Silk rather hopefully, "we will have discovered the secret of the universe."

In the summer of 1984 a workshop on galaxy formation was held at the Institute for Theoretical Physics in Santa Barbara. It drew a crowd of cold dark matter enthusiasts, including Davis, White, Bond, Szalay, Turner, Peebles, and a contingent from Cambridge, England, all intent on gang tackling the problem of making galaxies and the rest of the structure in the universe out of the fashionable new particles. The result was a document that Rees called "the cold dark matter manifesto," full of complicated graphs, in which four astrophysicists claimed that they had succeeded at last in explaining the existence and properties of galaxies from first principles and cold dark something—a triumph of the new cosmology.

Two of the prime authors of this manifesto were a pair of physicists, Primack and Blumenthal, at the University of California, Santa Cruz. Perched in a redwood grove on a hill overlooking the Pacific, UC Santa Cruz looks more like a national park than a college campus; one expects rangers and Boy Scouts, rather than the somewhat counterculturally clad student body and faculty to pour out the doors of the small rustic wooden-clad buildings. MBAs are not a conspicuous element on campus. At Santa Cruz you can get a doctorate in history of consciousness, and students a few years ago waged a campaign to adopt the banana slug as the school mascot.

Since its founding in the sixties, one of those rustic two-story buildings has housed the headquarters of the University of California's Lick Observatory, which operates a 120-inch telescope, among others, on Mount Hamilton outside San Jose. Lick has always been in the forefront of galaxy research; it was Lick, after all, that had helped compile redshifts for Hubble and Humason, and where the prodigious Shane-Wirtanen galaxy counts analyzed by Peebles had been done. Sandra Faber, another of the manifesto's authors and a protégé of Rubin's, had become a sort of guru of dark matter and an ace observational cosmologist. A collection of dark matter enthusiasts and theorists had built up around her.

One mild, damp winter day I drove from San Francisco down to Santa Cruz to hear Primack and Blumenthal explain their calculations on the genesis of a galaxy. It was the sort of day a banana slug would love. Fog shrouded the coast. Fat raindrops dripped like bombs from the redwood canopy. Blumenthal was waiting for me in his office. Reddish blond, bearded, and rangy, he had the sleepy mien of a mathematician, which had been his original specialty. He looked like the sort of man who has spent his whole life in corduroys. Blumenthal proceeded to explain how he and Primack had gone about rigorously reinventing the formation of galaxies.

"There is evidence of dark matter," he began professorially, pacing back and forth in front of a big blackboard that was soon to be the focus of intense scrutiny. "It has been growing in recent years. Meanwhile, standard big bang cosmologies, without exotic cold dark matter, just don't make it as far as galaxy formation is concerned." He reviewed the old Peeblesian theory of galaxy formation, in which small bits of ordinary matter—mostly hydrogen gas clouds—came together and aggregated into larger and larger chunks as the universe expanded. In such a universe, he explained, lumps of every size—from a few suns' worth to trillions upon trillions of stars—should eventually condense. Any astronomer worth his salt knew that that was wrong. Galaxies and clusters and other cosmic structures seemed to come in certain mysteriously prescribed ranges of mass and size. For example, normal galaxies had masses ranging between 10^{10} and 10^{12} solar masses; globular clusters and dwarf galaxies were typically 10^5 solar masses; a large cluster of galaxies was worth about 10^{15} solar masses, but never 10^{16}. In other words, out of all the possibilities, all the conceivable outcomes of cosmic evolution, nature favored a few magic numbers. Primack and Blumenthal and the other prospective galaxy builders had to ask themselves why.

The answer, they concluded, must have something to do with

the details of the dynamics and interaction of the light and dark matter in the universe. The first step, then, was to determine how much of each there was.

The Santa Cruz consortium, he says, proceeded to review and reanalyze all the astronomical data regarding missing mass and dark matter—the galaxy rotation curves and the cluster redshifts and the Virgo infall of the local group—to recalculate the mass-to-light ratios of various galaxies and clusters.[2] This time, to represent the luminous matter, they counted not just starlight but other kinds of radiation as well. Infrared heat radiation from dust clouds, and X-ray and radio emissions from hot clouds of gas in intergalactic space. And they got a different answer than Davis and Tonry had. The mass-to-light ratio did *not* increase with scale.

On every scale for which reliable data were available—from the local lanes of the Milky Way to the richest clusters of galaxies—they concluded, therefore, that the ratio of dark matter to light matter was the same; it was about ten to one. If the dark matter was indeed some exotic elementary particle, then this ratio, ten wimps for every ordinary proton, was a universal constant, a legacy of the big bang. "Particle physics," said Blumenthal grandly, "has to explain that number."

Just then Primack burst into the room, and what had begun as a laid-back tutorial turned into a stereo physics broadcast. Primack, who it became immediately obvious was the spark plug behind this theoretical extravaganza, was slightly thickset, about six feet tall with smooth features and thick brown hair. When he talked about physics, he smiled like a well-fed cat, with broad cheeks and his eyes twinkly slits behind glasses. He spoke with the self-assurance of a kid who has always known that he was the smartest one in his class—and in fact always was.

Primack was a theoretical particle physicist who also had a strong interest in public policy. He had spent a year in the Congressional Office of Technology Assessment, where he had met his wife, Nancy Abrams, a cabaret singer and lawyer who specialized in scientific mediation. Primack reminded me a little of Schramm; along with assurance he radiated ambition. Valedictorian of the Princeton class of 1966, he had earned a Ph.D. on the golden hills of SLAC and spent time as a Harvard junior fellow before landing in foggy Santa Cruz where he led a group to do the particle physics of cosmology—or was it the cosmology of particle physics? He was another convert to God's own particle accelerator, the big bang.

Primack and Blumenthal talked over and under each other, usually interrupting each other's sentences—"But the main point really is . . ."

2. The mass-to-light ratio is just the mass of an object in suns, divided by its luminosity in units of solar luminosity. Thus the mass-to-light ratio of our sun is defined as 1.0. The mass-to-light ratio of the Milky Way galaxy is about 50.

or "What he's trying to say . . ."—whenever confusion crossed my face. As it did often as I tried to assimilate this modern version of a fairy tale in which the smallest, lowliest thing in the universe, a quantum fluctuation, grows up to become a lordly galaxy.

The first element in this story, as Blumenthal had pointed out, had been the determination of a reliable recipe for the physical universe: 10 percent ordinary matter and 90 percent wimps or some other form of cold dark matter.

The next chapter was to ask how this ten-to-ninety broth was spread across the hot, rapidly expanding space of the big bang. As Peebles and the others had surmised years before, the distribution of matter and energy could not have been completely uniform, or there would be no galaxies today. There must have been small—minutely small—lumps and bumps in the embryonic universe. The difference between the eighties version and the old Rees-White seventies version of dark matter theory was in their assumptions about the quality and origin of this bumpiness. According to inflation, the modern version, the lumpiness would arise from quantum uncertainties in the mystical Higgs field during the very early period when the universe was inflating. At the end of the inflation, when the Higgs field finally decayed into ordinary matter and radiation, the quantum fluctuations would imprint a pattern of minute lumps—hot spots or density variations—on the universe.

This lumpiness would have the distinctive property of looking the same no matter what scale was used to examine it. A map or photograph of the early universe on any scale would show similar features— regions of high and low density and, within them, pockets of even higher and lower density that in turn were pocked with high and low regions, and so on. The proverbial bumps on top of bumps on top of bumps.[3]

This graininess was seemingly as subtle as it was fine. The apparent uniformity of the cosmic microwave background radiation implied that when the universe was a few hundred thousand years old, its densest parts were still only about 1 part in 10,000 denser than its most diffuse spaces. When I heard that I remembered that it is often pointed out that the earth is smoother and rounder than a billiard ball—a fact of small comfort to a trekker in the Himalayas. From such slight differences huge effects, effects as glorious as a Tibet or a galaxy or a Virgo cluster, can grow. I doubt that even the most audacious of the quantum theorists in the twenties imagined that the uncertainty principle could have a mani-

3. The modern mathematical term for a system that exhibits this kind of scale-free behavior is fractal. A common example is the coastline of Maine, which looks equally jagged whether viewed from the moon, from an airplane flying overhead, or skipping rock to rock over tidepools in some cove. IBM mathematician Benoit Mandelbrot was a pioneer in the study and popularization of fractals and their application to a wide range of physical problems.

festation as solid as a galaxy, but that was what Primack and Blumenthal were telling me.

As the universe expanded and cooled, the fate of any lump, Primack and Blumenthal explained, was a constant negotiation between the forces trying to smear the lump out and its own feeble gravity trying to pull it together. Initially inflation wrenched it apart, and radiation during the hot big bang era kept it apart. Slowly though, gravity gained a handhold. The denser a lump was, the more strongly it resisted being expanded. Eventually, for any size lump there was a turnaround time when it stopped expanding and started to contract.

Armed with a detailed statistical knowledge of the quantum fluctuations that started it all, Blumenthal and Primack could track the history of the whole community of lumps that was the universe and explicate the dance of light and dark matter that produced the visible structure of the cosmos. The story went on like this.

Gravity could rein in clouds of so-called cold dark matter and compress them much more easily than it could clouds of neutrinos, but it could not cool them off and cause them to condense past a certain point; the particles of dark matter were in a sense buoyed by the energies they had carried out of the big bang and from falling together. The same was not true of the ordinary matter, which could radiate away its energy. The result was that when a blob of universal broth at last began to contract, the ordinary stuff, cooling as it fell, would fall inward a little faster than the dark matter. Ordinary matter would rain toward the center of the blob.

The outcome of this collapse, they explained, depended on whether the dark matter fell together faster than the regular matter could cool and settle out at the center of the cloud. This depended on the mass of the cloud. Small ones would collapse quickly, leaving the dark and light matter no time to separate; the resulting cloud would wind up getting swallowed into a bigger cloud. On the other hand a very massive cloud would take a long time to pull itself together; in the meantime, the ordinary matter would cool off and condense into localized concentrations—somewhat like the small quartets of talkers at a large cocktail party—form stars, and light up. That was why a mass of 10^{15} suns wound up as a thousand individual pools of starlight—a cluster of galaxies—rather than one giant supergalaxy.

When Primack and Blumenthal ran the lumpy, expanding universe through this theoretical gauntlet of cooling and clumping, it turned out that the first lumps to puddle luminous matter and light up had masses of between 100 million and 1 trillion solar masses—that is, galaxies. Galaxies lay on the inside of Primack and Blumenthal's so-called cooling curve—a dense graph that took most of a morning's education to read. So by this epoch in cosmic history, galaxies bounded about like icebergs, 90 percent invisible, little puddles of light encrusted with darkness. "The mass range of galaxies popped out," Blumenthal explained proudly, "we didn't put it in."

Probing, calculating further, they found another feature of their mathematical cold dark matter universe that seemed to mimic the behavior of the real one. The densest lumps tended to group closely together, in regions where the background density was already high—they were bumps on the bigger, longer bumps—like tall mountains next to each other on a plateau. The galaxies produced under these conditions were tight, smooth, and very bright, and they had no spin. The reason for the latter was simple: under these congested conditions, the collapsing dark clouds constantly rubbed against each other, like people stuffed in an elevator and unable to turn around. This was in fact a description of elliptical galaxies, which astronomers since Hubble's time had known were found in the core of large clusters of galaxies.

The next densest lumps, as a class, were more spread out; they were, in effect, foothills or peaks on the flanks of the plateau. These came out of the calculations with their spins intact as spiral galaxies, and they were found where spiral galaxies were found in nature: on the outskirts of big clusters.

Blumenthal and Primack recalled that they were floating on air for about a week in the fall of 1984 over all this. "I claim this as a basic success of the theory," said Primack, purring. It was a spectacular victory for theoretical astrophysics and the cold dark matter picture of cosmology. It seemed, they explained, that galaxies were products of their dark matter genetics. Cold dark matter explained not only where galaxies came from, but why they look the way they do. Cold dark matter even explained the appearance of rich clusters of galaxies. The problem, if there was one, Primack said, his enthusiasm damping slightly, came when they tried to explain larger structures in the universe.

According to cold dark matter, superclusters shouldn't exist yet. There hadn't been enough time since the universe began for such large assemblages—clusters of clusters—to draw together, like developing nations amalgamating out of tribes, and form cohesive structures. On the flip side of the same coin, the same was true of giant voids. The observers kept bringing these things in, however, like dogs arriving at the theorists' back-door steps with half-dead snakes. There was the Pisces-Perseus chain of galaxies, which the Cornell radio astronomers had traced across the cosmos for 500 million light-years, and, most disturbingly, the Bootes void—more nothing than anybody had ever heard of. Were they real? More important, were they typical, or just statistical oddities?

"In our picture," Primack explained, "galaxies form throughout space, even in the voids. Galaxies should not be hard to find. We have to say there are galaxies in the voids."

By now the blackboard in Blumenthal's office was dense and unreadable. We dodged raindrops and slogged to the cafeteria to eat Mexican food for lunch.

Across the bay in Berkeley, where it was sunny and warm, I found Davis sitting in his new office full of computer printouts of simulated universes—clusters and swirls of dots—scowling like a tailor whose client has one leg longer than the other and won't stand up straight. There were neutrino universes, both with high and low values of omega, piled there with their too-tight knots and their neat voids; and cold dark matter universes that looked like shotgun sprays.

Having mastered the trick of simulating universes full of slightly massive neutrinos, Davis and his collaborators had been quick to apply their techniques to a universe of cold dark matter particles. The principle was the same—toss 32,000 imaginary points of mass around in a box representing the universe, and then see how they arrange themselves over simulated cosmic eons. Once again, they did simulations for both high- (omega equal to 1.0) and low- (omega equal to 0.2) density universes. This time, however, the initial pattern reflected the superior small-scale clumping power of cold dark matter.

There was both good news and bad news in the simulations. The good news was that even cold dark matter produced the frothy, stringy galaxy distribution that Zeldovich had posited for the pancakes and that the surveys were finding. The bad news was that none of them looked enough like the universe to suit Davis's quantitative physicist's eyeball. "The observers are fond of pointing at ever larger structures," he complained. "The field is lacking in good quantification. We need numbers, not pointing fingers at large-scale filaments."

It was the numbers, he explained, that killed cold dark matter. In the simulation of the omega equals 1.0 universe—the truth and beauty candidate—the suctioning of the voids was too violent, and the galaxy dots wound up with too much energy; as a result the computer universe was more turbulent than the real one, even though the two looked similar when frozen in a single moment. Meanwhile, in the open universe, the dot clusters and voids just didn't grow fast enough. He showed me a plot. What might have been clusters on it were blurred and indistinct; what might have been voids were not empty but sprinkled with a low density of dots.

"What does it mean?" I asked.

"If galaxies trace mass, the simulations aren't quite working," he concluded clinically. The simulations of cold dark matter that best matched the motions of galaxies in the observable universe confirmed what Primack had told me in Santa Cruz—that galaxies should form throughout space, even where astronomers in real life saw so-called voids. According to those simulations, there were no voids, only low-density regions. In other words if cold dark matter were correct, the "voids" astronomers did see were not empty, they were just dark. That meant that in some sense the structure of the universe as delineated by starlight was an illusion, a few strokes of greasepaint on the bumpy face of a clown.

"We have to bite the bullet," said Davis, not mournfully, "break assumptions. If it doesn't work, give up on it. Try something else. It may be that mass doesn't have very much to do with galaxies. Light may be like snow on the mountains."

Davis painted a picture of space-time as an undulating, mountainous landscape of dark matter. There were big mountains and little mountains, and mountains on mountains. Mountains everywhere. There were mountains in the valleys as well as on the plateaus, but only the highest peaks got snow. Only the densest lumps of matter in the densest regions lit up and became recognized as galaxies. Filling in behind and around them was a backdrop of dark clouds, failed galaxies. Looking only at the "snow," astronomers missed most of the mass of the universe, which was now in the form of dim or failed galaxies.

This idea—that galaxies only achieved luminosity in special high-density areas—was called biased galaxy formation; it went along with cold dark matter like hot fudge and vanilla ice cream. Using it, Davis and White could clean their charts of dots that would not light up, perhaps, and re-create the large-scale structures of the sort that the surveys saw. At the same time these invisible galaxies would add a background mass to the universe.

"The general consensus," Davis concluded, "is that if light traces mass then the universe is open, but if light doesn't trace mass, then omega of 1.0 is easy."

He stood up behind his hopelessly cluttered gray desk and rested his knuckles on the computer sheets in front of him. "We don't have to know what the cold dark particles are. That's as it is for astronomers. I'm not all that comfortable telling physicists that the photino has to be 10 giga electron volts. Let the physicists decide." He smirked.

"Cosmology can decide simple things, like is it hot, cold, so-so? Cold dark matter is a simple picture; it could be wrong, it could be more complex. There could be more parameters. Hot dark matter [neutrinos] had no parameters—it was a falsifiable theory, and it was wrong."

By the end of 1984 cold dark matter was the cosmologist's leading candidate for the formation of the galaxies and other structures in the universe. Not surprisingly, it had its critics; surprisingly, among them was Peebles, whose style seemed to have evolved into tossing out ideas and then shooting at people as they tried to chase them. He had registered an early (and obscurely written) disclaimer in his first paper, wondering if cold dark matter could produce large-scale structure, and he disdained the idea of biased galaxy formation. How did it work? he kept asking. How did a galaxy over here poison it for one over there?

That was the situation when Huchra and Geller embarked on a vast expansion of the CFA redshift survey. Geller is a handsome woman, with a corona of wavy auburn hair and a quick laugh, who projects a slightly disorganized air behind oversize glasses perched on the end of her nose. She had grown up in Morristown, New Jersey, not far from Guth. Her father was a crystallographer, and she had gotten interested in math at an early age. As an undergraduate during the late sixties at the University of California, Berkeley, where her father had moved, she started out majoring in math, but switched to physics. She recalled being advised against astronomy: "Why be in a field where you can only look?"

In 1970 she moved to Princeton where, without ever having taken an astronomy course, she became a graduate student of Peebles. Undergraduate Princeton had just turned coed, but the physics department was still a difficult place, Geller found, to be a woman. She would be, in fact, only the second woman in the history of Princeton to get a Ph.D. in physics, although the graduate school had long since been coed. "I had a lot of psychological difficulties. Faculty members would tell me they didn't think women belonged there."

She found Peebles professionally inspiring, but personally remote. "I learned how to approach big science problems," she said. "He had a program for research. I was impressed. He had a tendency to look at the foundations of a field. He did a lot to make cosmology a physical science. He organized it. Of course, he didn't do anything himself."

Inspiring as Peebles's vision was, the painful aspects of Princeton predominated for Geller. She couldn't get "charged up" about any particular idea in astrophysics. After a while, "I just wanted to get out. I did the minimum to get through." She and Peebles disagreed on what her thesis topic should be, and Geller wound up working with Ostriker instead.

When I asked Peebles, one day, walking from lunch, why Geller had had such a hard time at Princeton, he turned and grinned mirthlessly. "Graduate school isn't supposed to be fun, is it? She wrote a good thesis."

In the fall of 1974 she arrived at Harvard on a postdoctoral fellowship, still disgruntled with astronomy, wondering if she had done the right thing. Geller was not impressed with Harvard, and when Davis arrived shortly thereafter, the two of them ran an extragalactic seminar together, which they essentially wound up teaching. Davis was working with Peebles on their complicated BBGKY hierarchy. Geller thought about getting into observational astronomy. When Davis and the others began to organize the redshift survey, she stayed aloof, except to offer encouragement. "I was not paying much attention," she explained, "I was thinking of getting out of the field. The work I was doing was not interesting. Other people were getting better appointments. I was distressed."

In 1979, she went to Cambridge, England, and spent two years reading and searching her soul. She returned to Harvard as an assistant professor just as the first results of the redshift survey were coming out.

"The results of the first survey were not so striking," she told me, a judgment that seems not to have endeared her to Davis. "They were consistent with what people had thought. I would say there were no great revisions in people's views. Nobody looked and said, 'My God!' So Peebles was right about the correlation function. Big deal."

In 1984 she and Huchra, after much debate among the redshift group (including Davis), decided to extend the survey. By then Davis, the originator of the survey, had quit and gone to Berkeley. In the intervening years, as the redshift team puttered around, trying to decide what to do next, his place had been taken by Geller.

The original survey, as discussed earlier, looked at all the galaxies in the northern sky brighter than magnitude 14.5, and using their redshifts as an indication of relative distance, they constructed a three-dimensional model of the local universe. Geller and Huchra proposed to extend the survey another magnitude, to all galaxies brighter than 15.5. That sounds like a small jump, but increased the number of galaxies involved from about 2,400 in the first survey to about 15,000 and would double the depth of the universe sampled from about 300 million light-years to about 600 million.

The first survey team had been guided by Zwicky's massive galaxy catalog, but at these depths even Zwicky's compilation was suspect. Geller and Huchra had to photograph sections of sky and scan the plates for the locations of fuzzy nebulae. They used a plate density scanner (PDS)— one of those demon automated devices that Sandage despised—at Yale for the task, and the scheduling of time on the machine grated on Geller.

The PDS, she explained, was supposed to be a regional facility. But the only time the Harvard-Smithsonian crew got to use it was on weekends. One fall, Geller complained, they were scheduled on the machine every weekend that Yale had a home football game, and they couldn't even get hotel rooms in New Haven.

For purposes of the survey, Geller and Huchra sectioned the sky like a grapefruit into North-South strips, about 6 degrees wide. Each slice, they figured, contained as many galaxies as they could observe in a single season on Mount Hopkins.

By now the CFA team had the mechanics of the actual observing down pat. Latham and the others had continued to improve the spectrograph. The first slice, centered on the Virgo and Coma clusters, was observed in the spring of 1985. Late that fall, the computers burbled out the first maps under the watchful eyes of Geller and her graduate student Valerie de Lapparent. The known universe, the part that had been reliably mapped, was suddenly doubled and filled in, and what had been a sketch, suggestive but vague, solidified into a portrait. A portrait of how the universe was built and perhaps a revealing portrait as well of what mysterious dark stuff it was made out of.

The two clusters were providentially at the center of the map, and they made a little stick figure of a man. Around them were arrayed arcs and circles and curving sheets dense with galaxies, surrounding bubbles of nothing 50 million light-years or more across. What some other observers called filaments were actually slices of sheets. When she saw it, the first thing Geller thought of was suds. She later told *Time* magazine that she likened the local cosmic structure to a sink full of dishwater.

Geller thought she had at last found her "My God!" As far as she could tell, the data bore no resemblance to any of the standard theories of galaxy formation. The great curving sheets of galaxies were different from Zeldovichian pancakes—the walls were thinner and rounder than pancake theory predicted. The voids and sheets were bigger—in some cases almost to the scale of the survey itself—than cold dark matter could produce. In other words, nothing worked.

When the first report, by Geller, Huchra, and de Lapparent, who was working for her Ph.D. at the University of Paris, was ready, so was the publicity machine. Harvard and the Smithsonian pulled out all the stops. The results were debuted before 1500 astronomers in Houston at a meeting of the American Astronomical Society, in January 1986, complete with a film of the wedge of universe in question, its arcs of galaxies displayed from various perspectives. Geller got onto morning television, whisked away in a limo at five A.M. to have makeup applied for what she claimed was the first time.

As it happened, the Houston meeting conflicted with a cosmology ski meeting in Aspen, so the majority of the people to whom it was most directly relevant first saw Geller's map with the little Virgo-Coma man in the middle in *Time*.

When the second 6-degree wedge of sky was completed, I went to Harvard to visit Geller. We sat in a basement computer room with Paul Kurtz, a bearded, sandal-shod programmer, and looked at computer animations of three-dimensional representations of their slices of universe.

They looked like wedges of transparent grapefruit with galaxies for seeds. Geller and Kurtz gazed—one got the impression they would be happy to do this every day—turning the images over on the screen like jewelers admiring the faces of a crystal, while the voids and sheets rotated slowly past, presenting themselves, lining up, and losing themselves again against the dust of galaxies. "It's like a drug," Geller said, sighing.

Geller noted that ten years earlier, when she was working for Peebles, the redshifts of less than 1,000 galaxies were known; now the number was somewhere around 30,000. "This field," she declared, "has been so data poor. Now it's getting data rich. It's remarkable to me how poorly we understand the universe at low redshift. The field is young, but it's not that young. When so many discoveries can be made in such a short time about the nearby universe, it shows we don't know anything."

She added, "One of the most frightening things is that the biggest things we see are the biggest things we *could* see in this survey." To her the whole notion of the isotropy and homogeneity of the universe was perhaps still up for grabs. "Extrapolations of the universe are like explaining the earth from a map of Rhode Island," she said.

Upstairs in her office, I asked her about the standard theories that cosmologists subscribed to, including inflation, in light of her results. She had the typical, perhaps self-serving, observer's skepticism about the theorists. To her, none of the theories was consistent with the Harvard observations. Biased galaxy formation? "The evidence against galaxies tracing mass is not as strong as we'd like it to be."

Omega is 1.0? "We can't prove it is not 1.0, but we can't prove that it is 1.0. I'm not impressed by the flatness problem—omega is going to be one number. We don't know the cosmological model; there are very few definite tests."

She went on, "Not that inflation is not an elegant theory. It's just not the way to reach conclusions about a measurable quantity. People have stopped paying attention to measurement. Einstein got it right by pure thought, but usually experimental results precede theory.

"When I hear about this dark matter," she complained, "it sounds like the ether. What is it? Where is it, this stuff that explains everything?"

If there were any cosmologists who still doubted that large structures and organizations of galaxies and large voids—whether real or apparent—existed in the universe, the new CFA results swept those doubts away. For years afterward it was impossible to be at any cosmological meeting without seeing during somebody's talk that little stick-man map of Coma and Virgo. How revolutionary you thought the results were tended to depend on how firmly you had believed in voids and superclusters before. Few cosmologists were ready to agree with Geller that they had overthrown all the theories of galaxy formation. To Davis, there were no great revolutions in the new data. Turner cautioned that it was misleading to conclude too much from looking at pictures—that was after all the astronomer's disease. Cosmology needed numbers to be physics.

Peebles, however, found much to admire in the new redshift data. He had counted himself a nonbeliever in voids. The sharpness of the sheets also surprised, and, of course, the fact that they were curved. If there were pancakes, they were all warped.

One model that gained in popularity from the new CFA results was the idea that galaxy sheets and voids had been created by primordial explosions. Princeton's Ostriker along with Len Cowie, who later moved to the University of Hawaii, had invented it in 1981, borrowing from theories of star formation. Stars, it was thought, are born when a cloud of gas and dust is triggered to collapse by the shock wave of a nearby supernova explosion. Suppose, they said, the same thing could happen on a galactic scale. Suppose there was a series of supernova explosions in the first galaxy to form in some region of space. The explosions could push all the uncondensed gas out of the galaxy, essentially killing it in its crib, and blow a bubble in the surrounding primeval intergalactic gas. On the surface of the bubble, and especially where it collided with neighboring bubbles, the gas would be compacted and fragment into galaxies.

A nice feature of the Ostriker-Cowie mechanism was that it provided a natural way to separate ordinary and dark matter. An explosion—the push and shove of atoms—is an electromagnetic event. The blast wave would leave undisturbed whatever dark wimp particles were mixed in with the hydrogen and helium in space. Imagine the Piazza San Marco on a warm spring day, thronged with tourists. In the center of the square a policeman suddenly shouts in Italian the order to evacuate the square. The Italians dutifully scurry to the sides of the square while all the Germans, Americans, and Japanese, who didn't understand the message, are still standing around in the middle pointing Nikons at each other. In a similar way, ordinary matter hears the command to evacuate the bubble, but the dark matter doesn't.

The only problem, as Ostriker cheerily admitted, was that his explosions could only make bubbles 10 or 20 million light-years across—not 100 or 200 million light-years, which was the size of a typical CFA bubble.

Cosmic strings were another idea that came to the forefront of fashion. As pointed out previously, strings were one of the possible kinds of scars that could be left in space-time from the primordial agony of symmetry breaking of forces and the freezing of the Higgs field—along with monopoles and domain walls. They were really thin tubes of false vacuum, with masses of 10^{16} tons per inch. Born infinitely long, they would fly through space twinging like rubber bands, cutting each other and forming loops that would eventually shrink and decay by gravitational radiation.

Since strings were such fascinating objects to hypothesize about, a considerable body of research had built up before a use was found for them. Perhaps massive strings were responsible for carving the primordial matter into clumps to make galaxies and clusters of galaxies. The main proponent of string theory in galaxy formation was another young Englishman, Neil Turok. Turok was tall and good looking with dark, curly hair and a stylish disdain for fashion. He wore sneakers and long-sleeve dress shirts with the sleeves buttoned and the collar unbuttoned. Turok had been a student at Imperial College in London under Thomas Kibble, who had categorized the kinds of space-time scars that could result from symmetry breaking back in 1974. He exuded scientific confidence.

While visiting at Santa Barbara, Turok and Andy Albrecht made model computer universes of strings and then measured the correlation function of the loops—just as Peebles had done for the galaxies ten years before—and got the right answer. Loops and galaxies had the same correlation function. They also had the all-important property of looking the same on every scale.

This was fairly amazing because, as Turok explained triumphantly, there were no fudge factors in cosmic string physics that you could adjust to make the universe come out right. "There's nothing to fiddle with in string theory," he claimed. "It's either right or wrong."

One of the partisans of string theory was Schramm, who had never gotten over his infatuation for neutrinos—now known as hot dark matter. Schramm figured that maybe with a boost from cosmic strings, a neutrino universe would work at making believable galaxies after all. "No single candidate solves all the dark matter problems," crowed Schramm, "we have to invoke the tooth fairy twice." Schramm pushed strings as hard as he pushed everything else in his life. Chicago and Fermilab became a haven for string theorists.

The cold dark matter theorists, Davis and White in particular, continued to try their case. Cold dark matter worked too well on small scales to be cast aside. As for large scales, it didn't deserve to die because of a bunch of pictures. "You can show pictures until you're blue in the face," Davis argued, "but you have to do quantitative statistics—densities and correlation functions."

Technology was in fact conspiring with particle physics to over-determine the cosmological problem. A passel of groups headed south to conduct redshift surveys in the relatively unplumbed skies there. Unfortunately the skies were so unplumbed that there weren't even any good catalogs of galaxies down there. In the new age it was both easy and boring to generate reams of data to confuse the model makers. Many of the surveys were carried out in what is called the transit mode with the telescope fixed. Instead of following the motion of stars and galaxies mechanically across the sky, the new astronomers found it easier to program their CCDs to follow the galaxies electronically, transferring the accumulated signals from pixel to pixel as the galaxies wheeled. No sitting with an aching bladder with your balls wrapped around the cold of the universe. Prime-focus cages grew cobwebs. It was alleged that on the next generation of telescopes, observers wouldn't even have to go up on the mountain; they could run the telescope by satellite link from their urban offices. A young generation of astronomers didn't miss the discomfort they had never known. They ran the computers. They churned out maps of froth and deserts of light. The deeper they looked, the bigger the holes, the longer the filaments. Everybody was wondering how far it would go. Just beyond the last scale, as soon as we survey a little deeper, they kept saying, things will even out.

Then again, maybe not. That was the message David Koo and Richard Kron were bringing back from fifteen years of measuring redshifts of galaxies in just a few patches of sky about the size of the moon. They used photographs and CCD images through the Kitt Peak 4-meter telescope that reached fifteen times fainter, thirty times farther into space and time, than the Harvard survey. In effect Koo and Kron, who began their work as graduate students at the University of Arizona, were sticking a skewer halfway through the observable universe and counting the galaxies at different depths. At that faintness and distance the number of galaxies multiplied astronomically. Koo and Kron estimated that there were more than 10,000 galaxies in each of the fields; instead of measuring the redshifts of all of them, they resorted to random sampling of the fainter galaxies. On their skewer were massive clumps and vast voids hundreds of millions of light-years in extent.

Szalay, who knew large-scale structure when he sees it, joined Koo and Kron to help analyze the data. He found he liked observing. "It was like a giant video arcade," he reported after a run at Kitt Peak. He also discovered why it irritated observers when theorists like himself casually asked why they couldn't have a few thousand more redshifts.

It was a tough time for theorists, who were increasingly caught between a rock and a hard place, between the rich froth of today and the bland soup of yesterday. They were inundated with data about the modern universe, the surveys told you about the distribution of matter at a redshift of maybe 0.1, a few hundred million light-years out. Quasars revealed the universe at a redshift of 3, when it was quarter its present size and galaxies were forming. The microwave background revealed the distribution of matter and energy at a redshift of 1000, when hydrogen atoms were forming and the universe was a few thousand years old.

For twenty years, since the background radiation was discovered, radio astronomers had been running their fingers over that smooth opaque cell of vision, searching for the minute variations from which galaxies and clusters must have grown. And finding none. Down through the decimal fractions the upper limits went. By 1985 the microwave background was known to be uniform to better than a ten-thousandth of a degree (after the variation due to our galaxy's motion was removed). The smoother the early universe became, and the rougher and more contrasty the modern universe became, the more draconian became the measures theorists had to resort to in order to bridge the gap between past and present. According to hot dark matter, there were no such things as galaxies; according to cold dark matter, there were no such things as superclusters. No wonder theorists felt they were in a box.

Dark matter became orthodoxy, or at least respectable, finally, if a date had to be given, in the spring of 1985 when an International Astronomical Union symposium on the subject was convened at Princeton. The assembled astronomers agreed that there was no way out. Turner spun them a witty artistic rundown on the possible dark matter culprits and their roles in cosmic history. The official proceedings were illustrated exclusively with reproductions of his viewgraph drawings.

During a panel discussion at the end, Peebles asked who wanted to vote on the value of omega. The majority was up for it. "Write down their names," Avishai Dekel, a heavyset theorist from Israel, suggested menacingly. Rubin shouted "No!"

For the record, of those who chose a value (fifty-nine brave souls), twenty-eight pegged omega between .999 and 1.001—they believed in inflation. Two voted for a blatantly closed universe. Seventy-one voted "don't know."

By early 1986 the cosmologists had split into so many factions that the field was beginning to resemble the Italian parliament. None of the Chinese-menu hybrids seemed to be able to get a purchase on the truth of how galaxy clusters and voids formed, and they were rich in opposition. At a summer galaxy workshop Bond, the irreverent young Canadian theorist, listed fourteen different models on a blackboard and then shot them all down. He had time left over to take questions.

The more frenzied the theorizing the more cynical Huchra became. "I have theorists running up to me once a week with a new simulation." He made his voice go childish imitating them. " 'Here. Did I make it yet?' "

" 'No.' "

You had to feel that there was a certain amount of observational pride and turf on the line. "John and I are both real skeptics," Geller said. "It's what keeps us working together."

Neither strings nor any other theory seemed to make Peebles, the schoolmaster, happy. In Aspen the string theorists sat in a circle around him on the lawn while he gave them what amounted to homework. They claimed to be grateful for the attention, then out of his earshot Turok confided that Peebles's objections were not all that troublesome and went off to play volleyball.

Peebles put forth no theory of his own. He circulated what he called "screeds" shooting down the so-called standard cosmological model. One of them began with a quotation from Ira Gershwin: "It ain't necessarily so."

I asked him at one point if he didn't feel like the father of all this activity. "Oh, no!" he protested. "Certainly not! The subject would be where it is without me, I went along for the ride. I've managed to keep on the crest of the wave like a surfer, rather than . . . that's probably a good analogy, isn't it? I've kept on the crest of the wave, but I haven't generated the wave. That's generated by the whole community, and by the circumstances discovered. Sorta fun out there—a little precarious—you keep getting splashed with water. Sometimes you fall down into the trough and hit the surfboard.

"I remember writing a very virulently sarcastic paper on sheets of galaxies just before Geller et al. came out. It was full of amusing examples of people being misled into thinking they saw sheets when they weren't there. For example, you know there was a time when people seriously considered the possibility that there was spiral structure in globular clusters."

Even before the new CFA results had come out, Peebles had seen confusion coming. During a meeting in Tucson Peebles and I talked about the deterioration of the cosmological models. In a noisy poster exhibition room, they were serving beer in pint-size wax cups, and we were drinking it. Peebles had been wearing the same plaid shirt and slacks with the same holey undershirt all week long. He looked like an aristocrat who had fallen into a slightly dissolute life.

Among the things he now doubted was the assumption of assumptions, that we live in the mathematically simplest and most beautiful universe. "What evidence is there that omega is 1.0?" he asked himself at one point, and then answered his own question:

"Zilchville. Beware of theorists bearing long arguments."

He pointed out that another way to make the universe flat and reconcile theory and observations was to reinvent the cosmological constant, the repulsive force that Einstein had posited and gratefully trashed when he found out the universe was expanding. Except for de Vaucouleurs, the cosmological constant had had no fans to speak of since 1929.

When I expressed dismay, Peebles smiled wearily. "You have to say, 'I'm not going to know ultimate truths.' " He sighed. "Ultimate truths went out of style at the turn of the century. Instead, we have to be satisfied with making small progress. We can't get the final answer.

"I don't care." He shrugged and took another sip of beer out of his big cup. "I've learned to live with that last statement. Let's face it, astronomy isn't astrology, but we do make progress." He paused and looked down. "The hardest part of the game is to show that we make progress. There's no lack of definite answers. That they don't form a coherent web is just the way it is."

20

DREAMTIME

The search for first principles—the ideas that make the universe fly—proceeded as boldly and fitfully as the quest for the identity of the dark particles inhabiting it.

In 1980 Stephen Hawking received what is probably the most distinguished recognition a physicist can receive, short of the Nobel Prize. He was elevated to the Lucasian Chair of Mathematics at Cambridge; Hawking's predecessors included Newton and Dirac. On the occasion of his inauguration he gave a speech on the state of physics in which he wondered aloud whether the end was in sight for theoretical physics.

Mindful that such predictions had been made before, he concluded nevertheless that the answer might well be yes. By the end of the century it was conceivable, he declared, that physicists would have at hand the ultimate theory of nature, one that described all the forces and particles and explained with no fudging around why the universe was the way that it was. Moreover, he argued, physicists might be supplanted by their computers before the final touches had been placed on the computations involved. "So maybe the end is in sight for theoretical physicists if not for theoretical physics," he concluded.

Hawking was being typically brazen, optimistic, and controversial. What did it mean to know the ultimate law? Was there a difference between that and God?

By the very early eighties the particle physicists had blasted their way to within a billionth of a second of the beginning of time. The cosmologists contemplating the signatures of strings or supersymmetric gizmos in the arrangements of the galaxies had imagined their way a billion billion billion times closer. The veneer of existence was getting very, very thin, but it was in that last little crack of time—where space foamed into chaos and the spheres rang with harmonies undreamed of and symmetries were enfolded more intricately than a rose, where nothing happened and everything was possible—that the secret of gravity and existence lay.

If there was an answer possible to the question that Sandage most venerated—Why is there something rather than nothing at all?—physicists hoped that it might be found in the elusive fabled quantum gravity, which would finally marry general relativity and quantum theory and perhaps stave off the singularity. The answer, if it existed, lay with that totally unified and symmetrical force that manifested itself at the first instant of time—an unreachable El Dorado of elegance. Was quantum gravity the word that made the universe fly?

In the early eighties such a theory did not exist. Hawking and a few others hoped that supersymmetry and its cousin supergravity would lead them to El Dorado, or as he put it once, "the mind of God," but the case was not proven. (In fact, supergravity soon failed.) Lacking even an idea of what the principle behind the ultimate theory, if it existed, might be, because most of the physical evidence of that era had been erased by the magic of inflation, physicists who tried to deal with the origin of the universe were like bumblers in a dream, not knowing the answers or even the right questions to ask. Strange ideas floated by, paradoxes and crazy ideas from the subconscious basement of physics, to be grabbed onto and ridden by physicists hoping to float to the conscious surface of observable reality.

Only a few wild men of physics ventured into that realm beyond testable theory, leaving their reputations behind, groping, as Wheeler put it, "for an idea for an idea."

The original wild man was Wheeler, the black hole inventor, himself. When Princeton retired him in 1978 after thirty-eight years, he traded his teacup for a Stetson, read up on Sam Houston and moved to the University of Texas in Austin, where he could have students again. He had a huge office and a new house in the Austin Hills with an indoor swimming pool, in which he swam a quarter mile daily. During the day he ran up and down the stairs to visit his graduate students and postdocs on churning little pistons of iron legs.

In what some would call his old age, Wheeler came to view himself as a central collector of ideas in physics. He mused about establishing what he called a "charismatic chain" of contacts between the great thinkers of the ages who had known each other. "Parmenides, Socrates, Plato, Aristotle, and so forth—each of them knew the one before. I would like to have some souvenir, some word or stone or monument that symbolizes the contact between the two. I'm resolved to have one. It has to go through the world of Babylon, the Middle East, the Arabs, and Spain, up on into France and England. At the modern end I would like to have Bohr." And perhaps at the end beyond the end, Bohr's student.

"If there's one thing in physics I feel responsible for, it's this perception of the overall picture of how everything fits together," he explained. "So many people are forced to specialize in one line or another, that especially a younger fellow can't afford to try and cover this water-front. Only an old fogy who can afford to make a fool of himself." He laughed.

Once, when a wealthy student of his back at Princeton had offered to commission the university a sculpture by Henry Moore, Wheeler had visited the sculptor and been impressed by his technique. Moore's workshop, Wheeler observed, was a warehouse of shapes. The shelves were crammed with bones, stones, and carvings of different sorts. When Moore was working on a piece, he would select half a dozen or so objects whose shapes he wanted to incorporate into the sculpture and set them out on a card table where they could work their way into his subconscious. Ever since then, Wheeler had thought of himself similarly, a sculptor trying to mold a theory from a few favorite ideas. He sprinkled his office with what he called "clues," drawings and metaphors that would keep his subconscious churning.

"The central issue is to find the plan of creation, period," he said, shaking a fist in the air when I asked him what he was doing. He proceeded to draw a cartoon on the blackboard of an elaborately cuffed hand juggling a half dozen balls inscribed with mystical-looking equations—each representing in fact an important principle in physics. "It amounts to taking half a dozen balls and tossing them up into the air all the time and seeing them come down in new patterns. I have to keep making the circuit of these ideas," he said, touching up the sleeve. He frowned and stared. "Let's see, have I left out any clues?"

To Wheeler the biggest clue was still black holes, which he now called "gates of time." It was not the blackness of their surfaces that interested him any more, it was the oblivion inside, at the center, the notion that space-time had to come to an end. The lesson of general relativity, he always preached, was that there was no "before" the big bang, and there would be no "after" after the big crunch. The universe created and now contains space and time—space and time did not create or contain the universe.

Did the universe then also create the laws of physics, which are based on space and time? Or did the laws of physics create the universe? Did the law exist somewhere, somehow, independently of space and time, independently of the universe itself?

"The gates of time tell us that physics must be built from a foundation of no physics," he liked to say in his sloganeering fashion. How was that possible?

As a young Turk of general relativity, Wheeler had shared the hope that all physics would eventually be explicable as curved space, that is, geometry. But after a lifetime of staring at the problems he suspected that the key to the irreducible mystery of the origin of law and the universe had to lie instead in quantum theory. Why did the uncertainty principle exist?

As the reader will recall, according to that paradoxical-sounding law, certain properties of matter or of any system—such as its waveness or particleness, its position or velocity—remained suspended undefined, in a limbo of possibility, until somebody measured it. One of the stranger consequences of this fact was that nothing in the universe could ever be considered truly empty—particles could arise from quantum fluctuations in empty space (and as we have seen, grow up to be galaxies). In the as-yet-undiscovered quantum gravity, random fluctuations in a primordial nothingness might give rise to space-time itself.

Wheeler had a name for this primordial potential; it was super-space, a sort of mathematical ensemble of all possible universes, all possible physics. In superspace there were incredibly dense universes that collapsed in five minutes, universes in which all stars were blue, in which all stars were lumps of iron, universes in which there were no stars. Universes with unicorns and magnetic monopoles. Every possibility was there. Most of these universes would be without life; they would be, in Wheeler's words, "stillborn."

It was in superspace, Wheeler maintained, that the central mystery of quantum mechanics, the mystery of the wave function, fully asserted itself. Just as the wave function of the electron was spread throughout space before it was pinned down by measurement, the wave function of the universe was spread throughout superspace, a superposition of all possible universes, before it was somehow "collapsed" by measurement. But there were no observers to do an experiment at the beginning of time; Sandage was billions of years in the future. So how or what collapsed the wave function in superspace? How did God choose our universe? Or—as Einstein wondered—did he have a choice?

One way out of the paradox was the Everett-Wheeler many worlds approach that Hawking had espoused. God did not choose; there were uncountable billions of parallel universes. Every possibility was real.

Another idea that enjoyed some controversial vogue was called the anthropic principle. It was a descendant of the large-number coincidences that had captivated Eddington, Dirac, and Dicke. The proponents of this view, Dicke and Carter among others, argued that the properties of the universe—in particular the values of several physical constants—had to be just so, or life as we know it would have been impossible. This was a totally hindsight approach to physics. Wheeler followed along in its footsteps for a while, but eventually had come to embrace an even more radical principle he called Genesis by Observership.

It was based on the uncertainty principle, which, he explained, somehow involved the observer in the creation of physical reality. "In some strange sense," Wheeler said, "the quantum principle tells us we are dealing with a participatory universe." How far could you take this principle? He wondered aloud if it could be a prescription for building the laws of physics out of nothing—a command to create, to counterpose the black hole's urge to destroy—a way to create both the universe and its laws at once. Why else would there be something as strange as the uncertainty principle in nature, anyway?

As an example of the weird power of quantum theory, he described a famous *Gedanken* known as the delayed double slit experiment, in which by a choice in the present an experimenter could apparently influence what had happened in the past: An electron flies down a very long tube toward a screen with two parallel slits in it. Far beyond the screen—light-years, if you want—lurks the physicist with a choice of two mutually exclusive experiments to perform. One would measure the momentum of the electron and tell which of the two slits it went through; the other would measure its position and record part of an interference pattern from the electron's having passed through both slits (as a wave can). (Over time, more electrons would build up the complete interference pattern.) In principle the physicist could decide at the last instant what to measure and thus whether the electron had passed through one or both slits long after it had passed the screen.

In this result Wheeler saw a small sliver of daylight to the big bang. Without an observer there was no physics, but with an observer there could even, in principle anyway, be retroactive physics. According to the idea of Genesis by Observership, then, the universe was a kind of self-excited circuit. "Do billions upon billions of elementary acts of observer-participancy add up to all that we call creation?" Wheeler asked rhetorically, and then answered himself. "No other way has ever been proposed to bring into being plan without plan, law without law, substance without substance."

In an article in *Science 81* he wrote, "The past is theory. It has no existence except in the records of the present. We are participators, at the microscopic level, in making that past as well as the present and the future."

One day when I was visiting him in Austin, Wheeler asked out of the blue, "Do I strike you as ruthless in talking about these problems? A friend called me ruthless because the word implies a little insensitivity to human problems and concerns. I don't relish the adjective. And I confess I'm not very much the one to sit around and shoot the bull with students and so on. Time is short. I'm afraid I've never given the time it would take to make really close friends. Someone you spend three hours a week with, say, as minimum commitment to maintaining a close relationship, too much in a hurry."

He sounded lonely.

Few physicists understood Wheeler's position, and even fewer agreed with him. The notion that quantum principles somehow had something to do with the creation of the universe, however, became a favorite idea of those who fished these speculative waters, especially after the advent of inflation theory showed that it was possible to grow the entire visible universe from virtually nothing—an infinitesimal patch of false vacuum.

Suddenly it didn't seem like a big jump any more, only a tiny quantum leap, from nothing to something that could become the universe. If all the properties of the universe, such as charge and momentum, balanced out, as Guth, who was a fan as well as a scholar of theories of nothing,[1] pointed out to me, no law of physics forbade the spontaneous appearance of a universe—or a quantum piece of one. "It is tempting to imagine creating the universe from literally nothing," he said. "Such ideas are speculation squared, but on some level they are probably right."

Nothing, some physicists implied, might be the ultimate symmetry, everywhere, everywhen the same—a sort of gridlock of perfection. Mostly we knew what nothing was not. It was not anything. But it was the possibility of everything. And perhaps such beauty, nothing, was unstable. And the result was every once in an eternity it twitched.

---

1. The first soul brave enough to suggest that the universe was indeed nothing was Ed Tryon, now a physicist at Hunter College in Manhattan. Tryon blurted it out during a seminar by Sciama, "Suppose the universe is just a quantum fluctuation." Everybody laughed. There was little in the response to indicate that eminent scientists would be accepting travel money to talk about this stuff. Tryon eventually published these notions in *Nature* in 1975 and was mostly ignored. Peebles and Dicke had mentioned his work in their famous 1979 paper about enigmas and conundrums.

One April Thursday afternoon in Cambridge, when I was visiting MIT, Guth announced that we had to attend a seminar at Harvard. He wanted to hear a young Russian émigré by the name of Alex Vilenkin from Tufts University describe a new theory of the origin of the universe, of how it could have emerged molelike from nothing. I drove my car while Guth rode his bicycle, and we met outside Lyman Hall, the old red brick physics building just outside of Harvard Square in the late afternoon.

Inside a dusty classroom lecture hall several curious physicists lounged. Green spring light fell through the windows. Vilenkin was dressed in a dark suit. He was tall and immigrant slender with wavy black hair and ruddy high cheeks that made him look much younger than his thirty-five years. He had the sardonic eyes of an exile and a formal reserve that belied adventurousness.

Vilenkin's version of the infant universe, accompanied by chalk renditions of warped space-time sheets and wormholes, was of a kind of metaphysical mole. According to his proposal a bubble of universe, space-time, had "tunneled" into a Wheeleresque superspace of possible space-times and then tunneled again into "real" space and time. Tunneling was a well-known quantum process by which the wave functions that represented particles, or whole systems—or the universe—could melt through classically impregnable barriers, like a baseball passing through a wall.

But from where had the universe tunneled into this realm of existence? In Vilenkin's words, "from nothing."

What else was there?

Once it had entered a real space-time, Vilenkin's tiny bubble, which formed a closed universe because that took the least amount of energy to create and was thus the most likely to arise, inflated and went through the standard expansion and evolution of the big bang.

Vilenkin was greeted mostly by confusion. Afterward he, Guth, and Sidney Coleman sat and had a conversation that Lewis Carroll might have enjoyed, about nothing. It was Guth and Coleman's first encounter with the mole theory of Genesis.

Coleman, pale and bearded, pondered what he had heard. "What is nothing?" he asked, pressing his fingers together and staring through them.

"Nothing," answered Vilenkin deliberately, "is no time, no space."

Coleman brooded like a rabbinical scholar about that for a while. "There is an epoch without time," he said finally as the shadows lengthened. "It is eternity. So we make a quantum leap from eternity into time."

Then, as good physicists did, they repaired to a Chinese restaurant.

Later I went back to Tufts to learn more about Vilenkin and his theories. His life, it turned out, had been as adventurous as his ideas. Born in the Soviet Union, Vilenkin had studied physics and received an undergraduate degree at Kharkov University in the Ukraine, but his professor couldn't get him into graduate school because Vilenkin was Jewish.

He studied and published on his own for five years. Then he was drafted into the army, where he worked as a lab assistant. When he got out all he could find was a job as a night watchman in a zoo. They gave him a rifle, which he didn't know how to shoot. The main qualification for that job was that he was not a drunkard. "There was a little booth," he explains, "where there was alcohol."

In 1976 he was able to emigrate to Italy. While he waited for a visa to the United States, he saw an ad for the State University of New York at Buffalo, so he applied and went there, a decision he now thinks maybe he rushed a little. In a year he had a Ph.D. in biophysics. Tufts hired him as a solid state physicist. "Nobody minded when I switched to cosmology."

Vilenkin worked on cosmic strings for a while and then switched to the ultimate question. While he was in his wandering days before emigration, doing solid state physics at Kharkov, he had met and hung out with a frustrated scientist named Piotr Fomin, who—like Tryon—had suggested that the universe could appear spontaneously without violating any conservation laws. Fomin tried to sell his notion to Zeldovich and his colleagues, but they thought he was crazy. Fomin did publish a paper in 1975 and received no more attention than did Tryon.

Vilenkin's rather intuitive notion was that the potential universe could appear as a little bubble, like foam in an ocean of nothingness, a sort of virtual universe. Most of these bubbles would collapse immediately back into nothingness, but the lucky ones would grow and become classic space-time manifolds. "When you boil water," he explained, "you need bubbles of a critical size." Then he added, perhaps less than helpfully, "In this case the water doesn't exist."

Vilenkin took the view that the laws of physics should not just describe what happened to the bubble once it appeared, but must somehow determine how and why it appeared at all. "The bubble pops into existence in accord with the laws of physics," Vilenkin declared one rainy day in his Tufts office. "The laws of physics are already there."

"Where?" I asked.

"In God's mind."

Was that, I wondered, the same as the vaunted nothing that Coleman had called eternity? Vilenkin has a sardonic taciturnity and a reluctance to get dragged into foolishness. "Our laws are not meaningful laws," he went on, "all those we have are deeply rooted in space, time, entropy, energy, momentum, and they all lose meaning without space and time."

I asked him if nothing was the ultimate symmetry. "In thinking about these things you develop mental pictures," he answered. "We are not at the stage where these words have any meaning." People, he acknowledged, were still getting used to the idea of nothing. "The initial reaction was that it was not science."

This, he went on to point out, was not a new conundrum. He reminded me that Saint Augustine in a famous passage had asked what God was doing before he created the world and then answered it. "The answer is simple." Vilenkin repeated. "Before God created the universe, there was no time. The question is meaningless."

I went home and looked it up. What Augustine had also joked first was that God was busy preparing hell for those who ask such questions.

Hawking had no fear of hell. In the eighties he tackled Augustine's question head on. Hawking was idolized in the media often as "the greatest physicist since Einstein," somewhat unfairly to the community of geniuses who had devised such wonders as quantum chromodynamics and grand unified theories and, more recently, the new discipline of chaos theory. But in one sense he clearly deserved the elder wild-haired one's mantle. Not since Einstein had anyone boldly proclaimed and embodied the faith that the universe was comprehensible—that it was ruled by laws that could be discovered and understood by humans.

As Hawking told *The New York Times Magazine:* "The whole history of human thought has been to try to understand what the universe was like."

As he contemplated the coming ultimate unified theory of particle physics, however, he was vaguely disgruntled. To explain the state of the universe or any other system at any given time, it was not enough to know the laws of physics, you also had to know the initial setup, what physicists called the boundary conditions. Consider a pool table full of billiard balls. Newton's laws of momentum and energy were enough to describe how the balls bounced and rolled and caromed off the cushion, but to predict where they would end up, you needed to know where they were at the beginning of play: racked in a tight triangle with the 1, 2, and 3 balls on the corners, the 8 in the center, and the cue ball coming across the green felt in a certain direction, speed, and maybe with a little English on it. That was the boundary condition in a pool game.

In cosmology, Hawking knew, the boundary condition of the universe was the big bang singularity, where gravity crushed space, time, and the known laws of physics out of existence—in Wheeler's view a chaos of unrequited measurements waiting for the word to fly. Hawking had spent most of his adult life dealing with singularities in their role as destroyers at the centers of black holes. It was in trying to cope with the anarchic possibilities that could come out of this cosmic wild card at the beginning of time that drove otherwise sane physicists to notions like the anthropic principle. Hawking had once called the anthropic principle a "counsel of despair."

Hawking believed that a theory of the universe that did not include an explanation for the initial condition—the racking of the pool balls, if you will—was not a complete theory. It amounted to saying "The universe is as it is because it was what it was."

He had two choices. He could say the universe was the way it was because that was just the way God or the singularity made it—a random, arbitrary feature of existence. And that would be the end of physics. Or he could dare to go beyond description and try to explain *why* the universe had been the way that it was at the beginning. Hawking wanted to go ahead; he would never be content with a physics that just described. For him, as for James Bond, the view would always pall. That meant groping for some law or principle that even covered the point where law ended. That meant assuming that, somehow, even the beginning of time was subject to some greater law.

If he wanted to get rid of unpredictability (and God) in the universe, Hawking had to (gulp) get rid of the singularities. He was happy to do it. Even a relatively benign singularity as the big bang appeared to have been, he figured, was too much for a universe supposedly ruled by law. Like dictatorships, singularities could go from benign to ugly. If the laws of physics could break down at the beginning of time, they could break down any time or anywhere.

Wasn't this a contradiction of everything he had worked on and done for the past twenty years? Yes and no. Hawking's reasoning went like this: Singularities were a consequence of the classic theory of general relativity, which said that too much mass in one place could squash itself to infinite density. In the as-yet-unrealized theory of quantum gravity that everybody hoped would one day supersede general relativity, it was possible that quantum effects would prevent singularities from existing. Then there would be no place, even the beginning of time, at which the laws of physics—whatever they were—did not hold.

Hawking called this the "no boundary Proposal": The boundary of the universe, in whatever the final theory turned out to be, would be that it had no boundary, no place where space-time and the laws of physics disappeared. According to Hawking's proposal, the universe was regular and lawful everywhere and everywhen; there was no place for God to poke his nose in.

This was not such a dumb assumption. What sort of demon amok was the so-called big bang singularity anyway? A rather tame demon, Hawking thought as he reviewed the situation. The fact was, as any Fermiland cosmologist could tell you, the universe seemed to get simpler and smoother and more elegant the further back in time it was traced—not uglier and wilder. The universe started out smooth and bland as the microwave background, the forces were all unified, the particles didn't have mass. The universe seemed to have started out as simply as it could. According to Hawking's proposal it had no choice; order ruled absolutely. Did the no boundary mean that the universe *had* no beginning after all? Yes and no. One way out of that puzzle was to think of time as a circle. Any place could be the beginning and if you went far enough around you came back to the beginning.

In 1982, Hawking mentioned his no boundary idea at a cosmology conference at the Vatican, where the big bang had been orthodoxy since the early 1950s. Afterward he met the Pope, who impressed him as being very keen on the big bang and approving of modern cosmology—as long as the cosmologists, chuckle, chuckle, didn't try to look beyond the big bang. Inwardly Hawking chuckled back.

Later that year Hawking went to Santa Barbara for a cosmology workshop and fell into a collaboration with James Hartle, a large, round, and slightly softened man, balding with an egghead look and a diffident air, who had worked with Hawking back in Cambridge on black holes. Hartle had also been thinking about quantum origins of the universe. Whereas Wheeler and Vilenkin just talked about the universe as a quantum fluctuation, Hawking and Hartle decided to actually compute the wave function of the universe, the way Bohr and his cohorts had calculated the wave function of the hydrogen atom sixty years before.

Mathematically, Hawking's no boundary proposal resembled superficially Einstein's solution to the problem of the boundary of the universe. Einstein had declared that space curved back on space, back to join itself, and thus had no boundary. Hawking and Hartle did something similar except that their calculations included a mathematical convenience called imaginary time. The distinguishing characteristic of imaginary, as opposed to real, time, was that one could move backward or forward in it, which meant that they could treat it mathematically as just another dimension of space. Their model of the universe then was a four-dimensional sphere. Hawking liked to start lectures by asking his translator to draw such a sphere on the blackboard.

Their quantum cosmology assumed a closed universe, although it was one that could take an infinite amount of time to close, just as the inflationary universe could be closed but indefinite. You could think of Hawking and Hartle's space-time as being like the earth, in which time represents the distance from the North Pole and the distance around a latitude band is the size of the universe. As you went from north to south, the universe expanded and then at the equator began to contract.

At the North Pole, zero latitude, time went to zero, but just as at the geographic North Pole, nothing weird happened; it was just another point. The law extended everywhere. Asking what went before the big bang, said Hawking, was like asking what was a mile north of the North Pole—it wasn't any place. Or any time.

Another way of thinking of this nonsingularity business was in an analogy with the atom. The classic theory of electrons orbiting a nucleus predicted that the electrons would eventually spiral right into the nucleus and all the atoms in the world would collapse. That did not happen; atoms are stable. The reason was the uncertainty principle, which smeared out the position of the electron, so that it could never be thought of as closer than a certain small distance from the nucleus. In a similar way, quantum fuzz might smear out the condensed or collapsing universe, preventing it from ever achieving infinite density. When it was small, the universe had no more precise attributes than an electron. Zero, however, is a precise number and so the uncertainty principle suggested that the universe could never be exactly zero, only very, very close.

The simplicity and smoothness of the big bang suggested to Hartle and Hawking that the universe had been born about as close to that quantum smear around zero as it could get—what physicists called the ground state.

Their answer, then, to the question of why is there something instead of nothing was that we are descended from as close to nothing as you can get. The infant universe was a bare whisper on the white clouds of eternity, a kiss so soft it could not ruffle a soap bubble. The universe had been born smooth and unruffled and then had gotten lumpier and worse and more disordered ever since; all its complexity came from history and the corruption of time.

Hawking and Hartle constructed an entire quantum mechanics of the universe analogous to the quantum mechanics that physicists applied to submicroscopic systems. In ordinary quantum mechanics the wave function represented all the possible histories of an electron, say, as it went from point A to B. The wave function for the universe was a description of how the universe evolved. It was a combination of all the possible ways the universe *could* evolve, all the possible geometries it could assume over time. Hawking had never stopped subscribing to the many-worlds interpretation of quantum theory. To him all these geometries did exist in one universe or another; they were all out there—wherever "there" was. This particular universe, any universe, was just another quantum effect.

"It is a rather audacious theory," Hawking told *Vanity Fair* in

1983. Hawking's style had changed. For years he had given lectures by projecting a written text overhead as he dragged his paralyzed mouth through a forty-five-minute talk. Then on one trip, as any scientist will do, he forgot his slides and had to extemporize, drafting one of his graduate students to translate as he went along.

Hawking found that he liked that way of doing things. It gave him more flexibility and, in effect, control. He began to cultivate a comedian's sense of timing. The adroit pause and then the added quip. The quick spin of the wheelchair and then stillness to punctuate his conclusion. He traveled the world preaching that there was no "north" north of the North Pole.

What Hawking had in mind for the particle theory of theories when he started out in quantum cosmology was something called supergravity, an extension of supersymmetry. Supergravity soon died in the same mathematical minefield of infinities and absurdities that had killed other attempts at quantum gravity. But in 1984 a strange new theory—new to most physicists—emerged from twenty years of obscurity, apparently and miraculously free of these demons, and swept through theoretical physics like a prairie fire. Some wags and most media types referred to it as the theory of everything.

The theory was called superstrings. According to it, elementary particles were not points, but incredibly tiny vibrating loops of "string."

Such a relatively small adjustment—from points to strings—introduced into physics a rich and beautiful mathematics that seemed as seductive as a siren singing on the rocks of dreamy speculation. String theory's founders were seeking an explanation for why quarks were never found alone in nature; the idea was that they were bound on the ends of strings—cut a string and get two new ends, new quarks. When quantum chromodynamics took over as the theory of quarks back in the seventies, John Schwarz, then a young research associate at Caltech, refused to give up the haunting music of strings and with a succession of collaborators—most recently Michael Green of Imperial College in London—pursued it deeper into physics as an explanation not of nucleons, but of all elementary particles.

They spent every summer at Aspen where Schwarz has a house. In 1984 while in Aspen they made a breakthrough and were able to show that the only two mathematically consistent theories possible using superstrings not only succeeded in solving quantum gravity without the usual anomalies and infinities but also seemed to include all the other forces of nature as well. Every summer the particle physicists put on a cabaret show in Aspen. One year Gell-Mann had stood up announcing he had discovered the theory of theories and men in white suits had rushed him off, by prearrangement. In 1984 Schwarz stood up and repeated the claim and was also hauled away. The assembled physicists laughed. They laughed for the last time.

Ed Witten, a young Princeton physicist legendary for his brilliance and quickness who rapidly became one of the ringleaders of the superstring revolution, called superstrings a piece of twenty-first-century physics that had fallen into the twentieth century, and would probably require twenty-second-century mathematics to understand. This is what the twenty-first century had to say about the universe: Space-time had ten dimensions, although, alas, you could only park your car in three of them. The world was made of little—if little is a word descriptive enough for entities smaller than geometry—loops of "string" flopping and wriggling like a trout in the bottom of a boat. The "super" part of the name came from the fact that supersymmetry, and thus all the particles cosmologists loved, came naturally out of the theory—at least in principle. In practice it was difficult to calculate anything about the real world from superstring theory.

Unlike cosmic strings, which were little tubes of primordial false vacuum energy, superstrings had no internal structure. They were irreducible, the way a geometric line is irreducible. They were not matter, they were not energy, they were not geometry—they were the elements out of which all those things were made.

You could think of the elementary particles as notes played on God's ten-dimensional guitar. All our ears could hear in this cold lonely universe in the era of galaxies and protons was the lowest bottom octave. All the particles known to physics (and a lot still unknown), including the ponderous Ws and Zs, the undiscovered photinos, belonged to this lowest octave, in which their masses were negligible (to a string's way of thinking). The next octave and every octave after that would raise the mass-energy of these notes by nineteen orders of magnitude.

There were an infinite number of octaves, theoretically, but to hear even the next highest, physicists would have to pluck a superstring with a Planck mass of energy, 10 billion billion giga electron volts. In the eighties, the most wild plans of accelerator builders fell about fifteen orders of magnitude short. So the full power of superstrings was reserved for a crack of a moment at the beginning of time, when God's accelerator was fully charged, a golden second when the symmetry of nature must have been more dazzling than a diamond with a thousand faces. Physics itself was somehow a relic of that ancient era.

As Witten told *Discover*, "It's beautiful, wonderful, majestic—and strange, if you like—but it's not weird." Witten—tall, soft spoken, with a high voice and a boxer's bent nose—spent a lot of time searching for the principle that animated all this miraculous elegance. He suspected that string theory, when it was eventually understood, would be deeper than general relativity, deeper than space and time. It might even be deeper, he thought, than quantum theory.

What he was sure of was that physics itself would never be the same. "Anybody who ever wished he was there in 1926," he told me, referring to the quantum revolution, "has another chance now." Five years after the superstring commotion started he was still saying the same thing.

Superstring cosmology, which may offer the chance to calculate what actually happened at the so-called singularity, also seemed to be a subject for the next century. The earliest glimpse of superstring physics had two implications for cosmology. The first, and perhaps more trivial (but fun), was that the most popular form of the theory allowed for the possibility of another species of matter, dubbed "shadow matter," with its own set of particles and forces, cohabiting the universe with our own material. Shadow matter would only interact with the regular universe gravitationally. *The dark matter could be shadow matter.* Kolb, Turner, and Fermilab postdoc David Seckel whipped out an impish paper that coined the term—they invented it on Kolb's blackboard in about three hours—and sent it to Schwarz with the inscription "Beware the dark side" scrawled across the top.

The other contribution was all those extra dimensions. Actually, by the mid-eighties, extra dimensions had become quite fashionable in physics; the most advanced version of supergravity had eleven. The reason we weren't aware of these extra dimensions, the story went, was because they were rolled up in tiny little balls only a Planck length—10^{-33} centimeter—in diameter, a notion invented in the thirties by Theodor Kaluza and Franz Klein. Another way to think of these dimensions was to imagine that every point in space-time was like a giant office building. Finding the point only got you to the front door; to find someone inside a whole new set of coordinates was needed: wing, floor, room number, third desk behind the water cooler.

One of the thorniest, hairiest, most twenty-first-century problems physics had to face was how space-time had come to have this rolled up quality. The blazing symmetry of the pure theory decreed that at one time, the beginning, all the dimensions had been equal (and probably small because the universe was small). In addition to the sequential unfolding of symmetries, then, superstrings added another feature to the prehistory of the universe—the infinitesimally tight curling on themselves of most of the dimensions of space-time.

Every morning Schwarz (Professor Schwarz now) went swimming in the Caltech gym and then came back to his office, ruddy and smiling like a new parent, to fan the superstring papers that had arrived in that day's mail. The big names dropped into the game—Gell-Mann,

Weinberg, Salam—and Schwarz, too long an outsider, kept careful track of them.

One name that did not appear on Schwarz's list of converts was that of Sheldon Glashow, the witty Harvard theorist who had been one of the fathers of grand unification. In leaving the realm of experiment so far behind and being guided instead solely by elegance, the superstring theorists, Glashow argued, had left science. What they were practicing was more like medieval scholasticism. Glashow mocked superstrings in talks and in an article in *Physics Today* titled "Desperately Seeking Superstrings." He kept predicting the imminent demise of the field. Schwarz kept checking off the dates, fanning newspapers on his desk, and noting that the field was still alive.

One name that everybody wondered about was that of Stephen Hawking. Everybody looked to see if Hawking would jump on the superstrings bandwagon, but he gave little notice to them. Hawking being Hawking, he was preoccupied at the moment with a more ancient mystery, time.

Time. It always seemed to weigh on Hawking. Time was different from the other dimensions, despite the attempts of Einstein to meld them all into one flexible integrated package. You could go up the mountain or down, you could go from stage left to stage right and back, but you could go only one way in time. The seconds add—they didn't subtract. Why, as Hawking asked, do we remember the past and not the future? It was as deep and old a question as science could dare to ask. Only a Hawking would dare to attempt an answer.

There were actually two kinds of time. One is the t that appears in physicists' equations, the clock time that keeps track of events; give me the time and I'll tell you the size of the universe, or where the pool ball is along its trajectory. This kind of time made no distinctions of direction. The universe expanded or contracted; the ball rolled forward or backward. Without doing damage to any of the godlike perspective of general relativity or electrodynamics, the great and eccentric Caltech physicist Richard Feynman—another of Wheeler's students—was able to conceive of antimatter as ordinary particles going backward in time.

The second kind of time was the experience of transience, the kind of time Heraclitus had in mind when he said you could never step in the same stream twice, the experience of being *in* time. It was the Greek expression for this idea, *panta rei,* that the Szalay brothers chose for the name of their rock band. The universe is in flux. The dream flows on; we are skidding out front, down the wave, nose first.

What put the one-way sign on time, Hawking argued, was entropy, the surge of things toward the lowest common denominator. The arrow of time points toward disorder. Teacups fall off the table and break;

they do not assemble themselves out of pieces and leap up on to tables. "Later" time is when the teacup is smashed, when the pool balls once perfectly racked are dispersed, when black holes are bigger.

Now it was no coincidence, Hawking figured, that our minds, the psychological arrow of time, ran in the same direction, knowing or remembering the past but not the future. It takes energy to store a memory, a piece of information, whether in a brain or on a silicon chip, which is dissipated as heat. The increase in disorder is always greater than the amount of information recorded. So the price of memory is entropy; it defines the past as a time of less disorder the same way as the second law of thermodynamics. Disorder increases with time because we measure time in the direction of disorder. "You cannot have a safer bet than that," he said.

But there was always another why. What did this have to do with the biggest arrow of time of them all, the expansion of the universe? Does entropy increase with time because the universe gets bigger with time, and what happens if the universe contracts? That was where quantum cosmology came in. Hawking thought that maybe he had gotten the goods on the old law of thermodynamics at last. Things got worse, he decided, because the universe is getting bigger.

He was inspired in his thoughts about all this by the behavior of the cosmological wave function. According to the calculations of Hawking, Hartle, and Hawking's student John Halliwell, time began at the smoothest of the smooth points at the North Pole of space-time. The wave function then began to ruffle and get ugly with new forces, galaxies, and insider trading along about our own era. As time goes by the wave function gets more complicated; what was once a simple spike decomposes into wave packets of different wavelengths and amplitudes. These waves get pumped up by the growth of the universe, they oscillate and produce matter, life gets complicated. In effect this growth and dispersal of waves is entropy, they argued. It starts low and bounds upward. The galaxies storm; gravity eats us.

What happened at the other pole of time, the South Pole of space-time, to which, in Hawking's model, we are ineluctably proceeding once, even after a nearly infinite number of years, we pass the equator? Hawking reasoned that the South Pole should be just like the North Pole, unruffled, the universe returned to its mysterious "ground" state. That meant that all this nastiness and complication and rambling, bad-mannered disorder had to fold itself back up into a neat, unassuming little smoosh. The smoosh at the end of time.

In short, once the universe started downhill, once we passed the equator, entropy would have to decrease. It was like saying that a pool game had to end with all the balls back in a triangular cluster on the mark; at some point in the game they would have to start hopping back out of the pockets. The slogan of physics students in the contracting universe might be "things get better."

But that was where the psychological arrow put its whammy. Physics students, Hawking realized, might be better off, but they wouldn't know it. Computers (and people) would lose memory while they sucked heat out of the room air and into their circuits. In the contracting universe we would remember the future, and thus have less and less to remember as time went on. So we would still seem to see teacups break apart and ice cubes melt, not the reverse.

In the summer of 1985 Hawking took a tour around the world, going on his own personal quest for the singularity, so to speak. He stopped for a week at Fermilab along the way.

Kolb and Turner were in charge of Hawking's entourage. Hawking had been disappointed with a similar tour at CERN, so they took pains to give him the works. They poked him into every nook and cranny of the facility where in a couple of years trillion-volt protons and antiprotons would smash together in an attempt to slide particle physics another notch closer to the big bang.

The scientific climax of his visit was a technical lecture to the Fermilab community. At the appointed hour, the cream of American particle physicists gathered in the lobby entrance to Fermilab's underground auditorium, like rumbling cattle. All except Hawking, who was stymied because there was no wheelchair ramp or elevator from the lobby to the proscenium.

Turner and Kolb looked at each other, and then Turner scooped Hawking up out of his wheelchair and carried him down the aisle. Kolb took the chair. Suddenly it was very quiet. Turner was amazed at how light Hawking was. Halfway down he panicked, remembering how Hawking hated to have attention called to his disability.

Only about twenty people, Kolb later told a reporter, understood Hawking's talk. The next day, however, Hawking ventured into Chicago to give a public lecture. He received a rock star reception. After all, he was the most famous physicist in the world; his picture had been on magazine covers. Crowds followed him everywhere. Vietnam vets in their own wheelchairs raised their fists in salute and shouted "Right on" to him as Hawking whirred by on the street or the campus. A standing-room-only crowd spilled out the doors to the rain.

Schramm introduced him as the occupier of Newton's Lucasian Chair of Mathematics and then made a pointed reference to Hawking's children, saying he could obviously do more than just physics. Hawking blushed.

"They say it's Newton's chair," he quipped when his turn came, "but obviously it's been changed." Looking vulnerable and tiny from afar in a spotlight, he wheeled smartly about the stage. "Why," he asked, turning to the audience, "is it that we remember events in the past but not in the future?"

Hawking concluded by pointing out that if time did in fact reverse during the collapse of the universe, it would also reverse inside the collapse of a black hole. "So if anyone really wants to know whether I am right," he said, "then go and jump in a black hole." Thunderous laughter.

Reversible time was better than exploding black holes. And even more controversial. Like the best of Hawking's work it was couched in cryptic and bold calculations way out there where most physicists were too busy or timid to go.

The notion that time would reverse when the universe contracted seemed wrong to Don Page, an old friend who had been a postdoc at Cambridge and a Hawking household caretaker during the exploding black hole days. Now at Penn State, Page, who described himself as a born-again Christian, is tall with a round face. Like everybody who knew Hawking well, Page treated neither him nor his ideas with any particular reverence. He went to Cambridge for the summer in 1985 and spent it arguing with Hawking about time and entropy.

It was true that the separate wave packets that represent everything interesting in the universe would come together in the end, Page granted. But, he argued, they came together with more energy than they had in the beginning, due to inflation and the general knocking around of things in the universe that would heat them up. Any particular wave packet that you followed through history got large and then small again, Page conceded, but there were always more wave packets being launched into history with more and more energy.

"It's like shooting arrows into the air at an increasing rate," Page explained. "The recollapsing universe is like the down-falling arrows, always outnumbered by the up-going ones. It's subtle, and you miss it if you follow only one arrow."

The end, he concluded, would not be the same as the beginning. Disorder would go right on mounting throughout the sunset years of the cosmos. You could say that we would continue to remember the past and mourn the greatness of the expanded universe as we slipped toward the ultimate homecoming.

Page and Hawking wound up writing opposing papers for the *Physical Review*. Hawking helped Page, who finished first, with his. Page then had the journal hold his publication so that Hawking's could accompany it. In the printed version, Hawking's paper led off the set. Hawking concluded by saying that Page had some interesting arguments and that he was probably right.

In August, Hawking took another trip to CERN. While he was there he came down with pneumonia, the biggest danger to people with ALS; you never know when the lungs are going to shut down. He was rushed home to England, his lungs filling with liquid.

The doctors performed a tracheotomy, in which a breathing tube is inserted through a hole in the neck right into the patient's windpipe so a machine can breathe for him. Around the physics world the word flashed that Hawking was at the end of his rope. At two separate times he stopped breathing. In the words of one of his nurses, "He died twice."

Hawking hung on, and eventually his lungs cleared out. The tracheotomy, however, had ended his talking days forever. He was soundless and depressed. For a while Hawking's only form of communication was blinking. Page thought that for the first time Hawking had lost the will to go on.

A few years earlier that would have been the end of his career and perhaps his life. Now, however, a company known as Speech Plus in Sunnyvale, California, made a computer-activated voice synthesizer for victims of just his sort of malady. In January he was fitted with the device, which could be built onto his wheelchair and which he could operate with two fingers. Hawking became IBM-compatible. Hartle sighed with relief when I asked him about it and said that at last he could understand what Hawking was saying. Deterioration always had its good side.

In the spring of 1986 the Nobel Academy invited a few elite physicists to a symposium on superstrings. Hawking went and spoke, of course, about quantum cosmology. Nobody knew quite what to expect. He began, "Please pardon my American accent."

Hawking returned to Chicago for the Windy City's version of the Texas Symposium that December, the one that ended in a food fight, presided over by the expansive Schramm. His travel contingent had necessarily grown; Hawking went with three nurses and a caretaker/assistant who took turns watching over him in shifts.

His new device changed the way other scientists responded to Hawking. Once, out of awe, they left him alone. Now, when Hawking appeared in public with his new voice synthesizer, the younger physicists crowded around him. They were like teenagers admiring a new car. Hawking was happy to show off his system. Built onto one arm of his wheelchair was a small liquid-crystal display. On the top half of the screen a cursor constantly scanned a list of words, letters, punctuations, and editing commands. Hawking could select any word or command by just clicking a hand pedal. His text would build up on the lower part of the screen. When he was done he would play it out through the synthesizer or just let the questioner read it. In practice people wound up just looking over his shoulder as he put his words together. The program had a vocabulary of about 2300 words and as he used them, learned likely combinations, that "black" might often be followed by "hole."

Hawking's portable synthesizer seemed to have a Russian accent. For ceremonial occasions, he used an IBM computer with a bit of Scottish burr. He could put prewritten speeches through it and everything could be saved, which made him potentially the man with the most complete record of his utterances since Richard Nixon.

The organizers had showcased him, in a reprise of his amazing time lecture of the previous year, in an evening lecture. Back in 1974, after he had proven that Bekenstein was even righter than he thought, Hawking had enlivened his talks for a while with a viewgraph or transparency that announced "I was wrong." In Chicago he waved the flag of error once again. He took back the idea that time would go backward and disorder decrease during the collapse of the universe.

In its place he asked why disorder increases in the direction of the expansion of the universe, which he said was the same as asking why we live in the expanding phase of the universe, in which all these arrows coincide, and not the contracting phase. He answered himself by appealing to the well-worn anthropic principle, the idea that the universe is as it is because it has to be conducive to life.

Because inflation implies that the universe is so endlessly huge, Hawking argued, it would be trillions upon trillions upon trillions of years before the universe passed its halfway mark and began to contract. By then all the stars would have burned out, perhaps all the black holes would have evaporated, and all the protons would have decayed. There would be little of the raw material for life.

The anthropic principle was not popular, and Hawking's use of it disgruntled several of the more senior cosmologists present, who nonetheless did not want their names used. When I asked Hawking about that later in his hotel room, he was unapologetic. "I always wanted to know where the universe came from," he stated. "I think most people do. It is only by going out in advance that you make progress."

It was a slow conversation. I would ask a question and then sat listening for a minute or two of clicking, anticipating a longer answer than invariably turned out to be the case. Hawking composed flawlessly in real time on the screen. He never erased. Readers who have done some writing on word processors will appreciate how frightening that can be. Hawking looked more relaxed than he had in previous conversations, perhaps because he no longer had to distort his face in the effort to speak. On the other hand, he seemed less inclined to joke around than he had in the past.

I asked him why he had felt the need to make a public recantation of his earlier theory in the same city in which he had proposed it. "It should happen more often," he answered. "I think a lot of people wouldn't have read the article in the *Physical Review*. I never do."

In the end what I wanted to know from Hawking is what I have always wanted to know from Hawking: Where we go when we die. Whether the singularity is real, whether there is some witness to a miracle being born in these geometric machinations, a doorway out of the trivial solution, the formula of law and dead meat. Twenty years ago we had feared singularities as destroyers. Now he seemed to be sealing them out: what looked like singularities to our imperfect science were only approximations. There was only law, he proposed, the universe itself is the cosmic censor. I asked in the most neutral, emotionless language possible: So what does happen when *a*, the radius of the universe, goes to zero or we get to the bottom of a black hole? Where are we, and what are we then? Was there a hippie killer clown in our future?

But my supplication for certainty would always be denied. Yes and no. "There are no singularities in the metric in the path integral," Hawking answered crisply in his Russian voice, "but there are singularities in the classical solutions that correspond to the wave function." In a way I felt as if he were running over my toes again in the elevator. Just because classical space and time broke down, he was saying, didn't mean that physics would. The answer, it seemed, lay still ahead in the creepy shadows of quantum gravity where neither physics nor Hawking was yet ready to go.

By the following spring quantum cosmology had attained sufficient stature to merit a workshop on its own, at Fermilab. Accordingly a mixture of superstring theorists and cosmologists, including Hawking and Zeldovich, flew in for a long weekend.

Zeldovich was there courtesy of glasnost and the Chernobyl nuclear power plant disaster, which had cost one of his repressive higher-ups his job. So Zeldovich came at last to the United States, stopping at the National Academy of Sciences in Washington, D.C., to be formally inducted into membership and receive a medal.

Half the cosmologists on the East Coast assembled for Zeldovich's speech and lunch afterward under a circus tent on the academy grounds. Wheeler was there, Schramm flew his own plane from Chicago, and Thorne traveled from California. Zeldovich knew some of them from meetings in Moscow, others were strangers. When Geller, of CFA redshift survey fame, went up to introduce herself, Zeldovich looked right through her with the icy blank stare he used on people he didn't know. Geller was nonplused.

Ostriker, Geller's old Princeton professor, saw what was going on. He happened to have a copy of Geller's famous redshift map with the Virgo-Coma little man in the center of his briefcase. He pulled it out and pointed back and forth between the name "M. Geller" on the bottom and Geller standing there.

Finally the light dawned, and Zeldovich slapped his head in dismay and delight.

Rubin, who had been standing nearby, commented that he was confused because it had never entered his mind that "M. Geller" could be a woman.

A big smile crossed Zeldovich's face, and his eyes twinkled. "Ah," he said, "you must be a *feminist*."

In Chicago, at the quantum cosmology workshop, the cultural gap between the superstring theorists and the astrophysical cosmologists extended to the dinner arrangements. One night the physicists all went to physicist Chris Hill's house; the astrophysicists, Hawking and Zeldovich included, went to Kolb's.

There Zeldovich entertained the group by showing them the Russian method of opening a bottle of wine: Bang the base of the bottle stoutly against the wall a few times until the force funneled by the shape of the bottle has pushed the cork slightly out. Then grab the cork in your teeth and yank. The method is best advised for the cheaper varieties. Then Zeldovich insisted they all go to a disco and dance.

One day he and Schramm skipped out of the workshop. Zeldovich had some money from the academy and wanted to shop. He bought a diamond ring for his wife and then went looking for computer supplies. Zeldovich had a Commodore or an Atari—Schramm couldn't remember which. They went to all the computer stores and couldn't find what they wanted.

Finally Schramm figured out where the East's leading cosmologist could get his computer needs serviced. They went to Toys "Я" Us.

The scientific sessions frustrated Zeldovich, devoted as they were to abstract, almost theological interpretations of quantum principles and Planck-era microphysics hopelessly beyond experimental reach. As Wheeler had suggested years earlier, for cosmologists the quantum principle had indeed become a command, an invitation, to create the universe. The uncertainty principle ruled out nothingness. On that cosmologists seemed to agree, but everything else—how it evolved, whether its laws were fixed or fluctuated like the weather—was up for grabs. Some accounts of that meeting make it sound as if the workshop were conducted in Latin. Zeldovich was speaking for more than himself—he was speaking for all the intuitive hard-nosed hands-on guys—when he kept asking over and over again, "But what can I measure?"

IV

THE LAST GENTLEMAN

The weight of this sad time we must obey;
Speak what we feel, not what we ought to say.
The oldest hath borne most: we that are young,
Shall never see so much, nor live so long.
                    —W. Shakespeare

21

SANDAGE IN EXILE

Sandage was one of those who had little taste for the concept of the universe as a quantum fluctuation or for some of the other elegant ideas, such as inflation, that were seducing his colleagues.

"What is a superstring?" he asked me one day. "Can you tell me what a superstring is? Can you explain broken symmetry?" Sandage's voice lilted with scorn across the cocktail table.

"What is the nature of nothing? Does that statement have any sense, or is it gobbledygook? The nature of nothing. What they say seems to be absolutely nonsense. They say the universe could create itself out of nothing, a self-causing entity, because they say you could do it with zero energy. You just combine negative energy and positive energy until you have a zero-sum game. Well, that's just using words to convince themselves they can earn a living." He frowned and reflected on what he had just said. "That's unfair," he muttered to himself. "That's unfair.

"It means there are virtual particles going in and out of existence. I can talk the same language but I don't understand the words, and the more I read *Nature* and these popular books and conferences, I can talk the language better year by year. But I understand less and less of the thing. It's fascinating, but I don't think it gets down to the nub. I think it's a philosophical leap that can't be made, that there's an edge to science beyond which the questions of *why* are outside the realm of science.

"Do you believe in grand unification theories?" he asked me. "Why? Because everyone you respect tells you to. You are seduced by its beauty. And because it is so beautiful, it has to be true. Now that's an okay way to go, that's what Einstein said.

"Why do you believe that's true any more than the existence of God, which is also a beautiful theory and also explains a great deal? It's a hypothesis that is checkable in its consequences. But yet you would reject that beautiful hypothesis."

I ventured that I didn't know what he meant by the existence of God in this context.

Sandage responded with an ingenuous look and said, "I don't know what people mean by grand unification, broken symmetry. What is broken symmetry?"

I ventured lamely that it was the less than perfect realization of a perfect principle.

"Like a love affair," he murmured dryly. "Don't look at me with such a disdainful look."

Outside, a tattered, abandoned volleyball net flapped in the wind while a November storm threw flecks of sand and Pacific against the picture windows of the Sea Lion, a beachfront bar and restaurant north of San Diego. Inside it was the prelunch hour, and the restaurant was nearly deserted. The staff had lit candles and was drinking hot chocolate.

Near the window Sandage was slumped over a reddish tea-colored Manhattan. He was dressed in a sweater, blue jeans, and well-polished loafers. A long gray strand of hair fell across his forehead. His eyes twinkled, then went cold, and his face sagged. "One ought to understand," he said in a low voice, "when the time to quit is."

For years he had been glum and bitter about the attacks on his Hubble constant work, but lately there had been a new source of gloom. Mount Wilson, thought Sandage, the old Valhalla where giants trod the halls and you were never more than ten feet from a gentleman, was dying.

The writing had been on the wall for the Mount Wilson–Caltech collaboration since the early 1960s. It was then that Mount Wilson's owners, the Carnegie Institution, seeking access to the relatively unexplored southern skies, proposed to build an observatory at Las Campanas in the dry, calm, dark foothills of the Chilean Andes. Caltech, nervous about getting embroiled in South American politics, declined. Carnegie went it alone. In 1977, the 100-inch Du Pont telescope, an especially fine reflector with an unusually wide field of view, was dedicated there. Augusto Pinochet Ugarte, the infamous dictator of Chile, attended.

The addition of yet another observatory had made the already cumbersome arrangements between Carnegie and Caltech formally unmanageable in the view of Schmidt, who became director of Hale Observatories, as the conglomerate was named, in 1979. Schmidt stayed only long enough to dissolve the marriage. Caltech kept Palomar and the Big Bear Solar Observatory. Carnegie got Mount Wilson and Las Campanas. Mount Wilson retained its representation on the Time Allocation Committee (TAC), which divvied up the 200-inch time; they would share in the scientific administration of the telescope, but not in its ownership.

For the first time in three-quarters of a century Carnegie did not own a piece of whatever was the largest working telescope in the world. "Allan Sandage will never speak to me again," Schmidt told me one night on the mountain. And he was right.

Sandage had vowed never to set foot on Palomar again. He claimed that he had expunged it from his memory. "It's interesting," he said. "My psyche tells me I've never been on the mountain. And yet my wife tells me, 'Allan, you spent all those nights away for twenty-five years.' "

The advantage of the divorce for Carnegie had supposedly been that it would free their resources to pursue projects at Las Campanas. The recession of the early sixties, however, crimped those resources. There was talk of closing Mount Wilson. The Mount Wilson instruments, it was said, were old and creaky, needed expensive retooling, and the lights from Los Angeles made deep sky work impossible.

Meanwhile, driven by new technology such as computer-controlled mirror supports, telescopes that would dwarf the 200-inch, long considered the Everest of the astronomical world, were appearing on drawing boards of observatories around the world. The University of California was raising money for a 10-meter reflector[1] with a segmented mirror, but the standard for this large-aperture dreaming was 8 meters (320 inches), which was the largest mirror that could be cast in one piece in the University of Arizona's experimental rotating furnace. By the mid-eighties no observatory was an observatory that was not queuing up by itself or in collaboration for an 8-meter telescope. The European Southern Observatory—a sort of astronomical CERN—was planning an array of four 8-meter telescopes at its own site in the Andes.

Another sign of the changing astronomical order was the imminent launch of NASA's Hubble space telescope. Astronomers had been awaiting the space telescope as they had no instrument since the Palomar giant itself. Although it only had a 96-inch-diameter mirror—a pygmy compared to ground-based telescopes being planned— by orbiting above

1. As of this writing it is now nearing completion, as a joint project of the University of California and Caltech, on the summit of Mauna Kea on the island of Hawaii, and is known as the Keck telescope. The University of California had received a deathbed bequest for some $38 million from the widow of a Bay Area insurance magnate, but that was not enough to build the whole telescope. They invited Caltech in on the action to help raise money. Whereupon a Caltech alumnus and board member, James Keck, arranged to put up $70 million from the foundation established by his father, the founder of Superior Oil.

the smears and blurs of the atmosphere the device would achieve unheard-of resolution, which would enable it to see stars and galaxies fifty times fainter than the 200-inch. It would be able to dissect the nuclei of distant galaxies, take spacecraft-quality pictures of the planets, and measure Cepheid variables as far as the Virgo cluster. Originally scheduled for launch from the space shuttle in 1986, the Hubble telescope cost more than a billion dollars and would require something like ten million lines of software to operate. In the new Space Telescope Science Institute on the Johns Hopkins campus, former Harvard X-ray astronomer Giacconi was assembling a 250-person staff to operate the telescope; at Johns Hopkins itself—no fools they—the astrophysics staff was doubling.

In Sandage's eyes, large-telescope envy had especially hit the younger generation of Mount Wilson, people like the instrument builder Shectman and Alan Dressler, a slender, curly-haired former student of Faber's at Santa Cruz with an olive complexion and a calm, meticulous manner. Only with a new giant telescope of their own could the Mount Wilson astronomers keep up in the search for primordial galaxies and carry on the very deep redshift surveys needed to delineate the large-scale structure of the universe.

For a while Sandage had taken Dressler as a new protégé. There was a story about it in astronomical circles. One night when Sandage and another young astronomer were on the mountain, the story went, a third astronomer got annoyed on the younger person's behalf that Sandage kept referring to him as "my student." Sandage replied, "I was Hubble's student and this is my student."

Dressler was present when I heard the story and later admitted, somewhat embarrassed, that he indeed was the "student" in question. He had prevailed on Sandage to use some of the new technology, such as an intensified television tube rather than plates.

One night on the way down the mountain, conversation turned to the bleak plight of Carnegie. What should they do?

Dressler allowed as how maybe they should close down Mount Wilson. That was the end of the conversation and the friendship. The junior staff members had no business meddling in these matters, he recalled Sandage saying, it was incredibly presumptuous. They rode home in silence.

Sandage resented all this talk of Mount Wilson's obsolescence. In fact, a committee of the American Astronomical Society (AAS) had concluded that Mount Wilson was still a superb site for studying stars. The same inversion layer that trapped LA in smog made for exceptionally steady and clear images. It was just because everybody wanted to do cosmology, Sandage said. There were no stellar astronomers left. Now, in their unholy rush to stampede to the forefront of modern cosmology, to erect enormous mirrors in the virgin Southern Hemisphere, the young generation was willing to shutter Mount Wilson, home, among others, of the 100-inch Hooker telescope, discoverer of the creation event.

Partly as a rear-guard action, Sandage had been using that telescope himself on a project to analyze the stars in the Milky Way's halo. Despite an outcry from the general astronomical community, editorials in the *Los Angeles Times,* an inspired but misguided effort by amateur astronomers to conscript the Hooker telescope for public viewing of Halley's Comet, and the AAS report, no satisfactory transfer of the observatory's management to any other institution could be worked out. Mount Wilson closed in July 1985. At the same time it was announced that Carnegie had secured funds to build an 8-meter telescope in Chile, in partnership with Johns Hopkins.

Sandage's outspoken efforts to preserve Mount Wilson put him at bitter odds with the younger astronomers like Dressler and Shectman, as well as with Carnegie management. His life continued to be pared away. He learned in a memo that his observer was being let go. His photographic assistant, without whom he could not complete one of his longtime dreams, a photographic atlas of all 1300 Shapley-Ames galaxies, took the hint and went off to work at the Space Telescope Science Institute. The dead zone in Sandage's heart extended now to Santa Barbara Street and all of Carnegie. He sent a paper to the *Astrophysical Journal* listing his affiliation as "Astronomy Section, Carnegie Institution" and it was printed that way.

Around Santa Barbara Street the whole imbroglio was considered just another of Sandage's moods. One of the paradoxes was that he was in fact continuing to pour out papers. He and Tammann photographed, mapped, and classified the entire contents of the Virgo cluster, for example. "Allan Sandage concentrates so incredibly fiercely," said Leonard Searle. "He's a marvelous scientist. He flings himself into things; he seems to be a man of passion. The rate that he's pouring out publications suggests to me that he's not very different from the Allan we've always known. He strikes me as being in great form scientifically."

Searle, a middle-aged British spectroscopist, had earned Sandage's scorn by remarking during the debate on closing Mount Wilson that "the past is a burden." Searle explained that he had meant that in the long run they couldn't afford to be swayed by sentiment in deciding the future of the observatory; at the same time there would be a sentimental price to be paid. The blunt truth was, Searle said: "If we were stuck with 100-inch aperture telescopes, we weren't going to be in the forefront."

Soon Sandage's only friend at Mount Wilson was an unlikely one: Chip Arp, to whom he had stopped speaking over the quasar controversy. As Sandage tells it, one day he looked around and said to Arp,

"You're the only person here I can talk to. They both decided they had always been friends, even when they weren't speaking.

Since the sixties Arp had been collecting the novel and unexplainable, particularly galaxies or quasars in suspicious patterns that did not seem to be at the distances indicated by their redshifts. In his own mind, Arp had collected enough so-called anomalous redshifts to have overthrown conventional big bang cosmology, but outside of a tiny unexpanding circle of supporters, he was ignored. Most astronomers regard his results as coincidences.

In the old days the presumption, says Arp, was that a senior member of the Palomar-Mount Wilson staff could count on a certain number of nights on the 200-inch per year, but now that era was drawing to a close, and it closed on Arp's toe. Time was too scarce to spend beating Arp's dead horse. His somewhat pro forma request for observing time was turned down by the TAC. According to a letter which was leaked to the press, Arp's research had been too sterile and unproductive. Arp received time on the 100-inch at Las Campanas and went to Chile to work, but his feelings were hurt.

The next year Arp declined to submit a proposal, on the grounds that it was a formality anyway. Everybody knew what his research was about, he said. It was like volunteering for suicide. Naturally, he got no observing time. In the ensuing brouhaha, Sandage was the only one who stood up for him.

Arp took a sabbatical and fled to the Max Planck Institute for Astrophysics near Munich, where the traditional European tolerance for American crazies was a balm. Faced with the choice of coming back after two years or resigning Mount Wilson, he took early retirement and became a cosmic expatriate.

"I'm still rather pained by the whole episode," he said in Munich. "One unique thing Carnegie had was the ability to entertain and test new ideas. Carnegie had always been a haven." His voice drifted off. At sixty, Arp still had the dashing looks and moustache of the fencer, but he was tired of fencing. He rattled off a list of cosmological anomalies—suspicious patterns of quasar redshifts, galaxies of differing redshifts that appear to be connected, rumors of anomalous data that his colleagues never publish. Arp had concluded that quasars were the nuclei of new galaxies being born in the centers of old galaxies and then ejected, galaxies reproducing like amoebas, a variation of a theory put forth long ago by the Armenian Ambartsumyan, Arp's work was dying in a conspiracy of indifference, he thought, the evidence that could vindicate him buried in other astronomer's file cabinets. "It drives me up a wall when colleagues just say I'm wrong. It's a scandal," he complained.

Most astronomers would agree that Arp was a victim of his own stubbornness. Some men have to throw themselves under the wheel of progress for it to roll. For every Schwarz and Green who eventually shook the tree of physics there were a hundred Arps sitting agonizing in a Munich apartment, looking, like Archimedes, for a place to stand so that he could tip over the world. In his cheery moments he still insisted, "There's going to be a fundamental breakthrough that's going to change astronomy. I don't know when but it's inevitable."

One night we ate cake in his sparsely furnished apartment in Munich's student quarter while Arp wondered if there was any point to going on with his research. Was anybody listening? "Do they want more?" he asked. It was beginning to occur to him that the astronomical establishment might never be converted from what he thought were its mistaken ways. "Maybe," he thought, "every field has its great moments and then falls. This is a low point for astronomy." It had taken a hundred years for Copernicus's idea that the earth moves around the sun to prevail (over the Church, among other things). What if it took astronomy a hundred years to recover from its present falsity and by then he was forgotten?

He recalled having read a popular book on cosmology not too long ago. It sounded authoritative, he remembered thinking to himself, and then wondered idly if it could be wrong.

Reliving the moment Arp grimaced. Suddenly, he recalls, he had remembered who he was. "I slapped myself on the forehead. It *is* wrong!" The Munich sky beyond his Bauhaus balcony erupted with fireworks. Arp hoisted his two-year-old daughter to his shoulder and took her outside to watch.

One afternoon in the summer of 1985 Sandage stood frowning in the heat-buckled asphalt parking lot behind the Santa Barbara Street offices and watched silently with the other astronomers as workmen hauled file cabinets from the basements of the building to be loaded in trucks and shipped to UCLA. Inside the cabinets, in long yellowed and ripped envelopes, resided Mount Wilson's vast collection of solar memorabilia: glass plates containing photographs and spectra of the sun. Hale, Mount Wilson's founder, had been a solar astronomer and the first permanent telescope on the mountain had been a solar telescope. Solar astronomy had been the heart of Mount Wilson; now they were carrying it away. In the middle of all this rather depressing ritual, George Preston, Mount Wilson's director, noticed a rainbow and pointed it out. Sandage was mortally offended.

As he found himself more and more at odds with his own establishment, Sandage had begun appearing sporadically at conferences again, edging back out into the world, clasping old comrades and sometimes old opponents on the shoulders. Some astronomers didn't even recognize him, although his work was as much a part of their lives as the constellations.

Sandage told me gleefully of sitting in a large meeting in Tucson early in 1985 and squirming through a long lecture by Aaronson, when he heard the three astronomers sitting directly in front of him talking.

"Did you hear about Sandage?"

"No, what?"

"He's grown a big bushy beard and become a born-again Baptist minister."

Sandage decided to take a sabbatical and find out what life was like outside Pasadena. In late August 1985, he drove a VW camper to the back door of Santa Barbara Street and hauled the essentials out of his oak-paneled office: his can of bullshit repellent, his ancient eyeball magnitude measuring machine hung with a clove of garlic, books, papers, and giant photographs of galaxies marked like a surgeon's X ray for dissection. He carted them down the carpeted hallway past the portraits of the patriarchs of cosmology, Mount Wilson's heroes, his mentors, loaded them in the van, and drove south down the freeway.

His first stop was UC San Diego, where Margaret Burbidge was head of the Center for Astrophysics and Space Sciences and where her husband, Geoffrey, had retreated after having left the directorship of Kitt Peak. Sandage said he looked forward to drinking with Geoffrey Burbidge, if nothing else. He rented a house near the beach in La Jolla where the campus is. During the week he lived alone, working in a borrowed office on campus. On weekends he went back up to Pasadena to see his wife.

By then he claimed that he had forgotten that he ever was a cosmologist. "You may remember in 1956 that Humason and Mayall came out with this big catalog, and I did the magnitudes, so they put my name on this paper of redshifts from Lick and Mount Wilson starting in 1938. And only about thirty-eight percent of the galaxies in the Shapley-Ames had redshifts. That was another feeling of responsibility. When Humason retired, I felt I had to complete the Hubble classification, which led to the *Hubble Atlas*, and then I felt I had to complete the Shapley-Ames redshifts, and then I felt I had to complete Hubble's extension of the Hubble diagram, and then I felt I had to calibrate things from Cepheids.

"So I don't care about cosmology. I've just been forced into it by finishing up all these projects that were Mount Wilson projects, and now I'm labeled a cosmologist. But I'm really a stellar astronomer.

"So the hell with cosmology. It leads to nothing but grief, leads to nothing but criticisms. That's fine—" he added quickly, unconvincingly, "it strengthens the soul.

"Cosmology is a boring burden of pride," he said sternly, then continued reciting in a subdued monotone. "Pride is the downfall of any human being. Pride is the mother of the other six deadly sins. One's goal should be to live without pride. Pride is what will kill you. Pride is what makes you mad when you get a referee's report back. You've got to humble yourself before the problem if you're ever going to be a good scientist."

As the universe of galaxies and stars became more dead numbers, like salt from an evaporated sea of mystery, Sandage remembered and mourned the wonder with which he had viewed the universe as a kid. That sense of wonder had vanished, he complained, after two weeks at Caltech. There followed thirty years of duty, inertia, and discontent. Astronomy, once an escape from the human morass, led back into it.

About the time he received his Ph.D., he told me, he remembered having asked his father what the purpose of life was and being disturbed at not getting an answer. His father didn't know, and that seemed ominous. The question Daddy couldn't answer had come back to haunt him. One day, he says, somebody told him, "The purpose of life is to glorify God."

"That sounded right," said Sandage. Sometime around 1980, Sandage said, he converted to Christianity. He would not divulge any more details. He didn't want to be a nihilist. Life was not a dreary accident. He repeated it, "Life is not a dreary accident."

Sandage traced his slow withdrawal from the public and other astronomers to the debacle of his blue galaxies paper in 1965, in which with no referee he had attempted to identify a thousand new quasars. The debacle, in turn, he attributed to the cutthroat competition in Pasadena, or as he sometimes called it, "Sodom and Gomorrah." He became wary of the media.

"You said that in the sixties I was in the *New York Times* all the time. I'm not aware of that, honestly. If that's true, it's a terrible statement. I think the consequences of that are what caused a fifteen-year disappearance from the world." His voice was low and serious.

"You didn't see anything from 1969 to about 1983. Nothing. I rebelled against this media stuff. I knew . . . I had . . . it was nice as a young man to be so quoted. Suddenly there came a time when it was really clear that I had to choose being a spokesman in the media or being a scientist, and I completely withdrew. So for fifteen years there was a withdrawal phase. I had all these plates, I had a foundation that was laid by Baade and Hubble. I had an education that was laid on me, all the advantages in the world. If I didn't take advantage of that I was crazy, wrong, didn't rise to that responsibility. I cannot explain it, but the world became too hard, too complicated—the bloody *New York Times*."

Eventually Sandage learned how to—as he put it—"give away the world." What, he asked himself, if he just went on with his research as best he could and didn't pay attention to what his adversaries said? It would be like an actor not reading the critics. "At least I can do astronomy without getting hurt," Sandage said. "De Vaucouleurs can't stop me from publishing. The secret of peace is giving away the world."

He stopped and glared at me. "I don't know why I'm talking to you now, except . . . I'll say it." His voice went cold and dry, his eyes steel curtains. "You guys are superficial compared to what I'm trying to do."

My blood ran cold.

"Now that's a very prideful statement in some sense, but never mind," he went on. "Having given the world away, now I can come out of isolation where I went because the world was just too complicated. If I've come out from fifteen years of isolation, it's because you guys don't mean anything to me any more."

We went inside to eat, and he showed me how to make galaxies in a coffee cup. You get the coffee swirling and then pour a blob of cream right in the middle. According to Sandage, the trick worked best with real cream, but the Sea Lion only had half-and-half. Our galaxies were flawed. Nevertheless Sandage was soon calling out the catalog numbers of the creations in our cups.

While the theoretical physicists were attempting to make the universe out of quantum magic, Sandage continued working on the foundations of the subject of the expanding universe: distances and the Hubble constant.

"Getting distances is terribly hard and you try everything you know how," Sandage said one afternoon. "You have to have faith some-how that it's all going to work out. You push down a road until you're blocked and then you back up. You start to cut another way through the jungle and you see the Emerald City up ahead. And whether we've packed our way to the yellow brick road with supernovae or whether we're still in the jungle I don't know. Just cutting down the trees with the machete is already pretty interesting, because along the way you see lots of inter-esting plants and animals. But I have faith that we're pretty soon going to strike the royal road to the city of Oz. I think we're there already, but people don't believe that the road they see is yellow yet. They still think it's a dirt road with a roadblock up ahead. Or maybe they think their road is yellow and we're on a dirt road. I don't know.

"So it's gotta be fun. The travel itself has gotta be fun. I don't think anybody should tell you that he's slogged his way through twenty-five years on a problem and there's only one reward at the end, and that's the value of the Hubble constant. That's a bunch of hooey. The reward is learning all the wonderful properties of the things that don't work."

Dark matter was one of those interesting things cosmology had discovered along the road. Sandage had assembled and squirreled away with him down to La Jolla all the new data on distances and velocities of galaxies near the local group. He wanted to see if the dark matter supposedly clumped around the local group galaxies would show up as a distortion in the Hubble expansion close to home. It was a clever idea but exceedingly difficult to do right. The theoretical calculations had been giving him fits.

One Monday, as he described how he had found a way to do the necessary mathematics, his eyes grew bright and his hands screwed together jars in the air above the lunch table. "So I got a good night's sleep last night and I don't know how to solve it quite yet, but I think there is a solution there, and that's already nine-tenths of the battle." He was bouncing up and down in his chair.

So much for not being a cosmologist.

On my last day in La Jolla Sandage took me to his borrowed office. "I'm going to show you all my goodies." He was on a high because his calculations seemed to be working out.

We rode an elevator up the modernistic and stark Space Sciences Building. Sandage's floor seemed deserted. Inside his door the desk and walls were littered with working prints of galaxies, M81 and NGC 300, with Cepheids and other variable stars and calibration standards marked in ink. In the corner stood his ancient magnitude measuring machine, with its Ichabod Crane magnifier hanging on the front. "This replaces the CCD, the PDS, the VAX," Sandage trumpeted. It was plastered with stickers; one said: You will always be surrounded by true friends. Another: Clearly the best. People foist aphorisms, old Blake and Bible quotations, on Sandage, perhaps because he responds.

On shelves opposite, in long boxes the size of library card catalogs were Hubble, Baade, and Humason's plates—the expanding universe in a box.

Next to them sat a large candy jar labeled "megapotent Bible vitamins." Inside were little plastic capsules with sayings inside them, à la Chinese fortune cookies. He offered me one. I read, "And because you are sons, God has sent forth the spirit of his Son into your hearts Galatians 4:6."

Sandage shrieked with laughter. "You thought you were in the office of a reasonable man," he boomed. "You clearly are not."

Sandage took another one. "Oh Lord, our Lord, how majestic is thy name, in all the earth who hast displayed thy splendor above the heavens Ps. 8:1."

"There are three hundred and sixty-five of these in here and when we get through," he explained, "we put them back and recycle them. There are some days when I don't feel like vitamins. I'm sure there are some days when you miss your vitamins too."

He pulled out the big photograph of NGC 300, a spiral galaxy about four times the distance of M31. "Isn't that spectacular? That's where it's at. That's how all of us spend our time, just looking at black marks on a photographic plate.

"Here's where it all began, with M81," he said, slipping a piece of glass about the size of a playing card out of a yellowed stiff envelope. He read the inscription. "Here's a plate taken by Milton Humason with the 200-inch, February 12, 1950. These plates are all in order now. The last ten plates taken were j59 by Arp and a 1974 plate taken by me. Here is M81 itself. You've seen it in all the books. There it is for real. The red arrow points to one of twenty-seven normal novae." The face of the galaxy was streaked with dark cracks of lines, all radiating inward. "Dust lanes," he sighed, "so fine. The dust lanes know where the center is."

He went to the desk and picked up the M81 notebook and flipped its pages. In historic hands were calibration lists, sequences of star magnitudes, variable stars, comparison stars. He read, "N1 was discovered on plate 109mh, it's 21.9 apparent magnitude on that plate, March 18. Plates taken on the twentieth and twenty-first, it hadn't changed, but then it wasn't visible when Humason went up again. Here's Humason. He was principal observer for many months in 1950. Then here's a plate by Hubble, a plate by Baade, Humason, Baade, Baade, then I start, November 1951, and then here's Humason, Baade, and all the rest of these."

The room was crowded with ghosts. I felt the hot breath of nirvana. Messier 81 had been abandoned in 1954 as too hard, and left to dance in the shadow of NGC 2403; now it was being taken up again. Sandage had plotted the light curves of two Cepheids in that crinkly spiral, now revealed to be even farther than its putative partner. "Yes, yes," he shouted, "that's where it's at. It's this curve that has to be determined. HOW DO YOU DO THAT?

"You do that by going to the telescope.

"Here it is, okay, that goes from 22.0 to 23.2. This is absolutely at the detection limit, and this is a magnitude brighter. Look at the five years from 1950 to 1954, they are all phased." The points made a smooth sawtooth up and down in intensity. "That's the best light curve. Certainly it works." Sandage claimed his magnitude measurements, based on eyeball estimates with his ancient garlic-laden apparatus, were good to two-tenths of a magnitude—about 10 percent in intensity. "In these crowded fields that's all you can do. The people who have not been trained to do that think, aha, I'm going to do an unbiased measurement by putting this plate in a PDS measuring machine and scanning and then taking out the background, and they always get the wrong answer. The eye is an incredible machine, because almost independent of the background you can, if that star were a variable," pointing to a dot practically buried in other dots along a clotted arm, "estimate the brightness of this compared to that.

"I was trained to do that as a junior in college. I could do that when I came out and Hubble needed somebody. It was that training that Hubble knew I had that landed me a job as his assistant. When he had his heart attack, you know, I was on the train, going down the road, just incredibly lucky. But I had that skill and I still have that skill. When people then do these automatic techniques and make vast claims, and they have galaxies instead of stars, well . . ." he trailed off, shaking his head.

He picked up a photograph scratched with ink marks around different stars. M81 in the flesh. "Now all these markings were put on by Hubble. This is Hubble's marking chart. We would bring the plates down to Hubble from 1950 to 1952 and he did the marking. And there are novae, irregular blue variables. . . . I transferred all this over to a working chart, I guess I started this about 1954. I gotta finish this up. This will be duck soup for the space telescope." His voice trailed off.

Then he brightened. "I want to show you something extremely interesting. All the galaxies I'm going to show you distances for are like this, they're test objects." What Sandage had done in the hours when we weren't drinking at the Sea Lion was to plot all the distances and velocities relative to the Milky Way of all the galaxies within 16 million light-years—out beyond M81. It was the culmination of thirty years of staring at black dots on plates and 14-hour nights with an iron bladder in the cage of the 200-inch.

It was getting late in La Jolla, in the early November afternoon gloom. The flu was sealing my throat. Sandage talked faster and faster. "Let me identify these for you: IC 1613, member of local group; Andromeda nebula; 6822, member of local group; M33, John Graham's NGC 300, which you've just seen, at a Cepheid distance." His finger started thunking datum points. "Okay, Cepheid distance, Cepheid distance, Cepheid, Cepheid, Cepheid, Cepheid, Cepheid, Cepheid, Cepheid," as if it were a Fort Knox stamp, this was the good stuff, primo distances. "Here's my Cepheid distance, Cepheid distance, I divined that, and I divined that," he said finishing pointing at a pair of outer points. "So most of those are individual points that we observe."

He handed me the graph. Here, right at home, rising out of the scruff of random movements of local group dwarfs into a trend line of velocity with distance, the Hubble law, was the expansion of the universe, the mortality marks on the door, the cosmic riddle. "That's the beginning of the expansion," Sandage said pointing, his voice rising excitedly. "There it is, it's beginning and these are extremely close. How close? Look, one megaparsec, Andromeda is only that far." The finger cruised. "Two megaparsecs. Virgo's way out there. ISN'T THAT INCREDIBLE?" My response was silence, for my voice was clogged with a sore, congealed throat.

He whispered, "Well, I think it is."

I went to Kitt Peak in Arizona to watch the astronomers watch Halley's Comet in 50-mile-an-hour winds, and then went home. A month later, a week or so before Christmas, I called Sandage on the phone.

He asked me how the comet was. I said windy. "Now you know how it is," he crowed.

It was always this windy on the mountain?

"If not in nature, then it is in the journals," he answered, laughing. We went on to talk about M101, the showy spiral galaxy the Pinwheel, whose distance was both the linchpin and the shakiest element in the chain of observations and assumptions that Sandage and Tammann had employed to measure the Hubble constant. Sandage had inferred the distance to M101 cleverly but indirectly. Recently, he knew, Aaronson had measured the distance to M101 directly using modern high-sensitivity electronic detectors, CCDs, to observe Cepheid variable stars in the galaxy. Sandage didn't know Aaronson's result, but he feared the worst, that Aaronson would say that M101 was closer than he and Tammann had concluded.

Sandage knew that I would see Aaronson in Arizona. He asked if I knew Aaronson's result. As it happened, I did, but I had been sworn to secrecy. Sandage reacted bitterly when I said I couldn't tell him.

"Yes, yes," he said lifelessly, naming a value that was too low, "and Sandage and Tammann are all washed up." I felt stuck; I couldn't say anything. "Well, thank you, good-bye," he said. There was silence, but the line stayed open.

He asked, "Should we fold up our tents and go away?" I told him I hoped that they wouldn't.

"Now I feel worse than before you called," he whined. This cat-and-mouse game went on, designed to make me feel guiltier and guiltier. It wasn't my fault that I talked to Aaronson more than he did. "You must know you're being used," he said dryly, comparing the secrecy to Watergate. "They are out to destroy Tammann and me, not to do science. De Vaucouleurs doesn't care what the answer is as long as it's not the same as Tammann and me. I mean it."

Then he mellowed and exonerated the Aaronson collaboration— as he always did at the end of his tirades. At least they had gone down into the arena and brought back new data. "Once you publish a finding chart, magnitudes, and a period, you've exposed yourself," he said. When, he wondered, would Aaronson reveal his secret? The winter meeting season was upon us. Was he going to publicly destroy Sandage and Tammann at the upcoming American Astronomical Society meeting in Houston? Perhaps at a small upcoming meeting of observational cosmologists in Kona, Hawaii? How did Aaronson seem? he asked.

Confident, I said.

His voice fell further and further into the low, tired register. He was exhausted, isolated, beaten down, beleaguered by critics. "I think one ought to know when the time to quit is," he said, sighing. He couldn't shake the doom, the sense of a cliff a few strides ahead. Tammann was going to Kona; Sandage was going; so should I. Blood might be spilled. He tried to get me to say it would be his.

Suddenly at the end of the conversation a cosmic sigh lifted out of him. His voice resonated like a resigned warrior's. "You've got to grab life by the shoulders, Dennis, and shake it."

22

THE DAY THE UNIVERSE STOPPED EXPANDING

A month later, in the middle of January 1986, Sandage and I were standing under palm trees on a tongue of lava outside Kona. Around us the Big Island's gray-brown landscape—frothing lava frozen into spikes and ridges—was cut by small swaths of cultivated green, isolated shopping centers, and the domino march of high-rises down the coastline, each equipped with its own circular drive, pool, and gift shop.

The Keauhou Beach Hotel was near the end of the chain, ten miles from town. An air of distressed real estate speculation hung about the place like the salt spray from a big breaker. The room phones didn't have the hotel's phone number on them, as if they had just been borrowed for the week and could be repossessed. The coffee shop had no orange juice and closed after breakfast. For lunch you either hiked to the shopping center a mile away or had greasy sandwiches from the pool bar. The dining room had four entrées on the menu and closed at nine, just before the bar. After sundown it was possible to sit in absolute solitude on the balcony over the floodlit tidal pools watching puffer fish wandering among the rocks.

All in all, even with these amenities, the Keauhou Beach Hotel seemed an unpromising site for history in the making. The rustling palms and the rivulets of surf dripping from the lava offered no hint that they were about to host a week of shouts, sneers, and a general shredding of cosmological theories that would leave the astronomers in a state of confusion that could only be described as cosmic. That was not what I had signed on for.

I had come to Kona to see Sandage squirm and ultimately be vindicated when and if Marc Aaronson announced that his measurement of variable stars in galaxy M101 agreed with the distance that Sandage and Tammann had conjured a dozen years earlier. The distance of one galaxy, albeit a spectacular one, might not seem like much on which to stake or celebrate a career, even if it was the linchpin of contradictory efforts to gauge the size and age of the universe. What, after all, did it have to do with the meaning of existence? Or, to put it another way, how did it bear witness to the glory of God or to the wholeness of the ultimate symmetry?

The fact is that, for Sandage, every galaxy, every distance was a linchpin. He had spent more than thirty years assembling them, for that was what beauty broke down to—a chain of nitty-gritty details, smudges on plates and the slope of a line on a graph. I was rooting for Sandage to be right, not because I liked 50 better than 100 for the Hubble constant, or even because I liked him better than his opponents; in fact I liked Sandage's critics as well as I liked him. I rooted for Sandage the way I couldn't help rooting for every loner who was stubborn, every gambler who ever pushed all his chips to the center of the table and smiled waiting for the last card, every system builder who tried to re-create the whole universe in his own image, every dreamer who ever went for broke, every obsessive who lost his mind and his boundaries and didn't look back, every Coppola or Faulkner or Hawking who ever tried to say it all in one glorious fireworks burst of music or poetry or cold mathematics so perfectly that there was nothing anyone could do except stand there and be convinced, even if for one instant, that *yes, yes, this is the way it is*.

Nature was under no obligation to be fair. There was no guarantee that if we sent a man to dangle in a cage for thirty-five years on a cold mountain that he would emerge with a piece of the Truth. But I wanted to believe that it could be so. The fact is, there were a million ways for Sandage to lose the cosmological game, and only one way for him to win. I came to Kona to see Sandage draw his ace.

The Kona conference had been organized by Tully, the astronomer of Tully-Fisher fame, who was then ensconced at the University of Hawaii. The official name of this meeting was Distances to Nearby Galaxies and Deviations from the Hubble Flow, but everybody knew it simply as Brent Tully's meeting. Until that day, neither the guest list nor the agenda had ventured outside his quizzical blond head.

Early on Sunday evening I found Sandage, a drink in one hand

and the Kona program in the other, his brow knitting and unknitting, pacing back and forth next to the seawall. Beyond the seawall, the abrupt tropical twilight was punching holes in the lava hillside and coloring tidal pools vermilion and mercury. From inside a thatched-roof pavilion in front of Sandage came the surflike roar of forty or so astronomers and physicists, the blue water sailors of the cosmos who kept their heads in the galaxies, rattling mai tais and daiquiris—cosmologists in full throat.

"This is going to be a gauntlet of blood," he muttered.

Now as Sandage stared in and everybody else stared out, Tully's game was obvious. This room was full of Sandage's enemies. The room was full of cosmologists. Tully had assembled all Sandage's rivals, all the extragalactic barons who never spoke to one another, without telling any of them. All those people who scanned lists of invited speakers and then sent their regrets were swigging tropical drinks in little clusters like feeding sharks and looking over their shoulders.

There were Aaronson and his crony from Caltech, Mould, a red-headed Australian with a sort of roosterish posture; Gérard de Vaucouleurs, who had been the first and was still the most vociferous critic of Sandage; and Sidney van den Bergh, the Canadian cosmologist who belonged to neither Hubble constant camp and thus often got commissioned to write the review articles about it. With a tanned shiny dome and a voluminous white aloha shirt van den Bergh looked like some Indian swami. Many of those under the age of thirty had never seen or met Allan Sandage, whose life was the bedrock of their profession. And hardly anybody had ever seen him in the presence of de Vaucouleurs.

The word spread like a low rustle of gulls on a beach. *Sandage is here. Sandage.*

"Well," Sandage shrugged, "we're all in this Gamow box," drawing a box in the air with his fingers. "The question is whether we can tunnel out"—his hands fluttered to his sides—"or we just annihilate each other." He took a deep breath and plunged in. The room parted.

He went over to de Vaucouleurs and draped a big arm around his shoulder. "Let's fight it out like honorable men," Sandage proposed dryly.

De Vaucouleurs was short and pale. Dressed in a coat and tie and dark sunglasses, a splash of gray in his hair, and attached to his attaché case, with his determined dignity he reminded me of a spiffed up little kid trying to stay clean while being teased by roughhousing boys. He'd never be one of the guys. He squirmed as if a muddy Saint Bernard had just jumped up on him, shrugged fatalistically, and then started complaining about a recent paper in which Sandage and Tammann had criticized him.

He lectured Sandage about putting criticism in print where it could be taken apart. Sandage, he complained, once suggested that de Vaucouleurs not publish some argument because they were the only two people in the world who could understand it. Sandage reminded de Vaucouleurs of his suggestion that Sandage retire.

"You've made some great contributions to the field, but . . ." de Vaucouleurs replied, smiling thinly. "We always have these meetings, very friendly, but you resist any change."

Sandage laughed, "I'm not resisting, I'm telling the truth."

"You say you know the truth. You said so in Lausanne."

"I was just joking with you," Sandage said soothingly, clapping the shorter man on the shoulder. "You don't know when to take me seriously. You're a serious man, Gérard," he said in a deep, rolling voice. "You take yourself very seriously."

Sandage strolled away back outside to the seawall. "We always have these friendly talks," de Vaucouleurs complained as Sandage retreated, "but we never get down to concrete details."

Outside Sandage's jollity disappeared. He scowled and whispered, "If I had known all these people would be here, I wouldn't have shown up. I hate these confrontations."

Later we joined Aaronson and Tully for dinner on a balcony overlooking the tide pools. Tully, unfazed by his responsibilities as organizer, seemed to be running the conference out of his back pocket. With Tully as a laconic lubricant, Sandage and Aaronson bantered nervously. "You're a nice guy," Sandage said to Aaronson. "Why are you mean to us?"

He chided Aaronson for professing to lack a gut feeling about the real value of the Hubble constant. Aaronson, who wore (as he does everywhere) jeans, a short-sleeved shirt, and leather sandals, seemed bemused. He had wavy brown hair and a thick moustache. His brown eyes twinkled mischievously behind thick glasses. He replied that it was more exciting to be a slow learner who doesn't yet know. "If you know," he asked Sandage mischievously, "why are you still doing it?" Aaronson said he didn't care about H_0, he was interested in stars.

"Me, too," chimed Sandage, but what did Aaronson think? There was only one number, that was Truth. You only make progress, he intoned, when you have an intuition, an idea you want to prove or disprove. "We all bend the data and present the story in the best light."

Tully made demurring noises.

"But the truth always comes out and fouls you up," Sandage rolled on, his voice rising in volume and pitch. "That's what you don't realize in the moment of ecstasy."

Tully replied that that was why his publications were so far apart.

"I don't work on galaxies any more," said Sandage flatly to the belief of no one. He kept pushing. He asked Aaronson where the great papers were today, the massive compendiums of data and analysis that nourished previous generations, if today's astronomers have so much power at their fingertips. Aaronson pointed out that a lot of today's papers did have voluminous data.

"If Walter Baade were alive today, wouldn't he use the best technology—CCDs?" Aaronson asked back.

Sandage snapped back, "He'd use what he understood."

Later while we watched fish swim around under the floodlights, Tully and I talked about the upcoming meeting. To him, Sandage was just a sideshow. "The Hubble constant is not as rich a question as the origin and nature of structure in the universe," said Tully, slouching next to the balcony railing.

It was about to be High Noon, Tully predicted, for cold dark matter. According to that theory, as described earlier, 90 percent of the mass of the universe resided in clouds of slow-moving exotic elementary particles left over from the big bang. Their gravitational bulk sculpted in turn the visible galaxies and clusters of galaxies. As 1986 began, cold dark matter was the dominant theory of the universe, but there were hints of trouble.

It had only been two weeks before, via *Time* magazine and the winter American Astronomical Society meeting, that Geller had announced the results of the new Harvard-Smithsonian redshift survey and the "soapsuds" universe with 150-million-light-year bubbles in the galaxy distribution. The cold dark matter advocates were caught wondering if their photinos could have gathered themselves into structures so large and sharply rounded. Tully thought not.

Looking at maps, however, was like looking at pictures. The hardcore theorists didn't trust them. Who knew if the voids were really empty or if the glittering sheets of galaxies were just greasepaint? Which, explained Tully, was where the second half of the conference title—Deviations from the Hubble Flow—came in. If the knots and sheets of galaxies were real and not just illusory and if the universe was as dense as the theorists wanted it to be, then these huge structures should jostle each other gravitationally and distort the Hubble expansion.

The bigger the lumps in the universe, the wilder and faster the galaxies would move, which meant that by mapping the so-called peculiar velocities of galaxies and clusters, cosmologists could pinpoint and weigh the concentrations of mass in the universe. They could get *numbers*—velocities and masses—not just pretty pictures.

The name of this phenomenon was "streaming." There was already some evidence of streaming in the universe. One example was the infall of the local group into the Virgo cluster, which had been predicted by de Vaucouleurs back in 1953, and confirmed by Huchra, Aaronson, and Mould twenty-five years later. Now, as electronic devices facilitated a flood of distance and redshift data, hints were appearing of streaming on larger scales—much larger scales. Rubin and Ford's controversial suggestion of a distortion of the Hubble expansion could be due to streaming. The most unequivocal data came from the cosmic microwave background radiation, which had "poles" a thousandth of a degree warmer and cooler than average. The pattern suggested that the local group was moving through the microwaves like a goldfish in an aquarium. Interestingly, the streaming implied by Rubin's data and by the microwaves were in different directions. In the early eighties, Aaronson, Huchra, and Mould had combined the microwave data with their own observations and concluded that the entire Virgo supercluster was being pulled in the general direction of Hydra and Centaurus at some 600 kilometers per second.

Since that time there had naturally been a lot of thought about what could be doing the pulling. There was a cluster of galaxies, about 100 million light-years away, along the Hydra-Centaurus border, but our view of it from earth was spoiled by Milky Way dust. Nevertheless, it didn't look like Hydra-Centaurus had enough oomph to attract Virgo on its own, unless it was harboring a gargantuan amount of dark matter. Tully didn't think that cold dark matter could produce such large structures.

The putative program was heavy with data. Tully had scheduled four rigorous days of observational talks leavened with only two-hour breaks for lunch and swimming. "Meetings are usually boring," he said, "because everybody knows what everybody is going to say. This time I don't know what anybody is going to say." He shrugged and looked down at the fish. I should have started being suspicious right there.

In the morning we reassembled in the pavilion to debate the Hubble constant. Next door, one rocky outcrop away from the hotel, was a sandy cove. Throughout the day, while the microphone went on and off and reflections from the surf washed out the overhead projector, women clad in French-cut bathing suits strolled along a narrow wall to and from the beach, glancing in at pasty, sweaty cosmologists. Maybe it was my imagination, but none of them looked back.

Judgment Day started out disastrously for Sandage. He came in late—I could tell he hadn't had any breakfast—just in time to give his talk on the local galaxies and the calculations he had been doing in San Diego. He was breezy and authoritative as he stepped outward from the Milky Way. Measuring the distance of a galaxy, he reminded his colleagues, was excruciatingly painful, impossible work. Some of the results being presented here for the first time represented more than thirty years of work; most of the astronomers who had made the original measurements in these galaxies were dead. Sandage's main point was that the universe was expanding and expanding smoothly. The galaxies right next door to the local group were flying away with military precision at speeds perfectly proportional to the distances he had measured, corrected, tweaked, and retweaked over the years. The Hubble constant was still 50. Moreover, the lack of any noticeable deceleration of galaxies in close to the local group meant that the local group was not massive enough to contain large amounts of cold dark matter.

Sandage smiled tightly and strolled off when his time was up. During coffee break, Ostriker, the short Princeton theorist whose normal serious demeanor was muted by a floppy brown sun hat, had Sandage in tow out by the seawall. He was explaining rapidly in a soft voice where Sandage had made a serious mistake. By forgetting to take into account that the Milky Way and M31 are revolving about each other, his calculations had resulted in a ludicrously low estimate for the mass of the local group. Sandage was standing with his head down, his gut and jowls sagging, looking gawky, like a schoolchild being punished. His hand was curled over the top of his head, tugging on the lock that ran across his forehead as he followed Ostriker's explanation. I could see the confidence draining out the soles of his shoes.

At lunchtime Aaronson and I went snorkeling in the cove next door. Sandage grabbed a lounge and a drink by the pool.

Afterward, it was de Vaucouleurs's turn. Still dressed in a coat and tie, he spoke in a plaintive French accent. De Vaucouleurs's contention, something he had been arguing for thirty years, was that the universe was not so nice and smooth as Sandage claimed. It was lumped up with big concentrations of galaxies whose gravitation would distort the Hubble expansion. Moreover, he didn't trust nature to make anything uniform, including the variable stars that Sandage and others preferred to use to calibrate their distances. "I'm all for primary distance indicators," he declared, "with a healthy distrust of any one individually." De Vaucouleurs had invented his own bag of tricks to survey the distance to a cluster of galaxies in Hercules and concluded unsurprisingly that the Hubble constant was between 90 and 100.

Someone asked if this Hubble constant, which implied the universe had been expanding for about 10 billion years, didn't conflict with the 14-billion-year ages of globular cluster stars.

"No, sir," de Vaucouleurs stiffened. "I refuse to mix religion and cosmology. Ages are one thing, and distances are another. Let's not discuss religion. If we knew the answer, what would astronomers of the twenty-first century do?"

Tammann protested that de Vaucouleurs's distance indicators

were gobbledygook. De Vaucouleurs replied that Sandage and Tammann's recent criticism of his work had been riddled with errors of its own; he asked for ten minutes the next day to defend himself. Listening to this exchange of barbs about luminosity indexes and tertiary indicators, I couldn't help wondering what Hubble would have thought, whether he would have been ashamed to have his magisterial subject reduced to such pettiness disguised as nitty-gritty.

By the time Aaronson got up to speak, the sun was in the west, out over the ocean. Sunlight bounced off the Pacific chop and flecked the podium. Unfamiliar tropical heat washed the tent. Aaronson sauntered forward, like a gunslinger, I thought, offhand and low-key, exuding an understated confidence. The audience shifted in their plastic seats. I looked around. Sandage had snuck in and was sitting toward the back.

Unlike the others, Aaronson didn't claim to know the Hubble constant, only that it could and would be measured once and for all with the space telescope. The Tully-Fisher method his group used, he stressed, had no magic, only simple Newtonian physics. And his group was without history and feuds, without an emotional or philosophical stake in the outcome. "In five years," he concluded, "we can hope, if the checks and counterchecks agree, to have the Hubble constant to within 10 percent, and we can get on with our lives."

He teased the crowd along; everyone knew that Aaronson had more important news. Finally, almost as an afterthought he mentioned that they had detected Cepheid variable stars in M101, the controversial galaxy on which Sandage had pinned so much. He paused, eyes twinkling. "If you force me," he said, "I can give you a distance."

In my mind's eye, I saw Sandage rise at the back of the room, like a defendant in the docket facing the jury with downcast eyes.

Aaronson answered in the most obscure way possible, by giving the distance relative to that of the large Magellanic cloud, a satellite galaxy of the Milky Way visible from the Southern Hemisphere that is the main source of information on the properties of the Cepheid variables. Nevertheless, everybody knew immediately what he meant. He put M101 at virtually the same distance as Sandage and Tammann, 22 million light-years.

And that was it, I thought. A sixteen-year-old stroke of faith, guts, and intuition had been at least partly vindicated. Not every story ended miserably; not every galaxy was a demon trickster; not every astronomer was out for blood. Science, science, the dispassionate interrogation of nature, science prevailed.

As soon as Aaronson stopped talking, Sandage was gone. I didn't see him until the next morning when he complained of a cold. He had fallen asleep with the air-conditioner on. But his step and his shoulders seemed lighter. He seemed to have forgotten that M101 had ever been in doubt. The worst was over. In fact, by God, he said, the field was coming together. He beamed, "Brent Tully is a master psychologist. It's marvelous to have everybody here talking together. And the long knives aren't out."

Sandage's mood quickly became the talk of the town. Everybody agreed that they had never seen him happier. But then again, many of them had never seen him at all, let alone with his cosmic straight man, de Vaucouleurs. Sandage and de Vaucouleurs spent the rest of the week nose to nose, jibe to jibe. Every time I turned around they were going at it, like an old married couple breaking up a dinner party. Eyes and ears were constantly swiveling like in the old E. F. Hutton commercials.

They clashed again on Wednesday, the day that de Vaucouleurs had asked for ten minutes to rebut some Sandage criticism. Ironically, it was Sandage's turn to be chairman that morning.

"I'm going to be very naughty to you this morning," de Vaucouleurs whispered as he pushed past Sandage to his seat.

"Well, I'm going to be naughty right back, because you're wrong," Sandage replied.

De Vaucouleurs had asked for ten minutes to refute a Sandage and Tammann paper that was critical of something the Frenchman called the lambda index, which was supposed to be a distance indicator for spiral galaxies. In their paper Sandage and Tammann had apparently misidentified a few galaxies. De Vaucouleurs seized on this fact and launched into a stern tirade illustrated by blowups from Sandage and Tammann's own paper—lists of galaxies in which he found typos, inconsistent classifications, fictitious galaxies, and even the same galaxy twice, listed as a different type each time. "A masterpiece of confusion," he said, "I've never seen so many mistakes in one table by a few people."

All the time Sandage, chairman of the session, was standing ten feet away with a tight smile on his face, amazed, he later told me, that his stomach wasn't churning. It looked as though "giving away the world" actually worked for him.

At the end of the session the astronomers trooped over to the pool to pose for the official conference group portrait. Sandage and de Vaucouleurs went to the middle of the front row. Sandage put his arm around de Vaucouleurs, and the other shrugged it off, squeaking, "I'm European, and we don't do these things." As the camera clicked, Sandage called out, "Smile, Gérard." In the resulting picture de Vaucouleurs looks unyielding, while Sandage looks a little deranged and off balance, clutching his stomach as if he were sick.

It seemed hard for me to believe that we had only been at Keauhou Beach for three days. Every exchange seemed weighted with history, every joke or insult seemed destined to cast shadows down the corridors of cosmology. I could feel the center of gravity shifting. If the past belonged to Sandage, it seemed the future belonged to Aaronson, shuffling quiet and irreverent between the confused astronomers, offering a little observational support here, a little criticism there, adding names to his coalition. Aaronson was becoming the new kingpin of the Hubble constant. He was in the midst, moreover, of putting together his space telescope collaboration, the group that would decide, with techniques everyone could understand, what the real size of the universe was. He and Mould joked about getting Sandage drunk and signing him up. "I just want to have the strongest possible proposal," Aaronson explained with a straight face.

Meanwhile the real cosmologists, the large-scale structure mappers and the cold dark matter theorists were circling each other like a gathering storm. Tully had invited a handful of theorists, most of whom came straight from a workshop in Aspen the week before, to give talks on the last day of the meeting. The joke was that Tully needed a certain percentage of foreign scientists present to qualify his NATO sponsorship. There was Alex Szalay, the Hungarian, on the last stop of another two-year stint in the United States; Avishai Dekel, a heavyset Israeli visiting at Santa Cruz whose shirt tails were always hanging out; and Amos Yahil, Sandage and Tammann's old one-time sidekick; Joel Primack, the smooth prince of Santa Cruz. Szalay was wearing the cosmologists' uniform: a Space Telescope Science Institute T-shirt, jeans, and dirty sneakers; he and his comrades—Nick Kaiser, the inventor of biased galaxy formation, an extroverted Brit with a taste for aloha shirts, and Bond, the Canadian then at Stanford—claimed they were mainly looking forward to swimming and sightseeing.

From the beach and the swimming pool, the theorists were bemused by Sandage and de Vaucouleurs fencing and faintly alarmed—how would the astronomers like it if the physicists couldn't agree on, say, the mass of the electron? Szalay and his friends were more concerned about the upcoming clash of cold dark matter and large-scale structure promised by Geller's new redshift survey. How empty were the voids, really? How real were the shells? How new was it all? How much was in the eye of the observer playing connect a dot with an inkblot sky? What, if anything, did this have to do with the endless inability of the observers to agree on a Hubble constant?

What were these people doing on the same lanai?

Enter David Burstein, a young astronomer at Arizona State, and an unlikely candidate for hero (or villain) of the hour. Burstein was husky, with a broad face accentuated by a narrow fringe of beard. Without a moustache, his upper lip looked big enough to play tennis on. He had been walking around all week with it curled up into a smug smirk. "What I'm going to say makes sense of everything that's been said in the last two days," he told me with a sweaty undergraduate earnestness. I couldn't imagine what could be making him so smug. Burstein had been a student of Rubin's, and had been involved in measuring rotation curves in spiral galaxies. Burstein, in other words, already knew trouble. As a postdoc at Santa Cruz he had become part of a large collaboration led by another Rubin student, Faber, that would become known as the Seven Samurai.[1] Burstein's secret was that the Samurai had gotten into some of the same old trouble that had set the Furies after Rubin when she and Ford uncovered the alleged distortion in the Hubble expansion.

In 1980 the Samurai had begun an ambitious study of elliptical galaxies that was eventually to involve ten telescopes on four continents. Their aim was to investigate a relation which Faber and Lick astronomer Roger Jackson had discovered between the masses, and thus luminosities, of elliptical galaxies and the random speeds of the stars inside them. The star speeds are measured from the blurring of spectral lines—the wider the blur, the faster the stars were buzzing around and the more massive and more luminous the galaxy must be. The most important use of the so-called Faber-Jackson relation was in determining the luminosities and thus distances of elliptical galaxies, but Faber and her colleagues thought that it might also help ascertain other properties of the galaxies. Over the next few years they observed 322 galaxies that were spread over half a billion light-years of space, obtaining in the bargain information about their distances and redshifts.

What could they do with the distances of 322 elliptical galaxies? On the suggestion of Donald Lynden-Bell, one of the group, they decided to map the locations and velocities of the galaxies relative to the microwave background, which forms a kind of universal reference frame. In a perfectly uniform expanding universe, with no gravitational jostling, each galaxy would be like the proverbial raisin, snug at rest in the microwave expanding cake, racing away from us at speeds directly proportional to their distances. Astronomers already knew that was not strictly true; the local group and probably Virgo as well were sailing through the microwaves toward Hydra-Centaurus. Such flows were a powerful way of deducing the concentrations of mass in the universe and seeing whether the apparent superclusters and voids on redshift maps were real inhomogeneities or just greasepaint, as it were, on a uniform background of dark matter. If you looked over a large enough part of the universe and subtracted the

---

1. For the record the Seven Samurai were Sandra Faber, David Burstein, Alan Dressler, Donald Lynden-Bell, Roger Davies, Roberto Terlevich, and Gary Wegner.

proper Hubble expansion velocities, Faber and her friends reasoned, you would expect to see a random pattern: some galaxies coming toward you and others going away.

The Faber-Jackson relation has a lot of statistical scatter in it, which meant that it took a lot of time and analysis, beginning in 1985, before the data began to mean anything. Burstein was the first of the Samurai to see that it did not mean what they had expected—no random jostling of groups of galaxies. Rather, it seemed that the entire region they had surveyed—from the Perseus cluster, 300 million light-years or so away to the north, to Hydra-Centaurus in the distant south—the entire neighborhood, a hundred thousand galaxies, a thousand trillion suns, was moving together as a unit. It was going south, *past* Hydra-Centaurus, at 700 kilometers per second—2 million miles an hour.

This was not large scale—this was hyperscale. And it was a number—not a picture. The Samurai survey included galaxies scattered throughout the entire local universe. It was as if somebody had taken most of the known universe and set it on an ice floe adrift in some mysterious current. Moreover, the direction of this current was roughly in the same direction as the infamous Rubin-Ford effect. This caused consternation in the group—one day they believed it, the next day they didn't. Finally they believed it. Faber and Burstein, both old associates and admirers of Rubin, had each called to tell her the news.

Then they had proceeded to button their lips. While the Harvard-Smithsonian redshift team was making films and lining up talk shows, the Seven Samurai issued no press releases. They just kept grinding data out of the computer. A week before the Kona meeting Faber gave a presentation to the Lick staff. Then they fanned out for the winter meeting season. Later Primack told me he had been equally amazed at the result—half the local universe in contrary motion—and the fact that they had been able to keep it so secret.

Now, in Kona, the sly Burstein prepared to give his talk, a talk he knew would upset cosmology. He knew that if he lived to be a hundred he would never get to drop another bombshell like this. Gossip usually spread too fast for there to be real news at such meetings. Smiling to himself, enjoying the tension, Burstein drew out the news slowly.

He began by describing their survey and then reviewing the results of the old disputed streaming motions, the flow to Hydra-Centaurus. The simplest models of that motion, he reminded us, predicted that the attraction is mutual, Hydra-Centaurus drawing toward us. He paused, looking at Yahil, sitting in the front row, and his voice dropped dramatically. "Hydra-Centaurus isn't coming toward us; it's moving away at 700 kilometers per second."

Yahil turned white.

"We are seeing bulk motion of the whole shebang," Burstein concluded. "The question is at what? Don't expect the answer here."

The afternoon dissolved in pandemonium. I felt that I was watching forty lava tubes simultaneously erupt. As chairman, de Vaucouleurs maintained that Burstein's time was up, but the crowd insisted that he continue.

Having control of the floor, so to speak, de Vaucouleurs himself was the first to respond. "This confirms what I've been saying for years." The universe was not some noiselessly expanding raisin pudding of perfectly distributed galaxies and little round clusters; it was ugly and lumpy with misshapen clots banging one another around. A tricky place.

Nonsense, responded Sandage, almost as quickly. There had always been a little velocity scatter in the Hubble diagram, and, while 700 kilometers per second was fast, it was a creep compared to the Hubble velocities of thousands of kilometers per second that the outermost of Burstein's galaxies had.

"This does not destroy cosmology," he said with all the authority that he could muster from decades of Mount Wilson training. "The currents are on a small scale."

When they finally let Burstein go, Sandage came waltzing up beaming and clapping me on the shoulders. "This is crescendoing into the best scientific meeting ever." He danced away down the seawall.

I was in a state of shock. Having seen Sandage draw his ace, I had been prepared to spend the rest of the vacation on the beach. I wasn't prepared to see the cosmologists hopping up and down and shouting like football players in the end zone. I wasn't prepared to see cosmology go flying off the deep end into confusion. But confusion was what the Samurai movement boded. A rule of thumb was that the larger and faster a chunk of universe is streaming, then the more massive must be whatever is doing the pulling on the other end. Burstein's effect implied a concentration of mass somewhere far in excess of what cold dark matter could produce.

If Burstein and his colleagues were right and the whole neighborhood was sliding toward Centaurus as fast as he said, the voids and chains of superclusters would not be illusions. The universe would truly be uneven and chaotic, bulked up in some places like a weightlifter on steroids. And that meant that cold dark matter, which had been the closest thing to a standard model in Fermiland for almost two years—nearly a graduate student generation—could be a dead duck. Cold dark matter had no power to tug around whole sections of the universe or to build structures of trillions of suns. If the Samurai were right, Kona would be remembered as the place where cold dark matter died.

In the morning it appeared that might be the case. More astronomers seemed eager to add nails to the coffin. There were more tales of galaxies adrift in the currents of space. The reported velocities of these cosmic ice floes kept climbing, up to 1000 kilometers per second. I could see the universe, suddenly a chaotic unpredictable place, lumping up before my eyes.

Whatever happened, I wondered, to the even smooth Hubble expansion that Sandage had kept finding during the fifties and sixties? Had he just not seen far enough? Had all the observations to date been confined to a single, perhaps atypical, "floe"? Would the Hubble constant go up beyond it? Was de Vaucouleurs right? And did that mean that standard Friedmann cosmology was doomed? I could feel dominoes tumbling. Not now, Lord, not after I had finally learned to read Primack and Blumenthal's cooling curves.

As for cold dark matter, I decided to seek out Yahil, who had become an expert on cold dark matter. He frowned and admitted that he was still in shock from the day before. "Cold dark matter is just a cute idea," he said. "Don't take anything particle physicists tell you as gospel. They are the most theologically promiscuous people in the universe. Because as soon as something breaks down they will invent something new. I have an advantage because I came from particle physics and I know how easy it is."

Halfway through the morning we boarded buses to ride up to the top of Mauna Kea, thirteen thousand feet uphill from Kona, by acclamation the prime observing site on the planet, and home to enough big mirrors to cover a basketball court. The road wound around and through fields of lava sharp as frozen wheat. I rode beside Tammann, taking notes, for about half the way, until I realized I was woozy, whether from the curves or the altitude I didn't know. When we reached the halfway station and stopped for lunch I could barely stand up and had no appetite.

Meanwhile Tammann moved a seat back to sit with his leader, Sandage. They bantered about modern astronomy. "Did I tell you I got a personal computer?" asked Sandage. "I'm not nearly as afraid of the things as I was. It's wonderful learning to open and close files; I've got a great word processor and I've got a paper already half-written. All I need is somebody to teach me. In a couple of days I could be reducing CCD frames. I'll go back to the telescope and be a *real astronomer*," he cackled. They started mugging at each other.

Tammann asked if they were going to propose anything for the space telescope. Sandage put on a sober face and his voice dropped. "Well, it's terrible because it will change our lives completely."

We were lifted, after lunch, to a cold, nearly airless brown and gray cinder wasteland. The cinder mounds of Mauna Kea were each crowned with glistening white domes. In the far distance a huge square hole in the ground marked the future site of Caltech's 10-meter Keck telescope. Above us the sky was a brilliant, headachingly blue. On the top of Mauna Kea half the oxygen in the atmosphere is below you. Each step was an adventure in dizziness.

Approaching the summit, Sandage instantly turned flinty-eyed professional. It was his first visit to Mauna Kea, and his first trip to an observatory in three years. "It's always exciting going to the mountain," he said wistfully. Sandage had brought a down parka, hat, and gloves to wear, while Tammann had a suit and topcoat and made sure to smoke when he got to the top.

The astronomers broke into small groups to tour the real estate. I went with Sandage; by now I was like a barnacle on his side. Every once in a while he stopped to lean on a wall and catch his breath.

We walked into the tallest dome, the one containing the 3.6-meter telescope built by Canada, France, and the University of Hawaii. Sandage perked up. "It's getting to be an interesting size." We were standing below the big mirror staring up at instruments hanging from its back, at the Cassegrain focus, thick cables running away. Sandage's eyes glittered. "This is a *wonderful* telescope."

That evening we all ate sushi in Hilo where the makers of the Canada-France-Hawaii telescope had sensibly built their scientific headquarters. Buoyed by the general air of goodwill, Aaronson and his comrades were still talking about making another effort to recruit Sandage on their space telescope team. Just then Sandage wandered over and told them that he and Tammann were going to propose a small space telescope project, to look for Cepheids in a couple of nearby galaxies that had hosted supernovae. It was nothing, he stressed, that would compete with their program. Aaronson said his offer for Sandage to join his team still held.

Sandage demurred. Shaking his head, he clapped a hand on Aaronson's shoulder. "We get along on a certain level," Sandage said levelly. "It's different when you sleep together than when you just say hello. You're a nice guy. But I'm a son of a bitch." He waltzed away.

As I staggered away, as if propelled by the intensity of that exchange, I caught Szalay's eye. He was leaning against the wall by himself with a toothpick in his mouth, thumbs hooked in his belt loops, trying to look tough. "Are you ready," he sneered, "for the death of cold dark matter?"

In truth I wasn't ready for the death of cold dark matter. I rode back to the hotel alone in the noisy bus, feeling disturbed. In three days, hardly anybody had even been to the beach. The atmosphere of alarm was contagious. The meeting had begun to take on the air of a runaway constitutional convention, where almost any crazy thing could happen next. No one was in charge. I expected at any moment that someone would leap up and reinstate the steady state theory. I felt suddenly as if I didn't know anything. Around me rose a din of chattering excited astronomers. Not one word sounded accessible.

In the morning Sandage stood up and asked for five minutes. "In view of everything that has been said," he began, "one can wonder if there is such a thing as a Hubble constant." He showed the classic Hubble diagram—the redshifts of giant elliptical galaxies chased all those years from the cage of Palomar into the night—the maligned and cosmologically impotent Hubble diagram. It was a straight line. Sandage noted that at its farthest reaches, peculiar velocities could be as high as 4500 kilometers per second and not disturb his graph.

"I want to emphasize there *is* a Hubble constant!" he declared ringingly.

The cosmologists stood up and cheered.

A danger point had been passed. Tammann declared later, waving his cigarette holder, "If Allan Sandage hadn't been present, the astronomers at the Kona conference would have thrown the expansion of the universe right out the window."

The slaughter on Keauhou Beach continued.

It was Tully, the laconic organizer himself, who finally breached the bounds of credibility, which had been slowly stretching, stretching all week, when he claimed his own mapping efforts revealed that almost all the known galaxies and clusters lay in a four-layered pancake 1.5 billion light-years across. Tully, who is given neither to braggadocio nor false modesty, slouched to the lectern, hands in his pocket, and described his new work as "the most important input into observational cosmology since the microwave background."

"Go for it, Brent," somebody yelled.

Tully's pancake was worth a thousand of the neutrino pancakes with which Zeldovich and Szalay, only five years before, had tried to build the universe. Tully's pancake was too big even for neutrinos. It was too big for the Kona conference to swallow, gorged as they already were on amazing and controversial observations. At last—the chance to be skeptical.

Now it was de Vaucouleurs's turn to feel jolly. After all, he had always known and argued that the universe was ugly and lumpy and not a theorist's playground. "I seem to be hearing voices from the past," he cracked. "Allan never believed in the local supercluster."

Sandage piped up, "I didn't say that, McCray did."

De Vaucouleurs answered, "But you told me the news."

Before and between sessions de Vaucouleurs would hold court in the hotel lobby, his briefcase popped open on a coffee table. He mused as we watched people go by on their way to the small beach between lava points, "Is this a vindication?" He seemed looser than earlier in the week. He had traded his tie for a cowboy hat. "Of course, Brent had to convince himself. This generation has been raised with the belief that the universe is homogeneous and isotropic, that there is no local supercluster."

So in the end, both de Vaucouleurs and Sandage were vindicated. They would each take away a victory but no tan. How these "victories" would be reconciled with each other was something I was waiting to see, but I was to be disappointed.

The theorists were dispirited. Ambushed by superscale structure they'd never heard of, they confined their afternoon of talks to discussions of general principles. Kaiser offered some suggestions as to how surprisingly easy it was to misinterpret redshift surveys. Collectively the theorists felt they had let the side down. By compiling such a parade of observational surprises, Tully had beaten them.

On the last night, Szalay, Bond, Kaiser, and I went to eat Vietnamese food at the local mall and then take the Szalays to the airport to begin their trip back to Hungary. "If the observers only realized how much we thirst after these measurements—velocities," Bond lamented. "But they don't hunger for some theory. They don't take it well to have some theorist tell them what to go out and measure. It's a very asymmetric situation. Some of them try to do it all, but it's pretty pitiful."

In the morning Sandage came dancing up to me one last time. "Smile, you're amid famous astronomers. I don't know if you've caught the excitement. This is the most exciting field in science." He put on a serious face. "Brent Tully has done it. There has been a lot of forgiveness at this meeting."

On both sides? I wondered.

"Yes." Sandage waxed enthusiastic to the bigger end. "They've gotten distance indicators in shape, down so good they're getting beyond the noise," he said. "And look what they're finding! This is the best scientific meeting I've ever been to.

"Am I enthusiastic? Well, yes. I'm just starting out in this field."

23

THE OTHER SIDE

Cosmology, the search for the secret of the origin and destiny of the universe, is an old pursuit, an ancient yearning, but it is not an old science. It was only in 1917 that Albert Einstein had first conceived of the universe as curved expanding space. Hubble's discovery that the nearby nebulae were galaxies rushing apart from each other came a decade later. Depending on which of those landmarks you take as the birth of cosmology, by the time of the Kona conference, the discipline was only sixty or seventy years old; its entire history to date could be compressed into a single human life span—Sandage's, for example. He was right to say that he was just starting out in the field. We were all just starting out.

When I left Kona, I felt cosmology had crossed some kind of dividing line. I had the sense of one phase ending and another beginning, perhaps not yet born. It was difficult to say what was ending—maybe the arrogance of youth that the answers were there for the picking, that the universe was easy. We were not young any more. All the easy answers were slipping away, and they would not return.

There had been no press, no public relations officers, barely a program for the Kona meeting, and no advance word of what was going to be presented, that is, no papers circulating. As a result hardly anyone knew what had been said in our hermetic, feverish little colony on the lava. I waited eagerly to hear the buzz and the uproar from the larger community of astronomers—I heard nothing. I spent the next three months traveling and talking to scientists in Europe and Australia, and nobody said or wrote anything about Kona, about megapancakes and rushing hunks of cosmos. After a while, I began to think it had all been a group mistake. In the snows of Russia, Kona seemed like a dream, a brief disturbing glimpse of cosmological monsters out there behind the curtain of verifiability. The curtain was closed now, but the nightmare of how crazy the universe might be remained. Maybe Peebles was right— it just never would make sense.

That was true of science, and certainly of cosmology. Someone was always building cathedrals, and some innocent was always coming along and spoiling things with a discordant observation. The job of being a scientist, it seemed to me, consisted mostly of constantly exposing yourself to the people who could do you the most harm, the ones who could see through you and pick you apart—your peers. It was the constant grinding of egos that produced the sawdust of progress. Sandage had spent the first half of his life building a cathedral of time and distance and the second half shoring it up against attack. The inflationary universe filled with cold dark matter was another such cathedral; whether the Great Attractor—as Samurai Alan Dressler called whatever was causing the putative movement of the heavens—would eventually knock it down remained to be seen. But the smart money was on confusion.

Astronomers were reluctant to jump on the bandwagons of Tully, the Samurai, and the other wild men of Kona. Confirming their whoppers would be long hard work; it required access to southern skies, for example, where what few catalogs of galaxies existed were incompatible with northern ones. In the absence of other supporting data, second-guessing the results of the Seven Samurai and their analysis of those observations became a small cottage industry. The Samurai themselves soon fell to bickering. Indeed the Samurai data, like the sky itself, were a kind of inkblot in which any cosmologist, it seemed, could see his own truth.

The reader might be forgiven at this point for thinking that Truth is constantly dangled in front of him or her and then yanked back. Why learn all this stuff in the first place, if it's always going to turn out to be wrong? Zeldovich had the answer one gloomy winter day in his Moscow office. He was talking about superstrings. "In physics, we have something very definitely obtained," he said, waving an index finger in front of my nose. "No success in ten-dimensional superstrings will undo what we already know. What is already taken by science will not go back."

Like Woody Allen, failing with better and better women, cosmology, too, would fail upward. Science, despite the wishes of people like Arp, would not go backward; tomorrow's newspaper headlines would be even harder to understand. Galaxies would still be quantum fluctuations; dark matter would still enfold the universe, but new and amazing things would keep getting dragged to the back door by the observers. New and wild ideas of ineffable beauty and logic would spring from theorists' heads. That was great; it was a good show. But was there ever going to be a chance to rest?

It was with those kinds of thoughts rattling around in my mind, vaguely anxious for solid ground, wondering where I could ever dare end this chronicle, that I drove through a steady, gray November drizzle to Baltimore to hear Allan Sandage give one of the last major addresses of his career.

Sandage had become a wanderer. From San Diego, he had gone to the University of Hawaii, then England, and then China before going to Baltimore as a visiting Johns Hopkins professor and visitor to the Space Telescope Science Institute. The Kona glow of fellowship had lasted about as long as it took for the next round of papers to get into print. "Astronomy is an impossible science. All you're getting is opinions, including mine," Sandage growled once when our paths crossed. "It is fifty, regardless of what they measure."

"A burning bush told you that?"

"A Baptist minister. He didn't reveal his source."

This occasion was what would have been Edwin Hubble's ninety-seventh birthday, an anniversary that the Space Telescope Science Institute had decided to honor with an annual lecture. According to NASA's original schedule, this inaugural lecture would have been delivered a month after the Hubble Space Telescope had been launched and would thus herald the new era of astronomy. When the space shuttle exploded, two weeks after Kona, the new era had been blown into the indefinite future. So Sandage's talk would instead mark the shadow between two eras—one not quite dead, the other not quite born.

When I got there, the future of American astronomy, the space telescope institute, was shrouded in indecision and low morale caused by the tragedy of the *Challenger*. It huddled against a steep hillside on one edge of Johns Hopkins. Across the campus hot pink neon letters on the top of the Baltimore Art Museum stood out from the fog, beckoning.

Inside the museum's plush postmodern auditorium, standing on a corner of the stage, fretting, was Sandage. Wearing a gray suit, he looked distinguished; his features were tight. He said he was nervous, but he appeared fit and focused. He complained that he hadn't eaten or drunk anything all day.

When his time came, Sandage slid forth on a foam of nervous jollity, with his old friends Hubble and Baade and Humason metaphorically by his side. "Today astronomers are making incredible, fantastic claims," he announced dryly. "What I want to do tonight is to attempt to tell you, the nonastronomers in the audience, something about those fantastic claims to see if you buy anything before this space telescope goes up.

"We don't know what the possibilities are of finally discovering the solution to the cosmological problem. It has been not only the dream of the space telescope to do that but the dream of essentially all mankind in some way or another from the time that writing began."

What followed was a curious speech—one of the best I have ever heard from Sandage or any other scientist—but one that contained almost no science, at least at the level I have been describing in this narrative: no rotation curves or Higgs field diagrams or large-scale structure statistics. He was there to remind us all of something that was easy to forget in the blizzardy march of technical data—that astronomers *do* make progress. The stars are impossibly old, but they are not eternal. They are impossibly faraway but not infinitely faraway. I found this oddly reassuring. The great truths of cosmology were not to be found in the mixing ratio of light to dark matter or even the value of the Hubble constant. Great science like great art returns to the obvious. Great truths have been carved from the domes of Palomar and the cerebellum: Stars are born and die. Galaxies are born and die. Atoms are born and die. Particles are born and die. The very forces of nature, and perhaps even the dimensions, are born and die.

If Sandage was wrong and cynical to belittle the superstring theorists, he was surely right that the main lesson of cosmology is that at the heart of existence is a mystery, a mystery that ultimately cannot be solved so much as savored. That was the task of cosmology, like mythology of old: to bear witness.

"Why is there something instead of nothing?" he asked rhetorically. "In science," he answered himself, "you can ask the questions 'what,' 'how,' and 'when.' The question 'why' belongs to the realm of philosophy. I'm a scientist tonight so that question will surely not be answered.

"Yes, the evidence for the Creation Event is there," he droned archly. "Yes, the value of the Hubble constant is 50. No, not everybody agrees. . . . Progress in this field only comes at the funerals for the astronomers."

And so, as if on cue, he called the roll of cosmologists, going all the way back: Socrates, Plato, Aristotle, Copernicus, Kepler, Lemaître. "All the cosmologists have gone on ahead. They know the answer. Now because they know doesn't mean we know."

Finally he got to his beloved Hubble, who, it seemed, knew everything from his perch beyond. "We live in a black hole, and there's no communication with the other world," Sandage went on. "We can only speculate if Hubble is really there and if he is looking down on us and seeing the preparations to determine his constant. If he really is there, he's smiling because the problem is much more complicated than we know."

Is the universe right-handed or left-handed? Hubble knew. "We are matter instead of antimatter, and Hubble knows"—it was becoming a mantra—"*why* we're matter and not antimatter."

He paused. "If any of this is really true," he concluded, "then I think in our lifetime the question of mankind that has been asked for ages, all those people on the other side know." Then he stopped and stood alone in the spotlight, basking in a standing ovation.

Sandage had become a man of his times. His voice carried none of the old Hubble aura of arrogance. During the talk he had shown pictures of the old cosmologists—Baade, Hubble, Shapley—and called our attention to the fact that none of them was smiling. Cosmology was a serious business, he said slyly. Of course he was right. Cosmology *is* serious business, and in our hearts we are nothing if not cosmologists, hanging in a cold cage sifting the ruthless jewels of existence. Someday Sandage's atoms and those of everyone in that lecture hall would be splashed across space, stripped, and recombined in some as-yet-unborn star, or slumped into a black hole. There would be nobody to remember him or her, this room, what was said here, or any of the grand conclusions of cosmology. The only sensible future the cosmos had was in our own heads. This was the image I wanted to retain of Sandage, a cosmic quipster, edgy as a razor, scat quick, and humorously dry as a lemon, standing on the lip of the stage as if it were the edge of eternity, one eye on the audience scanning for trouble, some hard fact, the other one gazing much farther.

A man with friends on the other side.

Meanwhile, what Sandage calls the "sweaty work" goes on. The theorists gathered in Aspen in July 1986 to see if their theories could be repaired and reconciled with the monsters of Kona by cosmic strings. I spent a week filling a notebook with calculations and arguments over whether those dense little twanging tubes of energy left over from the symmetry-breaking episodes of the big bang might help whip the primordial stuff (whatever *that* was, dark or light) into galaxies and superclusters. Peebles circulated screeds, crossed and uncrossed his legs across the patio where talks are held in the summer, and held court by the picnic tables. The jury was still out, and would be out for some time to come, on cosmic strings.

At the end of the week about ten of us, Peebles, Schramm, and Steigman among them, including wives and girlfriends, attended Maurice's cooking school. Maurice is a short cheerful Swiss-trained French chef who runs a small restaurant in one of the Aspen condominiums. For a certain fee, Maurice will let small groups sit in front of his butcher block table in the kitchen and watch as he tells stories and cooks dinner for them all night long and his assistant pours all the wine they thought they could drink. Perhaps I should have mentioned Maurice earlier, because it was obvious from the conversation as we arrived that Maurice had long been a great asset to the cosmological community in Aspen. Any hope of taking notes and keeping track of Maurice's activities soon faded. The evening became a blur of chopping knives, bowls of bright spices passed around for sniffing, heat, wine, poached salmon, lamb Wellington, sweetbreads braised in veal stock, saffron rice: in all, the four best French dinners I ever had, plus two desserts. For the record, everything was baked at 400°F in a gas oven that had rags stuffed in the holes where the dials and handle used to be. And every recipe started out with a quarter pound of butter.

Staggering out hours later under the starry vault, Schramm eyed the swimming pool, but after a moment's indecision decided not to take the plunge.

The next day Peebles and I hiked up to a pass near the Maroon Bells. It had been a rainy, cool summer, and the high meadows were choked with thick snowfields and unfordable rushing streams. Coming down Peebles attempted a shortcut between two switchbacks and managed a graceful long glissade down a snowbank on a forty-five-degree slope, smiling tightly into the wind, knees bent, weight nicely forward as if he were on a ski slope. Following him I slid the whole way on my rear end.

"I don't know how to summarize," he said at length when we rested. "There have been surprises certainly, the subject of cosmology has advanced because of the observations. I think the theory has been awfully flaky. I still don't think we have what I would call a clean theory for what's going on. The barest outline, the expanding universe, that's clean. Gravitational instability, that's surely there, but the details—why there are galaxies the size that they are and distributed the way they are—are awfully uncertain still. I can't get excited about any specific model. The observations have changed dramatically since I was a kid. It was just open, clean ground when I started working. You could imagine the wildest things. And your imagination wasn't wild enough for what we found.

"The field is moving toward a crisis now, in the sense that the constraints on the theories are getting stronger and stronger. The theories are becoming more and more elaborated, detailed, and, therefore, predictive. And I think soon we're going to see a situation in which a lot of theories seem to be wrong. I think, if we're lucky, one theory will survive and shine through as a good prospect. If not, we'll just cast our nets a little more broadly.

"There's got to be a crisis, so that's what it's all about. And I think it's happening at last in cosmology."

I was leaving the next day for Santa Cruz where a galaxy workshop was already in progress. The Santa Cruz meeting would be full of observers, so I asked Peebles if he had any advice or wisdom for them. He grinned and cocked his head back at the sky.

"Tell them," he said, "tell them—to look again."

It was foggy in Santa Cruz. The rhythm of life there—fog in the morning, sunshine at noon, the tease of starlight, and then fog again the next morning—seemed like the perfect metaphor for the Sisyphean journey of cosmology: from gloom to sunny understanding back to gloom again. The old split that developed in astronomy back in Hubble and Shapley's time—between theorists in the East spinning elegant ideas and observers in the West obsessed with instruments and efficiently utilizing every clear second of telescope time—still persisted, and I felt it tugging at me as I flew from the ethereal blackboards of Aspen to Santa Cruz. Who could speak better for God? The theoretical physicists and mathematical geometers who claimed to glimpse occasionally the secret Platonic symmetry beyond the world's chaos—to read His *mind*, in other words—or the astronomers and experimentalists who spent their lives staring at His handiworks?

At Santa Cruz, the participants were also concerned about the ability or inability of current cosmological theories to produce the spectacular structures that were being observed. I had the impression that if you passed a collection plate for the cosmic string theorists you would have a hard time raising bus fare back to Berkeley, the symbolic resting place of the abstract minded. There were dark mutterings about theorists who had never seen a real galaxy who would come in on the last day of a three-day meeting to explain everything.

Davis, White, and Frenk had prepared a kind of counterattack to the cosmological alarmism of the observers and the string theorists. It was based on cold dark matter, the most conservative and by now orthodox of theories. When they adjusted their computer simulations of the universe to answer the question- -what would an astronomer looking through a telescope at a narrow slice of that universe see?—the predictions of cold dark matter, they claimed, were not so far off the observational mark. Apparent voids and superclusters showed up on the simulated data.

White arrived in Santa Cruz with a cast on his ankle, having broken it doing an Irish folk dance. He gave a fiery and stubborn talk on the next-to-the-last day of the conference, arguing emphatically, writing it in capital letters on a viewgraph, YOU HAVE TO DO WHAT THE OBSERVERS DO. His main point was that cold dark matter's successes came in realms where the observations were the most reliable—the properties of galaxies and small groups—and its putative failures came in realms where the observations were the least reliable. On that basis alone, it was too soon to abandon the theory; cold dark matter deserved to be the standard model. He couldn't understand, White added, why anyone would want to study strings.

That night a crew of string theorists flew in from Aspen to give talks on the last day. I ran into Neil Turok and Andy Albrecht as they were unloading backpacks from the trunk of their rental car. They looked lanky and tan after weeks of volleyball in the Rockies. They were loose like confident athletes on the eve of a big game.

I asked them about strings. "It's not a religion," insisted Turok. "It's the truth," added Albrecht. They staggered giggling down the path with the packs.

Turok gave his talk the next morning, but he didn't get very far into it before he was interrupted by White, who wanted to know: "What are we having the strings solve if we already have to have hot dark matter?"

Primack, who was moderating, tried to explain that strings were a generic feature of modern physics. "They're not obviously crazy," he added.

"Then we're not compelled to invent them?" piped up Colin Norman, a rambling blond English theorist from the Space Telescope Science Institute, in a booming voice.

"From a particle physics standpoint, yes."

Norman grumbled some more, "We had one theory yesterday [White's] that did quite well."

"Colin is making a statement that amounts to astrophysical chauvinism," Primack announced suavely. "It's one universe."

I recalled, then, as we filed out into the misty sunshine, what Hawking had said on the steps of the Royal Society a long time ago, that there was no single unique universe, that according to the strange laws of quantum theory some observers might see it collapse, others see it go on forever. Maybe, I thought, we could do the same for these guys. There could be one universe for the cold dark matter theorists, another universe in which strings ruled, another in which there was a cosmological constant, and another in which the microwave background was grossly uneven. Perhaps, like the figurative electron flying between slits, whose properties have not yet been discerned or asked for because no physicist has yet put the final detector in place, the nature of the universe still lay in a limbo, awaiting the definitive experiment that would define the dark matter, crystallize the large-scale structure of the universe. According to Wheeler and Hawking we were all in some sort of crisscrossing dream, collectively conjuring ourselves and our surroundings out of the quantum muck of possibility.

One universe and a billion dreamers—almost but not entirely a hopeless collaboration. The cosmologists were all on the same boat. It wasn't such a big boat, nor was it really, even given the occasional food fight, such an unruly crew. As Bob Dylan said in "Talking World War III Blues," long ago when most of those present at this workshop had been undergraduates, "I'll let you be in my dreams, if I can be in yours."

The astronomers and cosmologists gathered one night in the dining hall of one of the Santa Cruz residential colleges for a cheese- and wine-tasting fest. Primack's wife, Nancy Abrams, the cabaret singer–songwriter cum lawyer, entertained, and many of the songs were her own compositions with their particular blend of dark folk-techno humor. One had been written riding the bus down from Mauna Kea during the Kona conference.

> In the town of Santa Cruz
> Worked astronomers who searched the sky
> And they told me what they found
> In three hundred nights of observing time
>> We all live in an expanding universe
>> Expanding universe, expanding universe
>> We all live in an expanding universe
>> Expanding universe, expanding universe
> And they measured the Hubble flow
> Toward Hydra-Centaurus and from Virgo
> And they found a wondrous thing
> H-nought jiggles like a spring
>> We all live in an expanding universe
>> Expanding universe, expanding universe . . .
> Now there are voids and filaments
> Peculiar velocities that don't make sense
> It's time for theorists to get tough
> Cold dark matter is not enough
>> We all live in an expanding universe
>> Expanding universe, expanding universe . . .

A hundred cosmologists raised their chardonnay glasses and sang along.

EPILOGUE

Marc Aaronson died in a freak accident in the dome of the 4-meter telescope at Kitt Peak National Observatory on the evening of April 30, 1987. While preparing to search for more Cepheid variable stars in the contested galaxy M101, he looked out a bulkhead door. A ladder hanging from the rotating dome clipped the door, closing it on Aaronson, and killing him instantly. He was thirty-seven.

Halton Arp, Geoffrey Burbidge, Fred Hoyle, and the latter's long-time partners in heresy, **Jayant Narlikar** and **Chandra Wickramasinghe**, published a long paper in *Nature* in August 1990, in which they raked the big bang model of the universe over the coals. Reviewing the controversial aspects of quasar and galaxy redshifts and the difficulties cosmologists are having reconciling the smooth microwave background with the present cosmic lumpiness, they argue, "The currently popular cosmological model is subject to many doubts based on observational data which suggest that, perhaps, there never was a Big Bang."

Sydney Coleman, the Harvard physicist who pioneered the idea of the false vacuum and befriended Alan Guth's inflation theory, has taken up the problem of why the elusive cosmological constant, which represents the vacuum energy density of today's universe, seems to be zero.

According to a proposal advanced by Coleman, submicroscopic worm-holes in the quantum spacetime foam that prevails on the tiniest scales in nature may regulate the vacuum energy, perhaps distributing it to an infinity of other parallel but otherwise empty universes.

The first all-sky maps of the **cosmic microwave background**, from measurements by the Cosmic Background Explorer satellite (COBE), were presented to the American Astronomical Society in January 1990. They show that the microwave background is uniform to less than 1 part in 10,000. Inflation theory predicts that temperature variations of about 1 part in 100,000 should occur over angular distances on the sky of a few degrees; future COBE maps should reach that precision.

Alan Dressler was on the cover of *Fortune* in 1990 as one of America's outstanding young scientists. He and **Sandra Faber** have accumulated more than 300 additional galaxy redshifts and distances in a continuing effort to study the Great Attractor. Their data, buttressed now by many other surveys (including one completed by Aaronson and his crew before his death), suggest that the large-scale motion detected by the Seven Samurai is real. The Attractor is no particular object. It is, according to Dressler, rather an entire region of space some 300 million light-years across in which the concentration of galaxies gradually rises; its center appears bland, lacking even any spectacular clusters. The Milky Way is near one edge of this region. By detailed studies of the galaxies within it, Dressler and Faber hope to be able to probe the distribution of dark matter.

Margaret Geller was awarded a five-year "genius" grant from the McArthur Foundation in the summer of 1990. Earlier that year she and **John Huchra** announced that the CfA redshift survey had discovered a chain of galaxies more than 500 million light-years long, which they dubbed the Great Wall. Some theorists immediately proclaimed that cold dark matter could never make structures that large. Advanced computer simulations by **Edmund Bertschinger** of MIT, however, suggested that cold dark matter has been underestimated, and that reports of its demise as a theory have been greatly exaggerated.

Jim Gunn, Maarten Schmidt, and **Donald Schneider** continue to analyze data from a survey begun in 1982 to find high-redshift quasars and define their distribution in time. In 1990 they found a quasar with a redshift of 4.7, which would make it the oldest and most distant object in the universe.

Alan Guth is now the Jerrold Zacharias Professor of Physics at MIT and a member of the National Academy of Sciences.

Stephen Hawking has separated from his wife, Jane, after twenty-five years of marriage. As of this writing, his book, *A Brief History* *of Time,* has been on bestseller lists for more than two years, and is soon to become a major motion picture. Recently, like Coleman, he has been studying wormholes. Hawking finally settled his long-time bet with **Kip Thorne**, the Caltech relativist and time travel expert, about whether or not Cygnus X-1 was a black hole; he conceded that it was. On a trip to California in June 1990 Hawking and an accomplice broke into Thorne's office while he was away, found the paper on which the bet was recorded, and Hawking (being unable to write) officially signed off on it with a thumbprint. Thorne has since begun to receive the British edition of *Penthouse* in the mail.

The **Hubble Space Telescope** was deployed from the space shuttle into a 281-mile-high orbit on April 24, 1990. Two months later NASA announced that due to an aberration in the primary mirror's shape the telescope could not be focused. The defect was traced to a faulty instrument used to test the mirror during its final polishing. Astronomers throughout the world were devastated. Although some spectroscopic and photometric observations can still be performed, the telescope's imaging instruments were severely crippled, and most cosmological investigations, such as deep-space surveys and remeasuring the Hubble constant, will be delayed until at least 1993, when a new camera with corrective optics can be delivered to the telescope.

Edward "Rocky" Kolb and **Michael Turner** published in 1990 *The Early Universe,* a textbook for graduate students in cosmology, with particular emphasis on big bang particle physics. It received glowing reviews.

In January 1989 most of the telescopes on **Mount Wilson** were reopened. They are being operated under a trial arrangement by the Mount Wilson Institute, a private nonprofit organization which, pending adequate funding, hopes to take permanent control of the mountaintop facilities and reopen the 100-inch Hooker telescope.

In 1990 **Jim Peebles** and **Joe Silk** published a "cosmic book" in *Nature,* in which they gave odds on different cosmological scenarios, based on the theories' abilities to explain various observations, weighted by the reliability of the observations themselves. In the resulting tabulation cold dark matter still came out as the best of a bad theoretical lot.

Roger Penrose had his own fling with the bestseller lists when he published *The Emperor's New Mind,* a masterly popular exposition on relativity and quantum theory masquerading as an attack on artificial intelligence. His most controversial suggestion was that the as-yet-undeveloped theory of quantum gravity might be key to understanding consciousness.

Vera Rubin is a staff member at the Department of Terrestrial Magnetism of the Carnegie Institution of Washington. She has been observing spiral galaxies in small groups and in the Virgo cluster in an effort to find out what happens to dark matter halos of galaxies that live in crowded neighborhoods. The early results, she says, are "not earthshaking"; galaxies in crowds do not seem to have individual halos.

Allan Sandage was awarded the 1991 Crafoord Prize—the astronomical equivalent of a Nobel, and worth $260,000—by the Swedish Academy of Sciences. He and **Gustav Tammann** recently published the ninth in their series of "Steps" papers, renewing the argument for a small Hubble constant. Sandage hopes to soon complete the *Carnegie Atlas,* a photographic compendium of the entire Shapley-Ames catalog of galaxies. Since his Hubble lecture, he says, he has become a convert to grand unification and inflation.

David Schramm and **Gary Steigman's** prediction from big bang helium production that no more than three or four neutrino types could exist in the universe was vindicated. In October 1989 physicists from SLAC and CERN held competing press conferences to announce that studies of the W and Z bosons, carriers of the weak force, indicated that there were only three families of neutrinos and elementary particles. Steigman is now a professor at Ohio State University. Big Bang Aviation has traded up to a twin-engine Cessna with a pressurized cabin.

Leonard Searle is the Director of the Carnegie Observatories.

Paul Steinhardt has authored a new theory called extended inflation. It purports to solve some of the problems encountered by earlier versions of inflation by incorporating a gravitational constant, g, which characterizes the strength of gravity, that changes with time during the early universe. Whether a time-varying gravitational constant solves more problems than it raises remains to be seen.

Alex Szalay hosted an IAU Symposium on Large-Scale Structure in Hungary on his birthday in 1987. He splits time between Eötvös University in Budapest and Johns Hopkins in Baltimore. He, **David Koo**, and **Richard Kron** are continuing their deep narrow "pencil beam" redshift survey. In early 1990 he and Koo, together with **Thomas Broadhurst** and **Richard Ellis**, who have performed similar surveys in the Southern Hemisphere, published a preliminary combined analysis of two pencil beams. Their data suggested that enormous clumps or walls of galaxies recur periodically throughout space every 400 million light-years or so. Commenting in *Nature* about these results, **Marc Davis** wrote in a widely quoted phrase that if the results were true, astronomers knew "less than zero" about the early universe.

John Wheeler has retired from the University of Texas and has moved to what he calls "an old fogey's home" in Princeton, where he commutes every day to an office at his old university. He published *A*

Journey into Gravity and Spacetime, a popular, lightly mathematical exposition of general relativity, in 1990.

Ed Witten was awarded a Fields Medal, the highest honor in mathematics, during the International Congress of Mathematicians in Kyoto, Japan, in August 1990.

Jakov Boris Zeldovich died of a heart attack in Moscow in November 1987. He was seventy-six.

INDEX